Sustainable Agriculture Reviews

Volume 10

Series Editor
Eric Lichtfouse

For further volumes:
http://www.springer.com/series/8380

*Other Books by Dr. Eric Lichtfouse**

Sustainable Agriculture
http://dx.doi.org/10.1007/978-90-481-2666-8

Sustainable Agriculture, Volume 2
http://dx.doi.org/10.1007/978-94-007-0394-0

Organic Farming, Pest Control and Remediation of Soil Pollutants
http://dx.doi.org/10.1007/978-1-4020-9654-9

Climate Change, Intercropping, Pest Control and Beneficial Microorganisms
http://dx.doi.org/10.1007/978-90-481-2716-0

Sociology, Organic Farming, Climate Change and Soil Science
http://dx.doi.org/10.1007/978-90-481-3333-8

Genetic Engineering, Biofertilisation, Soil Quality and Organic farming
http://dx.doi.org/10.1007/978-90-481-8741-6

Biodiversity, Biofuels, Agroforestry and Conservation Agriculture
http://dx.doi.org/10.1007/978-90-481-9513-8

Alternative Systems, Biotechnology, Drought Stress and Ecological Fertilisation
http://dx.doi.org/10.1007/978-94-007-0186-1

Genetics, Biofuels and Local Farming Systems
http://dx.doi.org/10.1007/978-94-007-1521-9

Agroecology and Strategies for Climate Change
http://dx.doi.org/10.1007/978-94-007-1905-7

Organic Fertilisation, Soil Quality and Human Health
http://dx.doi.org/10.1007/978-94-007-4113-3

Environmental Chemistry. Green Chemistry and Pollutants in Ecosystems
http://dx.doi.org/10.1007/b137751

Environmental Chemistry for a Sustainable World
Volume 1. Nanotechnology and Health Risk
http://dx.doi.org/10.1007/978-94-007-2442-6

Environmental Chemistry for a Sustainable World
Volume 2. Remediation of Air and Water Pollution
http://dx.doi.org/10.1007/978-94-007-2439-6

Rédiger pour être publié ! Conseils pratiques pour les scientifiques
http://dx.doi.org/10.1007/978-2-287-99396-1

Call for review articles
Authors wishing to publish a review article in *Sustainable Agriculture Reviews or Environmental Chemistry for a Sustainable World* should contact the Editor.
E-mail: Eric.Lichtfouse@dijon.inra.fr

* http://www.linkedin.com/pub/eric-lichtfouse/19/493/720, http://www.researcherid.com/rid/F-4759-2011

Eric Lichtfouse
Editor

Farming for Food and Water Security

Springer

Editor
Dr. Eric Lichtfouse
UMR1347 Agroécologie
AgroSup/INRA/uB
17, rue Sully
21000 Dijon, France

ISSN 2210-4410 ISSN 2210-4429 (electronic)
ISBN 978-94-007-4499-8 ISBN 978-94-007-4500-1 (eBook)
DOI 10.1007/978-94-007-4500-1
Springer Dordrecht Heidelberg New York London

Library of Congress Control Number: 2012942331

Printed on acid-free paper

Springer is part of Springer Science+Business Media (www.springer.com)

Contents

Public Goods and Farming

Catherine L. Gerrard, L.G. Smith, B. Pearce, S. Padel, R. Hitchings, M. Measures, and N. Cooper

Abstract There has recently been an increase in interest in the "public goods" that could be provided by a farm alongside its primary function of agricultural production. This paper reviews recent reports on the topic of public goods and, in particular, the public goods provided by agriculture and then goes on to discuss the development of a tool which can be used to assess the provision of public goods on a farm across a range of areas: soil management, biodiversity, landscape and heritage, water management, manure management and nutrients, energy and carbon, food security, agricultural systems diversity, social capital, farm business resilience, and animal health and welfare.

Keywords Public goods • Sustainability assessment • Tool • Organic farming • Field margin • Insect • Water management • Earthworm • Wind power • Air quality • Animal health • Social capital • Biodiversity • Food security • Landscape

1 Introduction

There has recently been an increase in interest in the beneficial co-products that could be provided by a farm alongside its primary function of agricultural production. In particular it is possible that in the future policy makers may wish to encourage the

C.L. Gerrard (✉) • L.G. Smith • B. Pearce • S. Padel • R. Hitchings
Organic Research Centre, Elm Farm, Hamstead Marshall, RG20 0HR, Newbury, Berkshire, UK
e-mail: catherine.g@organicresearchcentre.com

M. Measures
IOTA, Cow Hall, Newcastle, SY7 8PG Craven Arms, Shropshire, UK

N. Cooper
Natural England, Victoria House, London Square, Cross Lanes, GU1 1UJ Guildford, UK

E. Lichtfouse (ed.), *Farming for Food and Water Security*, Sustainable Agriculture Reviews 10, DOI 10.1007/978-94-007-4500-1_1,

provision of "public goods" such as ecosystem services and rural vitality as a reason for continuing to support farming through EU's Common Agricultural Policy. As discussed by Cooper et al. (2009), a public good must be non-excludable, i.e. available to all, and non-rival, meaning that its consumption by one individual does not diminish its availability to others. As such, benefits such as an improved environment or better water quality can be perceived to be public goods. Pure public goods are rare as most are non-excludable and non-rival only to a certain extent (Cooper et al. 2009; European Network for Rural Development 2011; RISE 2009). It is suggested that all the environmental services associated with land management show these public good characteristics to some extent (RISE 2009).

Public goods are not frequently encouraged by market forces and so it can be argued that, if public goods are deemed valuable to society as a whole, it is reasonable for governments to intervene via subsidies to ensure their provision. If this is the case then it will become important to have a means of assessing the provision of public goods on a range of farms to evaluate the impact of such subsidies and to ensure that they are delivering the desired outcome.

Given the current interest in this topic a tool was developed to assess the provision of a wide range of public goods covering economic, social and environmental aspects, i.e. carrying out triple bottom line reporting. This paper discusses public goods and farming, sustainability assessment tools and then presents the development of the public goods assessment tool and initial in-field testing on a pilot study of 40 organic farms in England.

In Sect. 2 recent reports on public goods and, in particular the provision of public goods by agriculture are reviewed. In Sect. 3 the development of the PG (Public Goods) tool is described. In Sect. 4 the review and discussion of the tool are briefly summarised and concluded upon.

2 Review of Public Goods from Farming and Sustainability Assessment Tools

2.1 *Public Goods and Farming*

The idea of a public good was first established in economic theory. An externality is defined as a by-product of a process that affects third parties e.g. pollution (RISE 2009) and a 'positive externality' may be said to be a 'public good' if it is non-excludable and non rival. Thus a public good should be a good that, if available to one person, cannot be withheld from others and if consumed by one person this does not reduce the amount available to others. As discussed by Cooper et al. (2009) and RISE (2009) pure public goods are very rare and public goods can show different levels of these two characteristics. The example that Cooper et al. (2009) and European Network for Rural Development (2011) give is that a cultural landscape may be regarded as a public good but that if too many people wish to enjoy it *in situ* then it may become congested and so its value will be decreased suggesting that it is not entirely non-rival. European Network for Rural Development (2011) gives a less technical definition of public goods as “things of benefit to the public which cannot be bought in the marketplace and for which there is no incentive to pay … but which are valued by society as a whole”. This suggests that policy-makers are moving from the technical economic theory definition of a pure public good to a looser definition which is more inclusive.

As discussed by European Network for Rural Development (2011), agriculture has, over many centuries transformed the environment in many parts of Europe, for instance reducing the amount of forest in the UK and replacing it with heath, crops and pasture. This has had both positive and negative environmental impacts depending on local soil and climate conditions and the farm management practices employed (European Network for Rural Development 2011). It is suggested that recent trends have been towards high intensity agriculture with high productivity of food and other direct agricultural products which has caused environmental damage although some farming systems such as extensive livestock systems continue to provide public goods (European Network for Rural Development 2011; RISE 2009).

RISE (2009) suggest that market failures are occurring with regards to the provision of public goods through agriculture. They point out that the majority of land in Europe is managed with around half of the land in the EU being used for agriculture and so farmers, alongside foresters, play a vital role in the provision of many public goods. They state that “Quite naturally farmers will respond to market signals for their food and saleable outputs, and pay less attention to the impacts of their activities where there are no markets. They will tend to provide fewer “goods” such as habitats, species and cultural landscapes, which no one pays for; and too many “bads” such as pollution of the atmosphere, soil and water as long as they are not required to pay the relevant full social or environmental costs.” As discussed by European Network for Rural Development (2011), the market failure with regards to public goods occurs because consumers cannot, in general, be persuaded to pay for them because they cannot be excluded from consumption if they do not pay. As a result producers of public goods have no incentive to continue to produce them as they are not rewarded for doing so via the market. It is suggested (Cooper et al. 2009; European Network for Rural Development 2011; RISE 2009) that if the market fails to reward the provision of public goods then it may fall to governments and the EU to guarantee that it continues by providing financial incentives through subsidies.

European Network for Rural Development (2011) suggest that farming has a key role in the provision of public goods as hundreds of years of farming have created cultural landscapes and farm-specific habitats which support animals and plants and also because farming plays an important role in food security which makes the preservation of natural resources a matter of increasing importance.

Cooper et al. (2009) suggest a number of public goods which they believe to be provided by agriculture. They classify these as environmental public goods (agricultural landscapes, farmland biodiversity, water quality and water availability, soil functionality, carbon storage and climate stability, greenhouse gas emissions, air quality, resilience to flooding and resilience to fire) and social public goods (rural vitality, food security, farm animal welfare and animal health). Similarly European Network for Rural Development (2011) lists the main, most widely recognised public

Fig. 1 One example of a traditional agricultural landscape in the UK, an upland landscape

goods from farming as: environmental (farmland biodiversity, agricultural landscapes, high water and air quality, water availability, the functionality of soils, climate stability, resilience to flooding and fire), food security (maintaining longer term capacity to produce food, husbandry of resources such as land, skills and essential infrastructure), rural vitality (social viability of rural communities), and farm animal health and welfare.

2.2 *Specific "Public Goods"*

As mentioned in Sect. 1 Cooper et al. (2009) and European Network for Rural Development (2011) identify a number of beneficial by-products of agriculture which they deem to be public goods. Some of these will be discussed in a little more detail here.

2.2.1 Agricultural Landscapes

As mentioned previously farming has had an influence on the landscape in Europe for many centuries. As such, agricultural landscapes have become valued as the "traditional" landscape of many areas (Fig. 1). Some landscapes are a "mosaic" of

Fig. 2 Field margin planted with wild flowers to provide habitats and food for insects and small mammals

agricultural land, woodland and unmanaged land, others may include very specific agricultural elements such as orchards, lemon gardens etc. (Cooper et al. 2009).

These landscapes can be seen as public goods (Brunstad et al. 2005; European Network for Rural Development 2011; Fleischer and Tsur 2000; RISE 2009) as most EU countries now have legislation to allow some degree of public access and because, unless very large numbers of people try to enjoy the landscape, it is generally non-rival – its enjoyment by one person is not prevented by its enjoyment by another (Cooper et al. 2009). In the UK, the landscape is more accessible in Scotland, which has no law of trespass, than in England but there is legislation to allow some public access (Countryside and Rights of Way Act 2000).

2.2.2 Biodiversity

Certain species and habitats (Fig. 2) are associated with agriculture, particularly more extensive forms of agriculture e.g. farmland birds (Chamberlain et al. 2000). In some cases these species may be reliant on agriculture as the habitats which it mimics may no longer exist except as a result of certain agricultural practices. However, agricultural intensification has been repeatedly linked to worldwide biodiversity decline of a wide range of taxa (Benton et al. 2002; Green et al. 2005). Agricultural management practices therefore have the potential to have a great deal of positive or negative impact on biodiversity. Since 1985, agri-environment schemes in the EU have aimed to reverse the detrimental impacts of agricultural

Fig. 3 Agricultural land near water systems requires to be managed carefully to ensure that there are no negative impacts on the water quality

intensification by encouraging farmers to adopt more environmentally-friendly practices (Ovenden et al. 1998).

Biodiversity can be deemed to be a public good (European Network for Rural Development 2011; RISE 2009) because it is difficult to restrict the benefits associated with it and one person's enjoyment of biodiversity does not impact on another person's enjoyment of it (Cooper et al. 2009). Mankind also benefits from the services biodiversity provides, for example, pollination of crops, and control of pests (Klein et al. 2007; Losey and Vaughan 2006).

2.2.3 Water Quality and Availability

Agriculture can have a large impact on water quality and availability (Fig. 3). Water is frequently abstracted from other water sources to irrigate farming land, especially cropping land (Pimental et al. 1997). In Europe agricultural abstraction makes up 24% of total water abstraction and, unlike water abstracted for electricity generation, much of the water abstracted does not return to a water body (European Environment Agency 2009). Run-off from agricultural soils may also contain fertilisers, manures, herbicides, pesticides and veterinary medicines which may

Fig. 4 The presence of earthworms in the soil will help to improve its health and structure

impact on the ecosystems of freshwater bodies and the marine environment (European Environment Agency 2011). Changes and improvements to agricultural management such as placing manure storage on hard standing, far removed from water sources, can greatly improve the quality of water (Cooper et al. 2009; Environment Agency 2007).

Water quality and availability probably shows a lower degree of both public good characteristics. For instance, most European countries are progressing towards a water pricing mechanism for public supply of water (European Environment Agency 2009) and it is possible for one user of water to use so much or dam a water source such that other users are impacted. However, the quality of water cannot be restricted to some users and not to others and so this can be regarded as non-rival and non-excludable (Cooper et al. 2009; RISE 2009).

2.2.4 Soil Functionality

Soil functionality (Fig. 4) can include the amount of organic matter in the soil, the susceptibility to erosion or compaction, level of contamination, soil biodiversity, etc. (EEA and JRC 2010). Agricultural management can impact on all of these in either negative or positive ways (Cooper et al. 2009; DEFRA 2009). Certain agricultural practices have been suggested as reducing the impact on the soil including, no tillage, reduced tillage, cover crops, and crop rotation (EEA and JRC 2010).

Fig. 5 Producing renewable energy on the farm – such as utilising wind power – can reduce emissions

Soil can be owned as part of the agricultural land and so can be regarded as private but the long term benefits of good quality soil are less excludable and so soil quality can be regarded as a public good (Cooper et al. 2009; RISE 2009).

2.2.5 Air Quality

Agriculture can be a source of air pollution in the form of emissions from farm machinery, contamination from spraying, emissions from manure, and odours and greenhouse gas emissions from livestock (Cooper et al. 2009; Quality Meat Scotland 2010; Weiske 2006; Williams et al. 2010). Ninety-four percent of Europe's ammonia emissions are from agriculture (European Environment Agency 2010). However, agricultural practices can be changed to reduce emissions and air pollution. Farmers can also contribute to reducing emissions by producing renewable energy on their farm (Fig. 5) which may be able to provide their own electricity and also provide electricity to others (European Environment Agency 2010).

Good air quality is a public good in that it cannot be rival or excludable (Cooper et al. 2009; European Network for Rural Development 2011; RISE 2009).

2.2.6 Rural Vitality

Rural vitality can encompass many aspects and farming can help to support some of these by, for example providing employment within the rural community so that people do not have to move away to find work and by being one means of enhancing the local economy (European Network for Rural Development 2011). Rural cooking and folk music can also be based on the farming heritage of a particular area (Cooper et al. 2009).

Although employment itself cannot be seen as non-rival or non-excludable the cultural and social benefits of a strong rural community can be seen as non-rival and non-excludable and therefore viewed as a public good (Cooper et al. 2009).

2.2.7 Food Security

Per the World Food Summit of 1996 definition, food security is "when all people, at all times, have physical and economic access to sufficient, safe, nutritious food to meet their dietary needs and food preferences for an active and healthy life" (World Health Organisation 2011). Agriculture, as a major food producer, contributes towards this aim.

Ensuring the long term ability of agricultural land to continue to produce food is seen as a public good (Brunstad et al. 2005; Cooper et al. 2009; European Network for Rural Development 2011).

2.2.8 Farm Animal Health and Welfare

There is increasing emphasis on the importance of farm animal welfare with pressure being put on the agricultural sector by consumers and animal welfare organisations (Fig. 6). For instance, perceived higher animal welfare is often given by consumers as a reason for buying organic livestock products (Aertsens et al. 2009; Zander and Hamm 2010). The RSPCA now certifies food under its "freedom foods" label indicating that a certain level of animal welfare was provided in the production of those foods (RSPCA 2010a, b, c, d, e). There is also an argument that, as well as being important from an ethical viewpoint farm animal health and welfare is a public good in that it contributes to the health and safety of consumers (Cooper et al. 2009).

2.2.9 Summary

Not all of the "public goods" discussed above are pure public goods but they are generally goods which have a degree of non-rival or non-excludable characteristics and which are becoming increasingly important to the public but are not fully rewarded by market forces. These are the type of public goods which may need financial and policy support in the future. However, to justify subsidies governments and the EU will need some means of assessing whether a farm is providing these goods or not, as not every farm provides these public goods to the same extent.

Fig. 6 Access to outdoor grazing can be seen as beneficial to animal welfare as it allows the animals to express natural behaviours

This is where an assessment tool may be required. Such tools will be discussed in general in Sect. 2.3 below and then in Sect. 3 a tool specifically designed to assess public goods provided by farms participating in an agri-environmental scheme will be introduced and discussed.

2.3 *Integrated Sustainability Assessment Tools*

One way of assessing a farm's performance in any of a number of areas is to create a tool which will allow a consultant, researcher or farmer to enter data which will then be processed and provide a score for the farm. Such tools exist for a variety of different purposes. Some tools assess specific areas such as carbon footprint e.g. CALM (carbon accounting for land managers) (CLA 2010), IMPACCT (integrated management options for agricultural climate change mitigation) (University of Hertfordshire 2011), and EASI (the energy, emissions, ecology and agricultural systems integration project) (Organic Research Centre 2011) which also assessed biodiversity as well as emissions and energy efficiency; mineral use e.g. MINAS (mineral accounting software tool) (Halberg et al. 2005a) used in the Netherlands; environmental performance e.g. the tool developed at Organic Centre

Wales (Fowler et al. 2004) and EMA (Environmental Management for Agriculture) (University of Hertfordshire 2006); socio-economic foot-printing surveys (Lobley et al. 2005).

More recently there has been a trend towards tools which assess farms across a wide range of areas such as economic, social and environmental factors i.e. triple bottom line accounting (Global Reporting Initiative 2010; ICAEW 2009). Triple bottom line reporting has been used by companies, especially public limited companies (PLCs), since the 1990s as a means of communicating the "sustainability" of their business to their shareholders and other stakeholders. Several agricultural sustainability tools have been developed including Suffolk Farm Sustainability Appraisal (Ridley and Woolley 2002), the Organic Systems Development Group Sustainability Audit (Lampkin et al. 2006; Measures 2004), the tool developed for a Defra project on "quality and environmental benchmarking for organic agriculture (Organic Research Centre 2010), RISE (response inducing sustainability evaluation) (Hani et al. 2003) and MOTIFS (monitoring tool for integrated farm sustainability) (De Mey et al. 2010; Meul et al. 2008). Many of these tools, in assessing sustainability, assess the provision of public goods such as clean air and water by the farm.

These tools make use of a selection of indicators to assess the farms' performance across the economic, social and environmental areas. The selection of appropriate indicators is of great importance when carrying out farm assessment/benchmarking and should be according to the specific needs for information related to the objective (Halberg et al. 2005a). Halberg et al. (2005b) also highlight the fact that the indicators that are chosen should be able to be affected by the farmer's choices or strategies.

RISE (Hani et al. 2003) uses the sustainability indicators: energy, emission potential, water use, soil use, biodiversity and plant protection, waste and residues, cash flow, farm income, investments, economy, and social situation. These 12 main indicators are calculated from 68 parameters which were chosen to reflect activities that the farmer can influence. The systems unit is the farm (with the exception of some bio-diversity activities) and household and other non-agricultural activities are excluded. The period covered is 1 year. To assess sustainability the tool uses two scores: state parameters vary from 0 (worst) to 100 (best) while driving force parameters vary from 0 (best) to 100 (worst). This approach mirrors the Driving State Response (DSR) framework (Organisation for Economic Co-operations and Development 1997). This sets a concept for the development of indicators for measuring sustainability and environmental impacts (Stolze et al. 2000) based on whether an indicator describes a Driving force (D, i.e. the cause), the State (S, or quality) of the environment, or a Response (R) to degraded environment in the form of policies and targets. Overall results from RISE are shown on a "polygon" (a radar diagram) showing all of the indicators, their state, their driving force and the degree of sustainability (calculated as S-D), thus the final score goes from –100 to 100. RISE has been successfully tested in a number of countries demonstrating that it is able to provide results for a wide range of farming systems.

MOTIFS assesses ecological, economic and social sustainability through ten themes: internal social sustainability, external social sustainability, disposable income, use of inputs, quality of natural resources, biodiversity, entrepreneurship, efficiency and productivity, profitability, and risk. The MOTIFS tool operates on three levels: level one is an overview of the farm's sustainability, level two focuses in on a specific sustainability dimension and its underlying themes and level three focuses into the individual indicator scores for a theme. During the course of its development this tool was trialled on 200 farms and as a result some critical success factors (CSFs) (Campbell et al. 1999) were identified which De Mey et al. (2010) suggest may extend to all tools of this type. These critical success factors are:

- Attitude of model users towards sustainability – a positive attitude increased interest in using the tool.
- Compatibility – compatibility with current data systems, especially terminology, increased uptake.
- User-friendliness – easy use increased interest. Farmers did not want to spend time trying to calculate values – workload, costs and implementation are important.
- Data availability – ease of access to required data. They identified a need for improvements to availability of data particularly with regard to social aspects.
- Transparency – farmers needed to understand results especially with new/complex topics.
- Data correctness – inaccuracy in reported data from farmers led to inaccurate results and caused farmers to lose confidence in the assessment.
- Communication aid – the authors found the tool useful as a starting point for discussions between farmers and between farmer and advisor.
- Complexity – They found that the assessor needed to explain complex issues straightforwardly.
- Organisation of discussion sessions – communication was affected by trust amongst participants. Groups should not exceed 12. Similar farms were grouped so that the results were comparable.
- Effectiveness – farmers were finding it a useful tool to get a comprehensive view of a farm's sustainability. They also appreciated the opportunity to share knowledge during discussion session.

The PG tool discussed below is similar to these sustainability tools in that it was developed to assess a farm over a range of economic, social and environmental goods that it may provide. As such, it also shares several of the critical success factors identified by De Mey et al. (2010). These will be discussed further in Sect. 3.

3 Public Good (PG) Tool

3.1 Introduction

Like the other integrated sustainability tools discussed above the PG tool assesses a farm across a range of factors. It differs from other tools in that it has been developed in relation to a particular agri-environment scheme in England (Organic Entry Level

Stewardship – OELS) thus allowing policy makers to ascertain whether their schemes/subsidies are having an impact and allowing farmers to see where the provision of public goods on their farm could be improved in the future.

The tool development involved the following stages which are described in greater detail below:

- Identification of the main public goods to be assessed.
- Decision as to how to assess these through a number of "activities" (developing questions and considering data requirements and constraints).
- Production of the tool as an excel spreadsheet.
- Pilot assessment of the tool on 40 organic farms.
- Analysis of pilot assessment data and feedback from the pilot assessment.

A variety of public goods, which may be provided by an agricultural enterprise were identified after a thorough literature review and a stakeholder meeting involving researchers, agricultural advisors and representatives from Natural England. These were: soil management, biodiversity, landscape and heritage, water management, manure management and nutrients, energy and carbon, food security, agricultural systems diversity, social capital, farm business resilience, and animal health and welfare. These areas were chosen to account for a range of benefits which may be provided by farming systems and are similar to those suggested by previous authors (BioBio 2009; Cooper et al. 2009; European Network for Rural Development 2011; Kuratorium fur Technik und Bauwesen in der Landwirtschaft 2009; National Institute of Statistics of Italy 2001; Organic Research Centre 2010; Organisation for Economic Co-operations and Development 1997).

Each area is assessed by asking questions based on a number of key "activities". Each activity has at least one corresponding question and these allow the advisor, who is assessing the farm, to evaluate the detailed ways in which the farm provides each public good. The activities (listed in Table 2) are tested via detailed questions.

The choice of activities was influenced by the desire for the data collected to be of a type that a farmer would have in their farm records already, i.e. not requiring any further surveys to be carried out. Care was also taken to maintain a mixture of quantitative and qualitative activities as quantitative data can be seen as less subjective but to measure areas such as social capital and animal health and welfare it is likely that some qualitative data will be required. It was also necessary to maintain a balance between obtaining sufficient detail to assess the spurs while keeping the assessment to a reasonable length of time. The PG Tool assessment takes 2–4 h to complete depending on the size and complexity of the farm and therefore does not ask for a commitment of time on the part of the farmer which he may be unable to make but does collect sufficient information to provide a reasonably in-depth analysis. It can be seen that these requirements for the tool mirror the critical success factors of user-friendliness and data availability identified by De Mey et al. (2010) with regards to the MOTIFS tool.

Each question is marked with scores between 1 (lowest mark – no benefit provided) and 5 (highest score). Some questions have a not applicable (n/a) option. This is the case where a situation may arise such that the farmer cannot possibly provide that benefit, for instance, a farmer who does not have dairy cows will not have any members of staff looking after them but should not be scored lower for failing to do so and therefore can choose n/a in this situation.

Some activities are assessed using several questions while others require only one. Where multiple questions are asked their scores are averaged and rounded to the nearest whole number to give the score for that activity. Thus an activity requiring several questions is not weighted more heavily than one requiring only a few or one question.

The scores for each area are obtained by averaging the scores for all its activities. These are then shown on a radar diagram allowing farmers to see in which areas they perform well and which areas might be improved. A bar chart showing the activities on each spur gives information on the scores of all activities so that if the farmer scored less well on a particular area they can then identify the specific activities to work on to improve the score in the future. The scoring system and results output were chosen to be as straightforward as possible so that farmers can see, at a glance, how their farm is performing rather than have to spend time understanding a more complex scoring system and data output. This reflects the critical success factor of transparency discussed by De Mey et al. (2010).

3.2 *Pilot of the Tool on 40 Organic Farms*

To assess the suitability and performance of the PG Tool in the field it was decided to carry out pilot assessments on 40 English organic farms and obtain feedback from the farmers and advisors involved in the pilot. The farms were chosen with the assistance of the eight advisors who carried out the pilot assessments and were therefore not representative of the population as a whole, as they are more likely to be interested in public goods provision, hence their willingness to participate in this project.

It was decided that the farms assessed would be chosen to cover a spread over the main robust farm types as defined by Defra for the Farm Business Survey (DEFRA 2010). Details were also recorded of ownership status of farm, number of years since conversion and number of years fully organic.

The 40 pilot assessments were then carried out and the results are summarised below. The advisors also provided feedback on the tool and the farmers were given feedback forms to allow them to rate the tool's performance and these comments are also discussed below. The aim of the pilot was to assess whether the tool was able to separate good practice on the individual spurs and activities from areas where further improvement would be advisable, to evaluate whether it was user-friendly and whether it was seen as valuable by farmers and advisors in evaluating the provision of public goods on a farm.

3.2.1 Descriptive Statistics

Table 1 shows the mean, median, minimum and maximum scores for each public good across all 40 farms in the pilot study along with the standard deviations of the scores for each area. The mean, minimum and maximum scores can also be seen in Fig. 7.

Table 1 Individual public goods showing their mean, median, minimum and maximum scores and the standard deviation across the areas

	Mean	Median	Minimum	Maximum	Standard deviation
Biodiversity	3.0	3.0	1.5	4.5	0.7
Landscape and heritage	3.9	4.0	2.5	5.0	0.6
Soil management	4.2	4.3	3.0	5.0	0.4
Water management	2.9	3.0	1.0	5.0	0.9
Nutrient management	3.7	3.8	2.8	4.5	0.4
Energy and carbon	3.0	3.0	1.7	4.3	0.7
Food security	3.4	3.3	2.2	4.5	0.6
Agricultural systems diversity	3.5	3.3	1.7	5.0	0.9
Social capital	3.0	3.2	1.8	4.7	0.7
Farm business resilience	3.9	4.0	2.5	5.0	0.6
Animal health and welfare	4.2	4.3	3.3	4.8	0.3

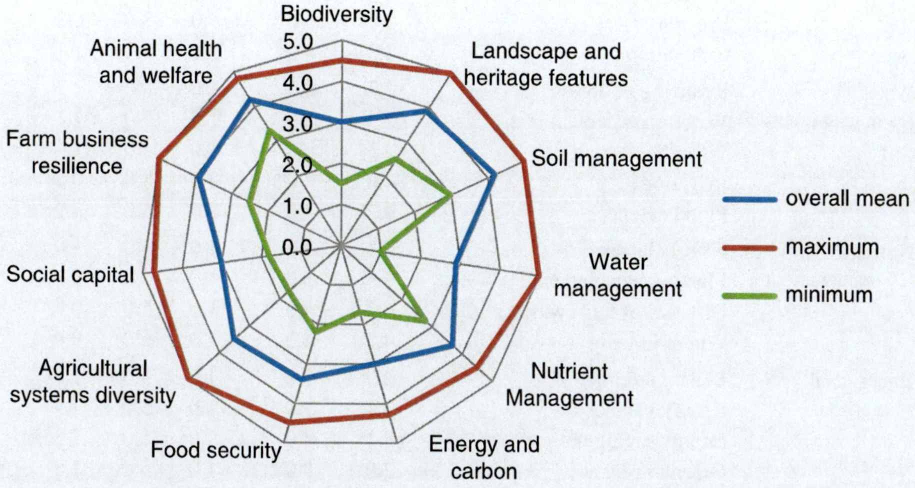

Fig. 7 Radar diagram showing the mean, maximum and minimum scores for each area for the pilot assessments on 40 organic farms

While there is some variation in the scores, no area gives a standard deviation of one or greater. However, there is greater variation in individual activities (Table 2). The highest scoring areas on average were animal health and welfare and soil management and the lowest scoring on average was water management.

The individual activities show a greater variation in scores as can be seen in Table 2. The greatest variation in scores was shown for the activities conservation plan under biodiversity, water management plan under water management (with standard deviations of 1.8) and on-farm processing in the agricultural systems diversity spur with a standard deviation of 2.0. The activities showing the least variation in scores were food quality certification, with a standard deviation of zero (as all of the farms in the pilot are organic and therefore score 5 for this), and the erosion activity in the soil management spur with a standard deviation of 0.2.

Table 2 Individual activities with their mean, median, minimum (min) and maximum (max) scores and the standard deviation (st. dev.) in those scores

Spurs	Activities	Mean	Median	Min	Max	St. dev.
Biodiversity	Agri-environmental participation	3.4	3.0	1.0	5.0	1.2
	BAP habitat and SINCs	2.0	2.0	1.0	5.0	1.2
	SSSI	4.4	5.0	2.0	5.0	1.0
	BAP and rare species	4.2	5.0	1.0	5.0	1.1
	Conservation plan	3.4	4.0	1.0	5.0	1.8
	Awards	1.5	1.0	1.0	5.0	1.2
	Habitats	3.2	3.0	2.0	4.0	0.6
Landscape and heritage	Historic features	3.8	4.0	2.0	5.0	0.8
	JCA and landscape features	4.5	5.0	2.0	5.0	0.7
	Management of boundaries	3.4	4.0	1.0	5.0	1.0
Soil management	Soil analysis	3.8	4.0	1.0	5.0	1.0
	Soil management	4.3	4.0	1.0	5.0	0.9
	Winter grazing	4.0	4.0	3.0	5.0	0.7
	Erosion	5.0	5.0	4.0	5.0	0.2
	Cultivation	3.8	4.0	2.0	5.0	1.1
Water management	Reducing pollution	2.7	2.0	1.0	5.0	1.1
	Water management plan	2.4	1.0	1.0	5.0	1.8
	Water harvesting	2.3	2.0	1.0	5.0	1.4
	Irrigation	4.0	4.0	3.0	5.0	0.8
	Flood defences	4.2	4.0	3.0	5.0	0.9
Nutrient management	NPK balance	3.1	3.0	1.0	4.0	1.0
	Manure management	3.3	3.0	1.0	5.0	0.8
	Disposal of farm waste	4.2	4.0	1.0	5.0	0.9
	Winter grazing	4.2	4.0	1.0	5.0	0.9
Energy and carbon	Benchmarking	2.9	3.0	1.0	5.0	1.2
	Energy balance	3.1	3.0	1.0	5.0	1.4
	Energy saving options	3.1	3.0	1.0	5.0	1.3
	Greenhouse gases	2.6	3.0	1.0	5.0	1.3
	Land use change	3.8	4.0	2.0	5.0	0.7
	Renewable energy	2.7	2.0	1.0	5.0	1.1
Food security	Total productivity	3.3	3.0	1.0	5.0	0.9
	Local food	3.6	5.0	1.0	5.0	1.7
	Off farm feed	4.1	4.0	1.0	5.0	1.0
	Food quality awards	2.1	1.0	1.0	5.0	1.7
	Food quality certification	5.0	5.0	5.0	5.0	0.0
	Production of fresh produce	2.2	1.5	1.0	5.0	1.3
Agricultural systems diversity	Cropland diversity	4.0	4.0	3.0	5.0	0.9
	Livestock diversity	3.4	4.0	1.0	5.0	1.1
	Marketing	3.7	4.0	1.0	5.0	1.3
	On-farm processing	2.8	1.0	1.0	5.0	2.0
Social capital	Employment	2.2	2.0	1.0	5.0	1.3
	Skills and knowledge	2.8	3.0	1.0	4.0	1.0
	Community engagement	2.5	2.0	1.0	5.0	1.4

(continued)

Table 2 (continued)

Spurs	Activities	Mean	Median	Min	Max	St. dev.
	CSR (corporate social responsibility) initiatives and accreditations	2.3	2.0	1.0	5.0	1.4
	Public access	3.7	4.0	1.0	5.0	1.2
	Human health issues	4.6	5.0	3.0	5.0	0.6
Farm business resilience	Financial viability	3.8	4.0	2.0	5.0	0.9
	Farm resilience	4.0	4.0	3.0	5.0	0.5
Animal health and welfare	Staff resources	3.8	4.0	3.0	5.0	0.6
	Health plan	4.6	5.0	2.0	5.0	0.8
	Animal health	4.3	4.0	3.0	5.0	0.6
	Ability to perform natural behaviours	4.4	4.0	3.0	5.0	0.6
	Housing	3.9	4.0	3.0	5.0	0.4
	Biosecurity	4.1	4.0	1.0	5.0	1.2

3.2.2 Discussion of Pilot Results

To investigate the variation in the scores within the pilot further it was considered whether the variation in the scores for the spurs is influenced by certain factors such as farm type, advisor carrying out the assessment, level of agri-environmental participation (OELS/Higher Level Stewardship – HLS), whether or not the farm was solely grassland, length of time the farm has been fully organic, or status of ownership (short-term tenant versus owner-occupier). A more detailed analysis was carried out across the spurs by the use of ANOVA or t-test. In particular, it was of interest to see whether the tool could identify any difference between farms in the OELS scheme and those in the HLS scheme as one of the aims of the tool, as discussed earlier was to allow policy makers to assess the impact of schemes, such as the agri-environmental schemes, on the provision of public goods on a farm.

The last two factors (tenancy/ownership status and length of time the farm has been fully organic) appeared to have little impact on the scores in the pilot assessment.

Level of agri-environmental participation had an impact on the biodiversity spur where membership of HLS had a significant (at the 5% level) chance of increasing the mean score suggesting that farms which hold HLS agreements do more to promote biodiversity.

The remaining factors investigated – farm type, whether or not the farm is grassland, and advisor – all showed significant differences on more than one spur. For farm type and whether or not the farm was solely grassland the same three spurs show significant results: energy and carbon (significant at the 1% level), food security (5%), and nutrient management (5%). For the advisor factor significant variations were again found for the spurs energy and carbon (1%) and food security (1%) (these were also significant spurs for the previous two factors), and additionally farm business resilience (1%), water management (5%) and social capital (5%).

These three factors are closely related. Grassland farms are generally livestock farms and so tend to be dairy or beef and sheep robust farm types. Most advisors specialise in a particular area of the country and/or certain types of farms and so the advisors involved in this pilot had each assessed only one or two robust types. It is, thus, possible to say that any or all of these three factors may have an influence on the results but it is impossible to say, from the analysis that has been carried out, whether the factors are independently significant or whether they interact with each other. To carry out an analysis which would give such information would require a larger data set with all advisors covering all types of farm and then would need the use of more sophisticated statistical techniques such as factorial ANOVA and was thus out of the scope of this pilot assessment.

3.2.3 Pilot Feedback

Another important aspect of the pilot assessment was to evaluate the reaction of advisors and farmers to the tool and to use their feedback to improve it where necessary and ensure that it will be a practical and useful tool in the field.

The advisors who were involved in the pilot provided feedback throughout the experience via e-mail and telephone calls. Additionally two conference calls were held to discuss the PG Tool, future development and to allow the advisors further opportunity to give feedback on the tool.

One advisor comment sums up the response "*Overall it was an interesting exercise and could be a useful tool with a bit of tweaking*"

Another advisor commented on farmer's reactions to the tool saying *"I would like to add that farmer's reaction was, on the whole, very positive. They were interested in the tool and its concept and entered into discussion very freely. The spider [radar] diagram was well received with interest not only in the high scores but also the low scores and the reason for them and how they could be improved."*

Direct farmer feedback was also generally positive. Of the 40 farms assessed 12 returned their feedback forms giving a 30% return rate. Of those, eight would recommend the tool to others in its current format and two more would recommend it once modified. It would also appear that the tool has generally increased farmers' understanding of public goods with nine of the farmers reporting a higher level of knowledge and understanding of public goods after the assessment than they had reported prior to it. The remaining three farmers already reported a knowledge and understanding level of either nine or ten out of ten and so would appear to have already been very knowledgeable prior to using the pilot.

Another area in which farmers scored the PG Tool highly was the opportunity to ask questions, eight rated this as excellent and the remaining four rated it as good. This suggests that the tool was valued as a communication aid as discussed by De Mey et al. (2010) and also mirrors the experience with the RISE tool where some farmers said that they felt that the interaction and discussion with the person carrying out the analysis was as useful as the actual figures produced (Grenz J (2011) personal communication).

The reporting format was also rated well with two excellent ratings, nine good and one fair. Lower ratings were obtained for the length of time taken to carry out the assessment which obtained four good ratings and eight fair ratings, one farmer commented that a farmer *"would need to be dedicated to return to it"*. The quality of questions received a mixed response with six farmers rating the quality of the questions as good, four as fair, one as fair/poor and one as poor/excellent (explaining that he felt that it was, *"early days – some great bits - some bits need work!"*). Further comments on the questions included *"some of the questions need refining in order to reflect properly the reality on the farm"* and *"I think it is good to be able to assess the public goods gained but some of the questions are a little vague and so don't always give a fair result as they don't give the whole picture but I suppose the tool would end up overcomplicated"*.

There was also a positive response to questions about the overall value of the PG Tool. With regards to value to their business seven farmers rated it as above average and three as high, one felt that it was too soon to tell and one rated it as below average. With regards to demonstrating the public goods obtained from farming to the wider community one thought it was of little use and one thought it was of no use, but four felt that it partly demonstrated this and six felt that it was a help.

4 Conclusion

Public goods are defined as goods which are non-rival and non-excludable. As such, there are very few pure public goods. However a range of goods such as clean air and water, biodiversity, landscape, rural vitality, food security etc. are seen as being on a scale of "publicness" as they tend to have the characteristic of being non-rival and non-excludable to varying degrees. Many of these public goods can be provided as a co-product of farming and there has recently been great interest in these as policy-makers view them as a potential reason for continued support of agriculture through subsidies.

Recent reports have suggested that some of the key public goods provided by agriculture may be agricultural landscapes, farmland biodiversity, water quality and water availability, soil functionality, carbon storage and climate stability, greenhouse gas emissions, air quality, resilience to flooding and resilience to fire, rural vitality, food security, farm animal welfare and animal health (Cooper et al. 2009; European Network for Rural Development 2011).

If such public goods are to be used in the future to justify support of farming through subsidies then it will become necessary to have a means of assessing the provision of public goods by individual farms. The PG Tool was developed with the aim of assessing the public goods provision on a farm so that the farmer can see where his farm provides high levels of public goods and where there is the potential for improvement. It is also the first tool developed in the UK which allows the provision of public goods during the course of a farm's involvement in an agri-environment scheme, such as Organic Entry Level Scheme, to be monitored so that policy-makers can see whether the scheme is succeeding in encouraging the provision of public goods.

The public goods assessed by the tool are very similar to those mentioned above as having been recently identified as the key public goods that could be provided by farming.

The questions used to assess the public goods provision on the farm use data that are available within the farm records thus not requiring further calculation to be carried out as part of the assessment. Quantitative data were used where available. Qualitative data were used for activities that were considered important by the stakeholders but for which few appropriate quantitative indicators are likely to exist on farm, such as some aspects of animal welfare and social capital.

The PG tool was tested on 40 organic farms and obtained positive feedback from farmers and advisors. The results showed that there were differences between farms depending on farm type, level of agri-environment participation and whether the farm was solely grassland.

Acknowledgements The authors would like to acknowledge the financial support of Defra and Natural England through their Organic Conversion Information Service (OCIS). The authors would also like to thank the advisors and farmers who were involved in the pilot for their invaluable help with this project.

References

Aertsens J, Verbeke W, Mondelaers K, Van Huylenbroeck G (2009) Personal determinants of organic food consumption: a review. Br Food J 10:1140–1167

Benton TG, Bryant DM, Cole L, Crick HQP (2002) Linking agricultural practice to insect and bird populations: a historical study over three decades. J Appl Ecol 39:673–687

BioBio (2009) WP2 Indicators for biodiversity in organic and low input farming systems. http://www.biobio-indicator.wur.nl/UK/Project+Work+Packages/WP2+Conceptal+foundations/

Brunstad RJ, Gaasland I, Vardal E (2005) Multifunctionality of agriculture: an inquiry into the complementarity between landscape preservation and food security. Eur Rev Agric Econ 32(4):469–488

Campbell D, Stonehouse G, Houston B (1999) Business strategy: an introduction. Butterworth Heinemann, Oxford

Chamberlain DE, Fuller RJ, Bunce RGH, Duckworth JC, Shrubb M (2000) Changes in the abundance of farmland birds in relation to the timing of agricultural intensification in England and Wales. J Appl Ecol 37:771–788

CLA (2010) http://www.cla.org.uk/Policy_Work/CALM_Calculator/. Accessed 18 Nov 2010

Cooper T, Hart K, Baldock D (2009) The provision of public goods through agriculture in the European Union, report prepared for DG Agriculture and Rural Development, Contract No 30-CE-023309/00-28. Institute for European Environmental Policy, London

Countryside and Rights of Way Act (2000) http://www.legislation.gov.uk/ukpga/2000/37/contents. Accessed 14 Sept 2011

De Mey K, D'Haene K, Marchand F, Meul M, Lauwers L (2010) Learning through stakeholder involvement in the implementation of MOTIFS, an integrated assessment model for sustainable farming in Flanders. Int J Agric Sustain 9:350–363

DEFRA (2009) Soil protection review 2010 DEFRA, London

DEFRA (2010) Definitions of terms used in farm business management

EEA and JRC (2010) The European environment state and outlook 2010: soil. Luxembourg

Environment Agency (2007) Waterwise on the Farm. Environment Agency, Bristol

European Environment Agency (2009) Water resources across Europe – confronting water scarcity and drought, EEA report 2/2009

European Environment Agency (2010) The European environment state and outlook 2010: air pollution. EEA, Copenhagen

European Environment Agency (2011) Hazardous substances in Europe's fresh and marine waters, EEA report 8/2011. EEA, Copenhagen

European Network for Rural Development (2011) Public Goods and Rural Development. EC: Agriculture and Rural Development, European Union

Fleischer A, Tsur Y (2000) Measuring the recreational value of agricultural landscape. Eur Rev Agric Econ 27(3):358–398

Fowler SM, Frost D and de Carle C (2004) Environmental and biodiversity impacts of organic farming in the hills and uplands of Wales. Organic Centre Wales report: [online] http://orgprints.org/10828/1/envimpactsuplands.pdf. Accessed 14 Sep 2011. Aberystwyth University, Aberystwyth

Global Reporting Initiative (2010) http://www.globalreporting.org/AboutGRI/. Accessed 2 Dec 2010

Green RE, Cornell SJ, Scharlemann JPW, Balmford A (2005) Farming and the fate of wild nature. Science 307:550–555

Halberg N, Van der Werf HM, Basset-Mens C, Dalgaard R, de Boer IJM (2005a) Environmental assessment tools for the evaluation and improvement of European livestock production systems. Livest Prod Sci 96:33–50

Halberg N, Verschuur G, Goodlass G (2005b) Farm level environmental indicators: are they useful? Overview of green accounting systems for European farms. Agric Ecosyst Environ 105:195–212

Hani F, Braga F, Stampfli A, Keller T, Fischer M and Porsche H (2003) RISE, a tool for holistic sustainability assessment at the farm level. Int Food Agribusiness Manag Rev 6(4):78–90

ICAEW (2009) ICAEW advanced stage audit and assurance manual. ICAEW, London

Klein A, Vaissière BE, Cane JH, Steffan-Dewenter I, Cunningham SA, Kremen C, Tscharntke T (2007) Importance of pollinators in changing landscapes for world crops. Proc R Soc Lond B Biol Sci 274(1608):303–313

Kuratorium fur Technik und Bauwesen in der Landwirtschaft (2009) *Bewertung der Nachhaltigkeit Landwirtschaftlicher Betriebe.* Darmstadt, Germany

Lampkin N, Fowler SM, Jackson A, JeffreysI, LobleyM, MeasuresM, PadelS, Reed M, Roderick S and Woodward L (2006) Sustainability assessment for organic farming – integrating financial, environmental, social and animal welfare benchmarking. In: Atkinson C, Ball B, Davies DHK, Rees R, Russell G, Stockdale EA, Watson CA, Walker R, Younie D (eds) Aspects of applied biology 79, what will organic farming deliver? COR 2006, pp 9–13

Lobley M, Reed M and Butler A (2005) The impact of organic farming on the rural economy in England, final report to DEFRA. CRR research report no 11. University of Exeter, Exeter

Losey JE, Vaughan M (2006) The economic value of ecological services provided by insects. Bioscience 56(3):331–323

Measures M (2004) Farm auditing for sustainability. British Grassland Society/Harper Adams University College, Warwickshire/Shropshire

Meul M, Van Passel S, Nevens F, Dessein J, Rogge E, Mulier A, Van Hauwermeiren A (2008) MOTIFS: a monitoring tool for integrated farm sustainability. Agron Sustain Dev 28(2):321–332

National Institute of Statistics of Italy (2001) Agri-environmental indicators to describe agricultural sustainability. STATISTICAL COMMISSION and COMMISSION OF THE ECONOMIC COMMISSION FOR EUROPE EUROPEAN COMMUNITIES CONFERENCE OF EUROPEAN STATISTICIANS EUROSTAT Joint ECE/Eurostat Work Session on Methodological Issues of Environment Statistics (Ottawa, Canada, 1-4 October 2001) WORKING PAPER No. 21 Ottowa

Organic Research Centre (2010) OF0348 Final Report to Defra – Quality and environmental benchmarking for organic agriculture. ORC, Newbury

Organic Research Centre (2011) http://www.organicresearchcentre.com/manage/authincludes/article_uploads/Research/Resources/Brief%20summary%20of%20EASI.pdf. Accessed 23 Mar 2011

Organisation for Economic Co-operations and Development (1997) Environmental indicators for agriculture. OECD, London

Ovenden GN, Swash ARH, Smallshire D (1998) Agri-environment schemes and their contribution to the conservation of biodiversity in England. J Appl Ecol 35:955–960

Pimental D, Houser J, Preiss E, White O, Fang H, Mesnick L, Barsky T, Tariche S, Schreck J, Alpert S (1997) Water resources: agriculture, the environment, and society. Bioscience 47(2):97–106

Quality Meat Scotland (2010) Estimating greenhouse gas emissions from Scottish livestock enterprises, 2009 production cycle: a discussion paper. Quality Meat Scotland, Edinburgh

Ridley N, Woolley N (2002) A sustainability audit for farms in Suffolk

RISE (2009) Public goods from private land, Brussels, Belgium. http://www.risefoundation.eu/index.php?option=com_content&view=article&id=55&Itemid=93. Accessed 14 Sep 2011

RSPCA (2010a) RSPCA welfare standards for beef cattle. RSPCA, Horsham

RSPCA (2010b) RSPCA welfare standards for dairy cattle. RSPCA, Horsham

RSPCA (2010c) RSPCA welfare standards for laying hens and pullets. RSPCA, Horsham

RSPCA (2010d) RSPCA welfare standards for pigs. RSPCA, Horsham

RSPCA (2010e) RSPCA welfare standards for sheep. RSPCA, Horsham

Stolze M, Piorr A, Häring A, Dabbert S (2000) The environmental impacts of organic farming in Europe. University of Hohenheim, Stuttgart

University of Hertfordshire (2006) http://sitem.herts.ac.uk/aeru/ema/old/index.htm. Accessed 1 Dec 2010

University of Hertfordshire (2011) www.herts.ac.uk/aeru/impacct. Accessed 24 Jan 2011

Weiske A (2006) Selection and specification of technical and management-based greenhouse gas mitigation measures in agricultural production for modelling. Report on 6th framework work programme project on Impact of Environmental Agreements on the CAP, MEACAP

Williams AG, Audsley E, Sandars DL (2010) Environmental burdens of producing bread wheat, oilseed rape and potatoes in England and Wales using simulation and system modelling. Int J Life Cycle Assess 15:855–868

World Health Organisation (2011) http://www.who.int/trade/glossary/story028/en/. Accessed 5 Sept 2011

Zander K, Hamm U (2010) Consumer preferences for additional ethical attributes of organic food. Food Qual Pref 21:495–503

Pesticides and Sustainable Agriculture

Patrick Drogui and Pierre Lafrance

Abstract The global perspective of this review is to present, in a particular way, many dimensions related to the use of pesticides and sustainable development in agriculture. Worldwide increasing use of pesticides induces high and very variable environmental pressures depending on the countries, their major crops and pesticide needs and the types of management of cropping systems. As an example, in 2003, the global market for pesticides by worldwide regions was the following: European Union: 28%; North America: 25%; South America: 18%; Asia and Africa: 30%. Between 1993 and 1998, the growth rate for synthetic chemical pesticides was: Latin America: 5.4%; Africa/Middle East: 5.1%; Western Europe: 4.6%; Eastern Europe: 4.4%; Asia/Oceania: 4.4%; North America: 4.0%. Mitigation procedures to minimize the environmental pressures due to the growing use of pesticides must use one or many elements of crop management strategies. These management strategies must be adapted to the specific combination of crops and pesticides. In 1997, at a global scale, the sales, in percentage of the total markets in billions of dollars, of some of the major crops were: fruits/nuts/vegetables: 21.0%; home/garden/ornamentals: 17%; cereals: 12.9%; maize: 8.1%; rice: 8.1%. At the same time, the sales for the main pesticides chemical groups were: herbicides: 47.6%; insecticides: 29.4%; fungicides: 17.5%. The reduction of pesticide losses in surface water is essential

---

Patrick Drogui is a professor at INRS-ETE. His research activities are focused on the development of innovative technologies for environmental applications, mainly to remove persistent organic pollutants from waters and wastewaters by using electro-technology and oxidative processes.

Pierre Lafrance is a professor at INRS-ETE. His research activities aim at understanding the mechanisms responsible for the contamination of surface and groundwater by agricultural pesticides. His studies include the evaluation of the impacts of various agricultural practices on the losses of pesticides into surface and groundwater.

P. Drogui (✉) • P. Lafrance
Institut National de la Recherche Scientifique, Centre-Eau Terre et Environnement (INRS-ETE), Université du Québec, 490, rue de la Couronne, Québec G1K 9A9, Canada
e-mail: patrick.drogui@ete.inrs.ca; pierre.lafrance@ete.inrs.ca

E. Lichtfouse (ed.), *Farming for Food and Water Security*, Sustainable Agriculture Reviews 10, DOI 10.1007/978-94-007-4500-1_2,

to limit toxicological concerns and the adverse effects of pesticides on natural organisms and human health. Also, the development and application of new biopesticides and transgenic crops, which could be tolerant to herbicides and killing weeds or resistant to insects, should be considered in sustainable development. The market for various biopesticides and transgenic crops was expected to expand by over 20% between 2000 and 2005. Despite their interest, they may be a source of unplanned social and environmental issues related to the ecology of crops, insect or plant resistance, and use of more efficient and toxic chemical pesticides. Water and wastewater treatment plants may also be used to alleviate the negative impacts, mainly on human health, of water contaminated by pesticides.

We reviewed four of the major goals linked to pesticides and sustainable development in agriculture. These goals are (i) Production, markets and worldwide uses of pesticides, including biopesticides and genetically modified crops, (ii) Rejection of pesticides in the environment and mitigation procedures, (iii) Impacts of pesticides in the environment and human health concerns, and (iv) Management and treatment of water contaminated by pesticides. The major advances and trends for these goals are the following. Production and use of pesticides will increase with the worldwide demographic evolution. Establishing a long-term tendency of the needs for synthetic chemical pesticides at a global scale may be hazardous, but by using demographic tendency, it is expected that in 2050 the world will need 70% more food than in 2010, especially in Africa, India and Asia. A significant growth of chemical pesticides use is unavoidable. The last 20-year period was characterized by the use of various types of biopesticides and genetically modified crops, namely plant-pesticides resistance to insects, i.e. *Bacillus thuringiensis* pesticidal protein or *Bt* crops, and herbicide-tolerant plants. This was a pivotal step in the development of modern agriculture and it induces a tremendous change in the use of pesticides by converting growers to intensive production of major crops. The next 20 years are expected to see a substantial increase in the use of genetically modified plants. This should contribute to the implementation of integrated pest management systems such as no-tillage practices in order to minimize the contamination of water by pesticides. Genetically modified plants can actually reduce some of the harmful side effects of insecticides. However, actual observations made until 2010 suggest that the rapid diffusion of genetically modified *Bacillus thuringiensis*-based crops will lead to pest resistance. A similar problem occurs with some herbicide-tolerant crops, i.e. for glyphosate. The spreading of these crops could result in the transfer of their genetic qualities to weeds, creating new generations of weed-resistance to herbicides and thus reduce crop yields. These paradoxes will possibly change the market of pesticides as well as the contamination of waters. Adapted management approaches will be needed in order to preserve the soil quality, integrity of eco-agrosystems and human health. Management and mitigation strategies to minimize pesticide losses in the environment will have to be more efficient in all countries no matter their development status. Human health, much rarely addressed, would probably be recognized as a key component in sustainable development. Efficient treatments are needed to remove pesticides from waters. Conventional methods such as chlorination have shown inefficiency to fully oxidize pesticides. Alternative methods such as the combination of ozone and hydrogen peroxide are more reliable to eliminate pesticides. The use of powdered activated carbon can be efficient for punctual contamination of raw water. The integration of a double layer filtration unit, such as granular activated carbon and sand filtration can constitute an interesting alternative method to remove pesticides in existing drinking water treatment plants. Advanced electrolytic-oxidation techniques are promising treatments to remove pesticides from waters.

Keywords Pesticides • Production • Agriculture, Management • Control • Heath • Impact • Contamination • Treatment • Transgenic plants • Chlorination • Ozonation • Advanced Oxidation Process • Membrane filtration

1 Introduction

Agri-pharmaceutical products or pesticides are widely used for agricultural purposes in many countries around the world; they are employed on a large scale for pre- and post-emergence weed control on crops, railways and orchards. Indeed, they are used for their toxicity to fight against pests and adventitious. They are responsible for numerous cases of groundwater and river contamination in the surroundings of treated areas. In order to face the global food production required by demographic expansion and specific cultural needs, both developed and developing countries need to maintain their crops yields. It could be expected that the difficulties in the growth of agriculture while maintaining sustainable development would affect in the future many countries all over the continents. With such a possible scenario, the traditional concept or image of duality in food production between

Source: Statistics Canada (2001)

developed and developing countries, basically based on advanced technologies for the protection and high crops production, could substantially change in the next decade. The following typical picture of pesticide spraying in the field represents the performance for crop production which is increasingly a worldwide issue. This means that whatever the specific needs for the countries and regions in the world are, such as speciality or intensive crops productions, the use of pesticides would remain a free-border concern.

Several studies carried out by the environmental protection agencies of various countries such as Canada, France, and the USA reveal the presence of many pesticides in ground and surface waters. The pesticides most often encountered in surface waters are herbicides, e.g. atrazine and metolachlor, insecticides, e.g. carbofuran and diazinon and fungicides, e.g. captan and chlorotalonil, which are usually found at concentration levels ranging from 0.01 to 10 $\mu g\ L^{-1}$ in streams draining cultivated watersheds. Globally, a very low proportion of these pesticides participate to the total dissolved organic carbon from pollutants (DOC < 1%) of a contaminated water, but their presence in water has to be taken into account due to their potential toxicity for humans, such as brain cancer, immune and reproductive system troubles and aquatic species, e.g. feminization, toxicity for fish and invertebrates. Modern legislation in all countries imposes environmental regulation and health quality standards that constantly become more restrictive. Different strategies need to be adopted by municipalities and industries to comply with various refractory pollutants, like pesticides, of which some compounds are included in the endocrine disruptors category. Most of these pesticides are stable and difficult to oxidize using traditional biological methods and require chemical or physico-chemical treatment. Advanced oxidation processes are used for the treatment of toxic residual wastewaters that are difficult to biodegrade. These processes, including, O_3/H_2O_2, UV/O_3, UV/H_2O_2, H_2O_2/Fe^{2+}, aim at producing hydroxyl radical in water, a very powerful oxidant capable of oxidizing a wide range of organic compounds with one or many double bonds such as pesticides. In spite of good oxidation of refractory organic compounds, the complexity of advanced oxidation processes, high chemical consumption and relatively higher treatment cost of these methods constitute major barriers in the field application. Alternative methods such as using activated carbon, notably in powdered form, can be efficient in times of peak pesticides contamination. The combinations of ozone and granular activated carbon or ultra-filtration and powdered activated carbon are more reliable to eliminate pesticides from water. Another alternate method could be electrochemical oxidation for water and wastewater treatments. The electrochemical method takes advantage of coupling chemistry, e.g. *in situ* generation of oxidant with electronic science, e.g. electron transfer.

The goal of this paper is to review the recent published literature regarding pesticide management in a world controlled by intense industrialization and growing demography, which have caused severe environmental pollution with adverse impacts on water, soil and atmosphere. Specifically, objectives are to: (i) describe the use of pesticides globally and in specific countries, (ii) focus on the release of pesticides in water and their impact on environment and human health, (iii) provide a database for the treatment and management of water contaminated by pesticides,

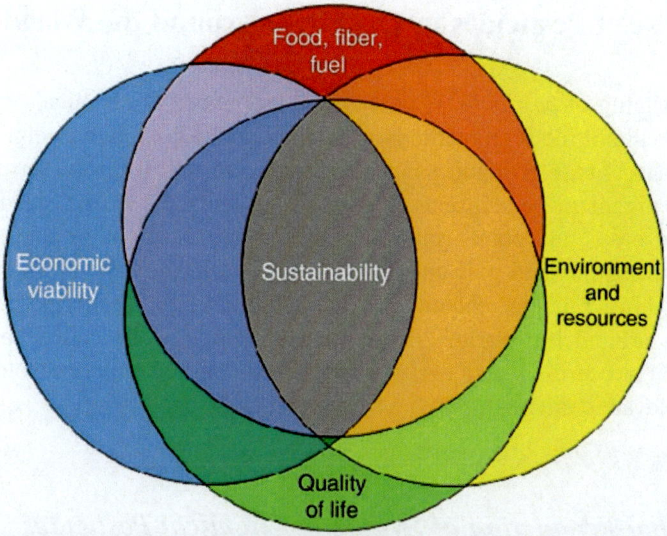

Fig. 1 Sustainability goals. The area where the four goals overlap represents the highest sustainability level (BANR 2010)

(iv) suggest new research orientations for the development of current and future management and pesticide control in the modern world.

Figure 1 shows that to be sustainable, a "global" or regional management system needs to balance four goals, illustrated by interlocking compartments. Our four abovementioned objectives are directly related to the compartments Environment and Resources and Quality of Life. Also, objectives indirectly address the compartment Economic Viability, thus contributing to attain a sustainable development.

Indeed, the extent of the use of pesticides in the world directly affects the components of agro-ecosystems and environmental health, including water quality in sources of drinking water, mainly in rural regions. If water contaminated by pesticides reaches private and municipal wells or streams used as drinking water supplies, then appropriate and efficient treatments are needed to remove these micropollutants that are likely to affect human health and quality of life. Pesticides may also directly reach storm or sanitary water systems. In such case, the inverse problems bring about similar goals. Wastewater treatment plants should be able to keep these contaminants from being rejected in streams, thus affecting the integrity of aquatic organisms and, ultimately, human quality of life. This occurs when pesticides are bio-accumulated and possibly bio-amplified in trophic levels from zooplankton to fish species, which are consumed after commercial fishing. The last objective is related to the economical viability of using strategies and techniques to reduce both use and losses of pesticides in the environment, as well as water and wastewater treatment costs. Ultimately, the losses of water contaminated by pesticides, perturbation and negative effects on aquatic and terrestrial fauna, and the costs needed to address the human health are part of this economic viability.

2 The Use of Pesticides in Countries Around the World

The first meaning of *pesticide* is "pest-killer" by the use of chemical or biological agents. We will not focus on methods such as inorganic compounds, e.g. arsenic, or antibacterial agents, e.g. liquid soaps. Similarly, we will limit the term "pest" to undesirable organisms, e.g. insects, nematodes, rodents and weeds causing damage to cultivated crops and plants, mainly in agriculture, horticulture and forestry or grass, e.g. turf grass and golf course. This approach allows avoiding the use of the generic term "biocide" encountered in some literature. The term "pesticide" is well-known around the world. "Biopesticides" may also be called "bio-rational products". Other terms likes "agri-chemicals" or "agri-pharmaceuticals" must be used in a relevant context.

2.1 Global Production of Synthetic Chemical Pesticides

Pesticides can be classified according to their target-organisms such as herbicides (weeds); insecticides (insects); fungicides (fungi or fungi spores); nematicides (parasitic nematodes); rodenticides (rodents); avicides (birds); algaecides (algae), etc. This classification according to their chemical group is also described in the literature. Indeed, atrazine is an herbicide from the group of chlorotriazine. Atrazine is widely used to stop pre- and post-emergence broadleaf and grassy weeds in major crops such as corn (*Zea mays*) and soybean (*Glycine max*).

A few words to consider the long term evolution of the US and other countries agricultural production over the last decades. The first and controversial generation of synthetic organic chemical pesticides was mainly associated with organochlorine compounds such as dichlorodiphenyltrichloroethane that appeared around 1940. Their appreciated insecticidal properties against paludism and other human diseases caused by parasites led to the era of synthetic organic chemical insecticides. Since 1948, synthetic pesticides have been used extensively because they provide many benefits to farmers and consumers. Fewer farmers on fewer farms produce more cultivated products. Major factors in the changing productivity patterns, either directly or indirectly, included the use of pesticides associated with more efficient agricultural practices. This combination, among others, allowed to continuously increase the U.S. crop productivity by a factor of near 2.5 over the 1948–2008 period. Establishing a long-term tendency of the needs for synthetic pesticides at a global scale may be hazardous, but by using demographic tendency it is expected that in 2050 the world will need 70% more food than in 2010, especially in Africa, India and Asia.

Table 1 indicates global pesticides sales as categorized by the major groups of chemicals, by crops and parts of the world in 1997. Pesticides represented here are mainly **synthetic organic compounds.** The global chemical pesticide market was about $31 billion in 1997. It was a mature market with a growth of about 1–2% per year (CLS 2000).

Table 1 Global chemical pesticide market (1997)

	Sales, billions of dollars	%
Product		
Herbicides	14.7	47.6
Insecticides	9.1	29.4
Fungicides	5.4	17.5
Others	1.7	5.5
Total	**30.9**	**100.0**
Crop		
Fruits, nuts, vegetables	6.5	21.0
Home and garden, turf, and ornamentals	5.25	17.0
Oil crops	1.75	5.7
Cotton	1.5	4.9
Cereals	4.0	12.9
Maize	2.5	8.1
Rice	2.5	8.1
Sugarbeet	1.0	3.2
Other	5.9	19.1
Total	**30.9**	**100.0**
Region		
North America	9.2	29.8
Western Europe	7.8	25.2
East Asia	7.1	23.0
Latin America	3.7	12.0
Rest of World	3.1	10.0
Worldwide total	**30.9**	**100.0**

CLS (2000)

These data are representative of the distribution of classical chemical pesticides sales by categories just a few years after the commercial use of **biopesticides**, such as transgenic pest-protected plants, started in the mid-1900s. This is thus a view of the situation before biopesticides (mainly transgenic crops) radically changed the sales and markets. The chemical group of herbicides was the greater sell with near 50% of the total market (Table 1). Within the classical chemical groups, insecticides are followed by a notable use of fungicides. Unsurprisingly, North America (near 30% of the global market), Western Europe (25%) and East Asia (23%) were the greatest consumers of pesticides. At the global scale, in 1997, the distribution of chemical pesticides by crops shows an integration of the major crops distributed around the world. Fruits and vegetables (21%) needed the highest amounts of pesticides. Indeed, these crops are distributed worldwide and require many applications of insecticides and fungicides. Costa Rica included in Latin America is an example of "limited" cropping areas which are dominated by the intensive culture of bananas that need a high pest control with these groups of pesticides. This country was, in the 2000s, the highest pesticide consumer in Latin America. It is thus useful to keep in mind that the release of pesticides in the environment is not only a function of the

Table 2 US chemical pesticide market by category (1997)

Product	Sales, billions of dollars	%
Herbicides	6.8	57.5
Insecticides	3.6	29.9
Fungicides and others	1.5	12.6
Total	**11.9**	**100.0**

Aspelin and Grube (1999) cited by CLS (2000)

global market, but also of the cultivated areas, the biogeography of main crops that can be very variable from a continent to another, the total pesticide application rates by crop, agricultural practices, policies for the conservation of resources and characteristics of the countries such as landscape, topography, pedology, climatology and hydrology.

Table 2 indicates U.S. pesticide sales as categorized by major groups of chemicals in 1997. The estimate for the U.S. is larger than others made for 1997 since it includes expenditures for non-agricultural pesticide uses (CLS 2000). This reflects application by owner/operators and custom/commercial applicators to industry, commercial and governmental facilities, land and homeowner applications to homes and gardens, including lawns. These pesticides have mostly been used since 1990 in and around urban areas for the maintenance of landscaping, e.g. lawns and golf courses. Their intensive use on restrictive land areas may be considered as a potential non-negligible source of water and wastewater contamination.

One must keep in mind that tendencies in the pesticide market may vary depending on the period of time and years considered (Fernandez-Cornejo and Just 2007). Table 3 shows the global growth rate of pesticides by regions between 1983 and 1998 during two intervals in this 15-years period. Between 1983 and 1993, the pesticide consumption and growth rate was much higher for the U.S. than between 1993 and 1998. Over these years, U.S. sales increased slowly in comparison with the 1983–1993 period. Following Yudelman et al. (1998), this change between the two periods may come from the difference between the real production and the expected and projected consumption for some regions.

The evolution of the U.S. expenditures in pesticides during the period of 1982–2001 is illustrated by Fig. 2 (Kiely et al. 2004). This typical pattern encountered for many developed countries should be viewed as a specific representation of a global evolution for all the market sectors. Also, these data do not take into account wood preservatives, specialty biocides, and chlorine/hypochlorites which are recorded in others databases. The temporal evolution illustrated by Fig. 2 is mostly due to the agricultural use of pesticides, mostly herbicides, with a growing need for herbicides after 1995 due to the introduction of transgenic crops and the increase of non-agricultural herbicides, e.g. from urban sources. Similar trend is observed in many developed countries since two decades.

In 2003, the global market for pesticides by worldwide regions was the following (Gorse and Dion 2007): European Union: 28%; North America: 25%; South America: 18%; Asia and Africa: 30%. The USA and China are the greatest pesticide consumers. It is projected that the supply of food in developing countries will have to consistently

Table 3 Global pesticide consumption, 1983–1998

Region	1983	1993	1998	Compound annual growth rate 1983–1993	1993–1998
	(US$ millions)			(%)	
North America	3,991	7,377	8,980	6.3	4.0
Latin America	1,258	2,307	3,000	6.3	5.4
Western Europe	5,847	7,173	9,000	2.1	4.6
Eastern Europe	2,898	2,571	3,190	−1.2	4.4
Africa/Mideast	942	1,258	1,610	2.9	5.1
Asia/Oceania	5,571	6,814	8,370	3.0	4.4
Total	20,507	27,500	34,150	3.0	4.4

Yuldeman et al. (1998)

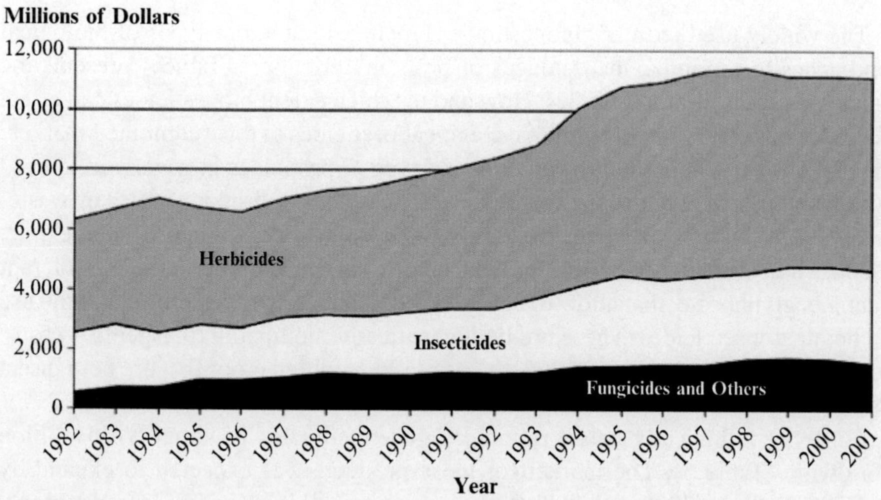

Fig. 2 Annual user expenditures on pesticides in the USA by pesticide type, 1982–2001 estimates, all market sectors

rise from the end of the 1990s to 2020 if the 6.5 billion population expected for Africa, Asia, and Latin America is reached, which would call for great global and regional efforts to protect crops and increase yields.

2.2 *Production of Biopesticides and Herbicide-Tolerant Crops*

The last 20-year period was characterized by field commercial uses for various types of biopesticides and genetically modified organisms, i.e. plants, which is a pivotal step in the development of modern agriculture. This induced a tremendous change in the use of pesticides by converting growers to intensive production of major crops at a large scale.

Table 4 Major categories of biopesticides

Product	Definition	Examples
Microbial pesticides	Microorganisms that operate as the active ingredient	Bacteria, fungi, viruses, virus coat proteins
Plant-pesticides	Substances that plants produce from genetic material that has been added to them	*Bacillus thuringiensis* pesticidal protein, potato leaf roll virus resistance gene produced in potato plants
Biochemical pesticides	Naturally occurring substances that control pests by nontoxic mechanisms	Pheromones, floral attractants and plant volatiles, natural insect-growth regulators, plant-growth regulators and herbicides

CLS (2000)

The widely used term of "biopesticides" includes numerous types of biological approaches to minimize the damages of pests in agriculture. Table 4 presents the major categories of biopesticides. How and to which extent biopesticides can affect the losses towards water of synthetic chemical pesticides in this "combined market" remains unsure. Changes in the amount and type of pesticides in water would need to be taken into account in the future for the treatment of drinking water or wastewater. From Table 4, we refer here only to one category of plant-pesticides, i.e. *Bacillus thuringiensis* pesticidal protein or *Bt* crop, and to one herbicide-tolerant plant, i.e. glyphosate, that allow to considerably reduce (and even eliminate) the use of chemical insecticides. These are both genetically modified – transgenic – crops. Commercial cultivation also includes genetically modified crops that are both insect resistant and herbicide tolerant.

In 1999, biologically based pesticides generated sales of about $700 million worldwide (Table 5). The market for these products was expected to expand by over 20% between 2000 and 2005. The most successful biopesticide is undoubtedly the transgenic crop *Bacillus thuringiensis*-based crop and over 40% of *Bacillus thuringiensis* sales are made in the USA. The rapid growth of genetically modified *Bacillus thuringiensis*-based crop is viewed as an alternative for competitive chemical products that are being banned or restricted. Herbicide-tolerant crops have also considerably modified the market at the global scale.

According to Yuldeman et al. (1998), biopesticides have under 0.45% of the market share of the multi-billion dollars agro-chemical market. Literature showed that the use of biopesticides in various countries represents less than a few percent of the markets. In 2008, genetically modified or transgenic crops were planted on a global area of 125 million hectares in 25 countries (Dymond and Hurr 2010).

Plant-pesticides and herbicide-tolerant plants considerably modified the choice and applications of conventional chemical pesticides. In agriculture, the genetically modified *Bacillus thuringiensis*-based crops include vegetables such as corn, e.g. *Zea mays*, soybean, e.g. *Glycine max* and other crops. This convinced growers to adopt systemic chemical herbicides such as glyphosate, e.g. a phosphonoglycine and glufosinate, e.g. a phosphorylated amino acid which are non-selective and

Table 5 Global biopesticide market (in millions of dollars)

	Year				
Market	1997	1998	1999	2004[a]	% change (1999–2004)
Microbial	65	66	67	72	7.5
Transgenic plants	405	429	455	610	34.1
Miscellaneous	180	184	188	208	10.6
Total	650	679	710	890	25.4

CLS (2000)
[a]Estimated

efficient in controlling many weed varieties. Glyphosate is currently the world's best selling herbicide, used in over 90 countries and on over 150 crops. The growing use of herbicide-tolerant plants has profoundly changed the herbicide use profile away from atrazine, metribuzin and alachlor to the benefits of glyphosate. This has the advantage of reducing losses of classical herbicides by runoff into drinking water sources.

The history of pesticides evolution in agriculture is marked by key developments in the production of new and efficient compounds or plants to increase crops yield which finally induce or lead to unexpected and new agricultural, environmental and health concerns. The following picture is a typical view of the tremendous efficacy of the growth of a contemporay herbicide-(glyphosate)-tolerant plant while weeds are eradicated. Nevertheless, the components of agro-ecosystems often showed their weakness to substantial perturbation as well as their capacities to adapt to major changes. The effect of both of these tendencies may result in unforeseen consequences.

Source: Unidentified

3 Rejection of Pesticides in the Environment and Mitigation Procedures

There are two primary sources of water contamination by pesticides: diffuse and punctual sources. Diffuse sources are characterized by the application of pesticides on soil at moderate to low rates. Moderate loads are then distributed over large areas in watersheds; pesticides are likely to be found in little amounts in streams but losses are widespread. Inversely, punctual contamination from industries and urban areas is characterized by a great or high application of pesticides on soil over "limited" areas. The loads are then high but initially affect restricted areas. There is no precise technical distinction between the types of contamination. Depending on the type of the concern addressed, one may view a high or dense amount of punctual sources widely distributed in large areas as a "diffuse" source. In all cases, the contamination of water by pesticides may affect the quality of the environment and human health.

3.1 Diffuse Source of Contamination: Agriculture

In agriculture, since pesticide concentrations found in streams are highly variable, often from 1 ng L^{-1} to 100 µg L^{-1}, it is difficult to assess the total amounts of dissolved and particle-sorbed pesticides (soil erosion) that are released and transported in the streams at the field and the watershed scale. Obviously, the sedimentation of soil-sorbed pesticides has an impact *in situ* near the release of pesticides in water, although dissolved pesticides are more likely to move to surface- and groundwater, which are pumped by users with or without appropriate drinking water treatment.

Pesticide releases are the results of their intrinsic properties and the extrinsic conditions that interact between them under variable spatiotemporal patterns. Pesticide properties will determine the compounds volatilization, e.g. vapour pressure, the soil adsorption capacity, e.g. solubility and soil organic carbon/water distribution coefficient or K_{oc} and the extent of chemical, e.g. hydrolysis, oxidation or biological degradation in soil, e.g. persistence and half-life time. These are the main factors that will affect the attenuation processes of pesticides in terms of their mobility with water and the concentrations available for the transport by runoff and drainage. The extrinsic conditions are, among others, the landscape, e.g. pedology, infiltration, slopes and hydrological networks, the distribution of cultures in the watershed, e.g. types of crops with or without inter-rows, tile drainage, natural drainage and ditches, the nature of pesticide treatments, the regional and local meteorology. The latter refer to the number and duration/intensity of rainfalls, delay between pesticide application(s) and earlier rainfalls following the pesticides applications.

3.1.1 Release of Agri-Chemicals into the Environment

The main source of diffuse stream contamination by pesticides is surface runoff. However, one should keep in mind that hypodermic and subsurface, e.g. below the root zone of plants, flows might contribute to export some pesticides. This pathway has not been extensively studied at a field or watershed scale for different pedological and topographical conditions. Another under documented pathway is the infiltration of water and pesticides in groundwater and their possible release in streams from tile drainage and eventually by natural drainage. On the other hand, numerous experiments have reported the effects of different agricultural practices, including pesticide treatments, on the losses of pesticides in controlled experimental plots of small and various cultivated areas, e.g. a few hundred of square meters. Many of these studies at the plot scale were made under natural rainfall or artificial rainfall, or under irrigation systems, mainly since the 1980s. Obviously, these studies do not reproduce the characteristics, the behaviour and the interactions of various conditions encountered at the watershed scale. Nevertheless, they are useful to better understand the pathways and mechanisms of pesticide losses under particular circumstance.

Quantitatively, a review of the literature generally showed that pesticide masses exported with surface runoff account for 1% or less of the total masses of pesticides applied in the field or in experimental plots (Wauchope 1978; Flury 1996). An unusual worst case reported by the review of Wauchope (1978) indicates a loss of atrazine and metolachlor of 2–4% of the masses applied under unfavourable conditions of slope and rainfalls. At the field scale (ten fields from 28 to 60 ha) in California, USA, Spencer et al. (1985) observed that, under irrigation water systems, losses of herbicides accounted for 1–2% of the masses applied, while the exported loads of insecticides were of less than 1%. In this study, tile drain effluents contributed in an insignificant amount to losses of the studied pesticides. Such a result is consistent with literature, which generally shows small concentrations in tile drainage. In some cases the volume of drainage water is higher than that of runoff water. Since pesticide concentrations are higher in runoff than in drainage, masses of exported pesticides are greater in runoff (Gaynor et al. 1995).

In Québec, Canada, a study compared the release of atrazine and metolachlor at the field-scale (from 1 to 9 ha) under natural rainfall using conventional and no-tillage practices for corn growth (Lafrance et al. 1997). This 2-year study was performed directly in the field using the machinery of the growers. During the two first rainfalls that followed herbicide applications using continuous monitoring during the rainfalls, the concentrations leaking from two fields by runoff were very variable and reached concentrations as high as 60–500 μg L^{-1} for no-tillage practice and 130–2,400 μg L^{-1} for conventional tillage. This suggests that soil conservation tillage (no-till) contributes in reducing pesticide concentrations leaving the fields with runoff water. This assumption was more realistic when applying a reasonable runoff coefficient for the region and comparing mass losses both by runoff and by drainage.

The study also confirmed that concentrations in tile drainage were low; under 6 μ L^{-1} for two fields, e.g. clay soil, but reaching 40–60 μg L^{-1} for other ones. The later concentrations observed in tile drainage were by far higher than Canadian criteria to maintain the quality of aquatic life. A combination of different slopes, soil textures, tillage practices and meteorological conditions may account for these results. In comparison with runoff, tile drainage does not greatly contribute to the overall pesticide masses exported to surface water. The patterns of herbicide concentrations in surface runoff and tile drainage during the rainfalls were also compared. It was shown that the rapid increase of herbicide concentrations in runoff followed the pattern of the cumulative quantity of water during the two rainfalls. A similar temporal evolution of herbicide concentrations in tile drainage was observed after a tile drain response from 6 to 12 h depending on soil texture. This showed at the field-scale the relative contributions of surface and drainage waters to the exportation of herbicides into streams.

One of the most important factors driving the transport of pesticides from the field to surface water is the delay between pesticide application and the first rainfall events that followed the application and their intensity (Lafrance et al. 2001). An exponential decay of pesticides concentrations in the runoff during the first and sometimes the second rainfall event that followed the application of atrazine and metolachlor was observed at the plot scale with no crop, e.g. bare soil, and under natural rainfall. Such a pattern in the decrease of concentrations during the first rainfall following application was sometimes similar to the one observed between a limited and successive number of rainfall events. By using the same protocol, Lafrance et al. (1993–1996) observed that the major losses of these herbicides occurred during the first rainfall following application; up to over 80% of the total herbicide losses during the period studied may occur during the first rainfall. This pattern was also observed in various other studies. Such a transient behaviour of the pesticide export in terms of pluviometry and concentrations peaks in the streams must be taken into account when evaluating impacts on aquatic environments.

In the next section, mitigation strategies to reduce pesticide losses into water will be presented. One of these measures is the use of vegetative buffer strips which can efficiently reduce surface runoff and thus both dissolved and particle-sorbed pesticides. Many studies have addressed this subject (Reichenberger et al. 2007). We present here a particular study made at the plot scale and under natural rainfall to illustrate the role of runoff and subsurface drainage in pesticide losses passing through those grassy buffer strips. In Québec, Canada, the study (Caron et al. 2007) was conducted to evaluate the efficiency of buffer strips, e.g. 5 m length, to attenuate losses of dissolved atrazine and metolachlor used in corn crops in both runoff and subsurface drainage, e.g. below the root zone; depth of 90 cm. This subsurface drainage system did not correspond to tile drainage and was placed just below the buffer strips; the objective was to compare the contribution of surface runoff passing through the strip and subsurface infiltration under the strip to pesticide losses. A control plot was used as a reference for buffer strips efficiency.

In 2004, the first three rainfall events that followed the herbicide application were monitored. Considering the total of the three rainfalls, the buffer strips reduced the dissolved masses of herbicide exported in runoff by 40–60% but increased by

800–1,100% the dissolved masses in subsurface drainage, i.e. depth 90 cm. In such conditions, 84% of the sediments were removed from runoff. The infiltration of water and herbicides in the grassed strips greatly increased the contribution of subsurface drainage to herbicide losses. However, compared with the control plot (no buffer strip), total losses of herbicides in runoff+subsurface water are considerably lower in the presence of grassed buffer strips. This study was completed in 2005 by using the same experimental design and conditions but with a different natural rainfall pattern. It was observed (Caron et al. 2010), that buffer strips reduced the total of mass losses of dissolved herbicides by 75–95%. Previous observations made in 2004 were confirmed in 2005 despite different rainfall patterns. One may note that in 2004 the relative contribution of runoff and subsurface drainage to the mean masses released for the three rainfalls was established for both atrazine and metolachlor. For both herbicides, surface runoff in the control plot accounted for over 95% of mass losses, indicating that low infiltration occurred in the absence of buffer strips. On the other hand, buffer strips slow down runoff and favour the infiltration of water and herbicides; the contribution of subsurface drainage was then of 40–45% for the mean masses of both herbicides released.

It is important to note that the "mass" of pesticides exported is a good indication of the percentage of applied pesticide released. Despite this useful information, environmental studies use pesticide concentration for comparison with any criterion, recommendation or standard for water use or environmental protection. For most agro-environments on a watershed scale, the mass of pesticides released from a field into water is diluted in the hydrological network, generally resulting in concentrations lower than those of the surface runoff leaving the field. This makes it difficult to evaluate the level of pesticide concentrations in rivers, which are a source of drinking water. Since the level is site- and time-specific, an approach other than experimental studies should be used to predict pesticide concentrations at the watershed scale; various modeling tools have been developed in the past decades. For example, Rousseau et al. (2011) used a semi-distributed hydrological model to evaluate these concentrations at any point of the hydrological network. As with experimental studies, the modelling of pesticide transport in streams has limitations. Nevertheless, this model was applied to six Canadian watersheds across the country, each having different characteristics, crops and pesticides. Results confirmed that the masses of pesticides reaching the six watersheds outlets represented less than 1% of the mass applied to cultivated areas. Rapid progress in pesticide transport models is expected, and such tools would be helpful for the planning and implementation of mitigation strategies in order to reduce pesticide losses to water.

3.1.2 Strategies for Controlling the Release of Agri-Pesticides in the Environment

Conventional intensive agriculture allowed increasing the yields of crops, but also generated problems of agro-ecosystem degradation. Among others, soil erosion and losses of soil quality as well as the export into water of nutriments, pesticides and other compounds or microorganisms, had a harmful impact on water quality. Since the 1960s, agricultural practices have been developed in order to improve the environmental performance of conventional agriculture, namely the Integrated Pest Management. The Integrated Pest Managements were developed to better preserve the environment while trying to maintain the industry's productivity and economical performance. Integrated Pest Management practices can be used in combination with other strategies that aim at reducing the application of pesticides as well as herbicide-tolerant weeds. More recently, Best Management Practices were developed at the field and watershed scales as a conservation program for agro-ecosystems including both resources, e.g. soil and water, and their wildlife.

BANR (2010) summarized these conservation practices. They are presented below and adapted to better show their impacts on the use or release of pesticides in water.

- **Crop rotation**, which involves the successive planting of different crops on the same lands in sequential seasons to avoid the build-up of pathogens and pests, as well as pest resistance, which often occurs when one species is continuously cropped.
- **Cover crops** as part of a crop rotation, which involves the planting of crop varieties that can potentially protect fields from soil erosion, reduce weeds and pesticide use.
- **Reduced-tillage** and **no-till practices**, in which a crop is planted directly into a seedbed not tilled. This minimizes soil disturbance and leaves plant residues on the surface of fields after harvest, increases pesticide retention in soil and reduces pesticide losses by runoff;
- **Integrated pest management**, which involves the strategic use of complementary practices—including cultural, mechanical, biological, ecological, and chemical control methods—to keep pest levels below critical economic thresholds.
- **Precision farming** practices, which combine detailed spatial information about soil conditions and indicators of crop performance to target fertilization and possibly pesticide applications.
- **Diversification of farm enterprises**, which helps increasing biodiversity, controlling pests and diseases, and reducing risks from climatic and market volatility.
- **Other agricultural conservation best management practices** includes the use of buffer or filter strips, grass waterways, riparian area access management, manure handling and management, wildlife habitat enhancement within agricultural landscapes, composting and increasing irrigation water use efficiency.
- The development of crops and animals that have **enhanced genetic resistance** to climatic extremes, pests and other threats, often using new genetic engineering tools.

These practices combine mechanical, cultural, chemical, biological, and other non-chemical methods of pest control. However, herbicides may affect sensitive plants in crop rotation. Also, weed populations may always develop a resistance following the repetitive application of the same herbicides, leading to decreased crop yields. Because growers expect to maintain their productivity, they will probably continue to use chemical herbicides as an important method of weed control. The integration of chemical and nonchemical techniques will increase the complexity of weed management. From the "ecological-based" practices to intensive "conventional" ones, there is a wide range of combined practices that would be site-specific and year-specific. The best combination of practices at a large scale has not been extensively documented yet regarding losses of pesticides into water. Nevertheless, Reichenberger et al. (2007) addressed in details the efficiency of some of these practices, including **best management practices** and especially grassed buffer strips. The buffer strips are a promising tool to protect water quality, provided that they are properly designed and have a sufficient length of vegetation. Indeed, in Québec, Canada, it was found at the plot scale that under moderate rainfall, a strip of 3 m had the same efficiency to reduce atrazine and metolachlor concentrations in runoff than a 6 and a 9 m buffer strips (Lafrance et al. 2001) The main three ways to control pests (see the conservation practices above) are to: (i) reduce the loads or application rates of pesticides (local applications where weeds are present; application on the seeded rows for inter-row crops; crop rotations, etc.); (ii) limit their transport out of the field, e.g. no-tillage that retain pesticides by adsorption on crop residues left on the ground and that reduce pesticide runoff; and (iii) implement vegetative barriers such as buffer strips that attenuate surface pathways which greatly contribute in the export of soil particles, water and pesticides into streams.

3.2 Punctual Sources of Contamination: Release of Pesticides from Urban Areas

Punctual sources of contamination may lead to direct or indirect exposure, e.g. skin contact, inhalation, ingestion, eating food containing pesticide residues. This is mainly observed in the continuum of synthetic pesticide production to pesticide transportation and distribution and finally spraying procedures, e.g. severe cases of aerial drift in small areas. Losses of pesticides that occur during their industrial manufacturing are likely to severely affect the vicinity of the production sites at both environmental and human health levels. However, and in most cases, real losses are often perceived to be underestimated by companies and it is arduous to quantify the amounts rejected in the environment and sewer systems. It was not the purpose here to make an historical inventory of punctual and episodic contaminations due to pesticide manufacturing and transport since many contamination cases are unique examples which must be viewed in their wide range of circumstances. We will thus only consider pesticide contamination sources in urban areas and their potential to reach sewer systems and water and wastewater treatment plants.

Few global or regional studies are found in the literature with regards to the release in water of chemical pesticides from urban areas. CLS (2000) addressed this possibly important concern. Residential pest control is applied by consumers for aesthetic or recreational purposes (Templeton et al. 1998). In 1995, about 12% of households in the USA dealt with lawn-care companies for fertilizer or pesticide application. In 2000, almost 80% of the expenditures for pesticides in the USA were for agricultural purposes (Kiely et al. 2004). However, two non-agricultural sectors, namely "industry/commercial/government" and "residential home and garden" appeared elusive in urban areas. In 2001, the first sector used 12%, 16% and 21% of the total herbicides, insecticides and fungicides, respectively, while the second sector used 10%, 41% and 6% respectively. Health effects on homeowners of pesticide exposure are very difficult to calculate. There are no statistics available to quantify pesticide intoxications in this sector (CLS 2000). Moreover, losses of pesticides from these urban activities in both storm and sanitary water systems remain unclear and need to be documented.

Todd (2010) reported a study that addressed the losses of pesticides in the streams draining mostly urban watersheds. In 2009, the sale and use of pesticides for cosmetic purposes were banned in Ontario, Canada. Pesticide concentrations were monitored in ten urban streams before (2008) and after (2009) the ban took effect to verify concentration changes after the ban. A total of 105 pesticides and their degradates were monitored. Among these watersheds (1.4–60 km^2), urban lands accounted for 35–97% of the surface of the watersheds. In 2008, 36 pesticides were detected at a concentration >1 $\mu g\ L^{-1}$. Two or more pesticides were found in all samples. Herbicides 2,4-dichlorophenoxyacetic acid, dicamba, glyphosate and methylchlorophenoxypropionic acid and the insecticide carbaryl were the dominant detected pesticides. In 2009, concentrations of 2,4-dichlorophenoxyacetic acid, dicamba, methylchlorophenoxypropionic acid, total phenoxy herbicides and total insecticides were significantly lower in comparison with 2008 and a decrease in carbaryl concentrations was noticeable. Pluviometry was similar in 2008 and 2009. This led to the cautious conclusion that the reduction in pesticide use after the cosmetic pesticides ban was responsible for the changes in stream concentrations. Water quality criteria for the protection of aquatic life were developed for over half the pesticides that were detected at a concentration >1 $ng\ L^{-1}$. Pesticide concentrations in urban stream waters rarely exceeded these criteria. In 2008, carbaryl exceeded a criterion in 12.5% of the samples, permethrin 4.2% and total phenoxy herbicides 3.4%. The only pesticide to exceed a criterion in 2009 was the insecticide permethrin. This suggests that the ban of cosmetic pesticides had a rapid effect on the quality of urban streams. One may also note that other studies have concluded that atmospheric deposition may be largely responsible for the presence of some pesticides and their degradates in urban streams, such as the widely used herbicide atrazine (Phillips and Bode, 2004 cited by Todd 2010). The major pathway for these herbicides to enter surface waters is runoff, especially in urban landscapes where impervious surfaces promote runoff and storm water systems increase the efficiency of drainage. The highest and lowest concentrations of pesticides in urban streams generally occur during storm flow and base flow, respectively (Phillips and Bode 2004 cited by Todd 2010). Most of the individual pesticides measured were below their respective criteria. However, some chemical mixtures may have additive, synergistic or antagonistic toxic effects. Pesticide mixtures at low concentrations are very often observed in surface water and their effects on aquatic life, which are of great environmental concern all around the world, are the subject of ongoing research.

4 Impacts of Pesticides in the Environment and Human Concerns

A considerable amount of literature has examined both toxicological and ecotoxicological (*in situ*) direct and indirect impacts of many pesticides on numerous aquatic and terrestrial organisms living in or near the cultivated regions of watersheds, e.g. diffuse source. The approach adopted here will be to consider some environmental and especially human health concerns related to pesticide contaminations. Indeed, the "ecosystem and eco-health" approach is based on human health, which is the result of many inter-linked factors (Lebel 2003). This approach was defined by the International Development Research Centre, a Canadian governmental agency which supports research in developing countries to promote growth and development. Briefly, human health is viewed as the most important indicator of the degradation of environmental quality. This means that sociological, ethnological, cultural, industrial, economical, environmental, ecological, biomedical, sanitary and other aspects are likely to disturb human health and world sustainable development. This approach is similar to the schematic definition of "sustainable development" presented in Fig. 1, except that the global concern here is: "May human health be preserved if the environment suffers from problems caused by a dysfunction of one or many disciplines mentioned above".

Yuldeman et al. (1998) addressed the wide spread presumption that chemical pesticides are harmful to human health and the environment. Organochlorines are highly toxic synthetic pesticides of the first generation. They are susceptible to adversely affect human health even today since their persistent residues are often present in aquatic and terrestrial environments. The use of some of them even today is known to have endocrine-disruptor properties and can cause cancer, birth defects, male sterility, genetic mutations and behavioral changes. Pesticides can also alter human health by causing allergies or breathing trouble or by affecting the liver, kidneys and nervous system. Air borne pesticides may fall down near their sources or may be submitted to long-range transport in Nordic regions and in the Arctic. In both cases, bio-concentration by organisms and bioamplification along the trophic levels are frequent following the predator-prey relationships. This is particularly remarkable in the Arctic where no pesticides are produced and where air borne insecticides are accumulated in fish, seals, whales, polar bears, and finally in mothers' breast milk fat at levels higher than for the mothers who live in developed countries and are submitted to more environmental pollution sources. Yuldeman et al. (1998) summarized reports related to the effect of pesticides on the endocrine system of both humans and wildlife.

A study by the World Health Organization (WHO 1972 cited by Yuldeman et al. 1998) estimated that 500,000 annual pesticide poisonings and about 5,000 deaths occurred globally (Farah 1994, cited by Yuldeman et al. 1998). In 1989, WHO and the United Nations Environment Program estimated that there were one million human pesticide poisonings annually, with about 20,000 deaths (WHO/UNEP 1989 cited by Yuldeman et al. 1998). It was suggested that occupational pesticide poisonings may affect 25 million people or 3% of the agricultural workforce worldwide each year, and may lead to three millions severe poisonings a year with 220,000 deaths (Jeyaratnam 1987, cited by Yuldeman et al. 1998). Poisonings from pesticides appear to be high in both developed and developing countries.

These data concerning poisoning and human deaths in the agricultural industry should be taken cautiously. Such data are arduous to collect and are often disclosed many years after the beginning of investigations. Also, it is tricky to distinguish all the causes of these human injuries, e.g. accidents, human errors, unadapted equipment, working conditions imposed by landowners or by the markets. Subsistence agriculture made without safety rules in poor regions of developing countries is an example of a lack of educational programs and social responsibility. Also, the economical needs in some of these developing countries, e.g. in Africa, led some governments to allow intensive agricultural activities that are entirely supervised by foreign companies coming from emergent countries. Even today, after 2010, the local population is relocated to areas of traditional production. In parallel, the totality of the industrial crop yields produced by these foreign industries is exported and leave local populations to subsist with poor cultivation lands and areas contaminated with pesticides. Such cases illustrate a relationship between political, economical, social, educational and environmental concerns which affect human health. Sustainable development includes many of these aspects. Disrupting one of them will probably lead to impacts on the environmental quality and thus on human health.

5 Management and Treatment of Water Contaminated by Pesticides

Modern legislation in several countries imposes environmental regulations and heath quality standards that steadily become more restrictive. The World Health Organization (WHO 1988) has set the recommended level of pesticides in drinking water varying from 0.03 to 300 μg L^{-1} depending on the type of pesticides. The Table 6 compares the guidelines from U.S.A., Canada, France and New Zealand with those recommended by World Health Organization. Guidelines proposed by World Health Organization are more restrictive for the following pesticides: aldrin/dieldrin, chlordane and cyanazine for which 0.03, 0.2 and 0.6 μg L^{-1} are respectively proposed. The highest concentration of 300 μg L^{-1} is proposed for bentazone, probably less toxic than the others. By comparison, Canadian Environment Quality Guidelines propose 0.7 μg L^{-1} for aldrin/dieldrin and 10 μg L^{-1} for cyanazine. For the herbicide atrazine, most often encountered in surface waters, a limit value of 2.0 μg L^{-1} is recommended by WHO, compared with 5.0, 3.0, 0.1 and 2.0 μg L^{-1} recommended by Canada, U.S.A., France and N. Zealand respectively.

It is worth noting that the standards imposed in each country take into account local realities. In all cases, water treatment plants should be adapted to efficiently remove pesticides to meet standard regulations. Tables 7 and 8 respectively compare some physicochemical and oxidation processes used for pesticide removal from water.

Table 6 Water quality guidelines for pesticides

	Countries				WHO[e]
Pesticides (μg L^{-1})	USA[a]	Canada[b]	France[c]	New Zealand[d]	
Alachlor	2.0	–		20	20
Aldicarb	7.0	9.0		10	10
Aldrin/dieldrin	–	0.7	0.03	0.03	0.03
Atrazine	3.0	5.0	0.1	2.0	2.0
Bentazone	–	–	–	400	300
Carbofuran	40	90	–	8.0	7.0
Chlordane	2.0	–	–	0.2	0.2
Cyanazine	–	10	–	0.7	0.6
Cyclodienes	–	–	–	–	–
2,4dichloro-phenoxyacetic	70	100	–	40	30
Dichlorodiphenyl-trichloroethane	–	–	–	–	2.0
Diquat	20	70	–	10	10
Diuron	–	150	–	20	–
Ethylene Dibromide	0.05	–	–	–	0.4–15
Fenoprop (2,4,5-TP)	50	–	–	10	9.0
Heptachlor	0.4	–	–	–	–
Hexachlorobenzene	1.0	–	0.01	1.0	1.0
Lindane	0.2	–	0.1	2.0	2.0
Methoxychlor	40	900	–	20	20
Metolachlor	–	50	–	10	10
Pendimethalin	–	–	–	20	20
Pentachlorophenol	1.0	60	–	10	9.0
Permethrin	–	–	–	20	20
Picloram	500	190	–	20	–
Simazine	4.0	10	0.1	2.0	2.0
Terbufos	–	1.0	–	–	–
Trifluralin	–	45	–	30	20

[a] EPA, U.S. Environmental Protection Agency (2000). Office of Water 4304. EPA 822-B-00-001
[b] Environment Canada (2001). www.ec.gc.ca/ceqg-rcqe/prot.htm
[c] European Union (1997). Interinstitutional file 95/0010 (SYN)
[d] Ministry of Health, New Zeeland (2000). www.moh.govt.nz
[e] WHO, World Health Organization. Document WHO/PEP/89.4 (1988). Food Chem. Toxicol. 38, S87–S90 (2000)

5.1 Conventional Techniques Used for Pesticide Removal

5.1.1 Coagulation-Flocculation

The coagulation-flocculation process may be defined as breaking the stability of dispersed or suspended colloids by neutralizing the potential energy of repulsion between charged particles using Al^{3+} or Fe^{3+} as coagulating agent, followed by floc formation. It is a set of operations to remove solids and colloids from raw water,

Table 7 Comparison of different physicochemical processes used for pesticide removal from water

Processes	Pesticides	Operating conditions	Removal rate (%)	References
Coagulation-flocculation	Alachlor	$[Alachlor]_0 = 45\ \mu g\ L^{-1}$; $[Alum] = 15\ mg\ L^{-1}$	4.0	Sheoran (2008)
	Metachlor	$[Metachlor]_0 = 34\ \mu g\ L^{-1}$; $[Alum] = 30\ mg\ L^{-1}$	11	Sheoran (2008)
	Dodine	$[Dodine]_0 = 250\ \mu g\ L^{-1}$; $[FeCl_3] = 10–100\ mg\ L^{-1}$	15	Kouras et al. (1995)
	Simazine	$[Simazine]_0 = 554\ ng\ L^{-1}$;$[Alum] = 10–40\ mg\ L^{-1}$ Al	20	Ormad et al. (2008)
	Atrazine	$[Atrazine]_0 = 551\ ng\ L^{-1}$;$[Alum] = 10–400\ mg\ L^{-1}$ Al	10–25	Ormad et al. (2008)
	Lindane	$[Lindane]_0 = 10\ \mu g\ L^{-1}$; $[FeCl_3] = 0.06–0.62$ mM Fe or Al $[Al(OH)_xCl_y(SO_4)_2]_n = 0.1\ mg\ L^{-1}$	8	Kouras et al. (1998)
Powdered activated carbon	Lindane	$[Lindane]_0 = 10\ \mu g\ L^{-1}$; $[PAC] = 20\ mgL^{-1}$	>99	Kouras et al. (1998)
	Simazine	$[Simazine]_0 = 554\ ng\ L^{-1}$; $[PAC] = 10\ mgL^{-1}$	55	Ormad et al. (2008)
	Atrazine	$[Atrazine]_0 = 551\ ng\ L^{-1}$; $[PAC] = 10\ mgL^{-1}$	55	Ormad et al. (2008)
Coagulation-flocculation Powdered activated carbon	Dodine	$[Dodine]_0 = 250\ \mu gL^{-1}$; $[FeCl_3] = 10–100\ mg\ L^{-1} + [PAC] = 100\ mg\ L^{-1}$	98	Kouras et al. (1995)
Granular activated carbon	Lindane	$[Lindane]_0 = 10\ mg\ L^{-1}$; $[GAC] = 62\ mg\ L^{-1}$	99.8	Sotelo et al. (2002a)
	Atrazine	$[Atrazine]_0 = 4\ \mu g\ L^{-1}$; flow rate = 90 $L\,h^{-1}$; GAC bed depth = 100 cm; column diameter = 14 cm	90	Knappe et al. (1997)
	Malathion	$[Malathion]_0 = 7\ \mu g\ L^{-1}$; $[GAC] = 50\ g\,L^{-1}$	90	Jusoh et al. (2011)
Membrane	Simazine	$[Simazine]_0 = 100–400\ ng\ L^{-1}$; Nanofiltration (NF70)	50–100	Agbekodo et al. (1996)
	Atrazine	$[Atrazine]_0 = 500–1000\ ng\ L^{-1}$; Nanofiltration (NF70)	50–100	Agbekodo et al. (1996)

Table 8 Comparison of chemical oxidation processes used for pesticide removal from water

Processes	Pesticides	Operating conditions	Removal rate (%)	References
Chlorination	Atrazine	$[\text{Atrazine}]_0 = 45$ ng L^{-1}; $[Cl_2] = 3.8$ mg L^{-1}; pH = 5.5	35	Westerhoff et al. (2005)
	Simazine	$[\text{Simazine}]_0 = 554$ ng L^{-1}; [NaClO] = 18 mg L^{-1} Cl_2	50	Ormad et al. (2008)
	Atrazine	$[\text{Atrazine}]_0 = 551$ ng L^{-1}; [NaClO] = 18 mg L^{-1} Cl_2	20	Ormad et al. (2008)
Ozone (O_3)	Simazine	$[\text{Simazine}]_0 = 554$ ng L^{-1}; $[O_3] = 4.3$ mg L^{-1}	65	Ormad et al. (2008)
	Atrazine	$[\text{Atrazine}]_0 = 551$ ng L^{-1}; $[O_3] = 4.3$ mg L^{-1}	50	Ormad et al. (2008)
	Aldrin	$[\text{Aldrin}]_0 = 500$ ng L^{-1}; $[O_3] = 4.3$ mg L^{-1}	85	Ormad et al. (2008)
Photolysis – UV light	Carbaryl	$[\text{Carbaryl}]_0 = 4$ mg L^{-1}; UV (250 nm) intensity (90 mW cm^{-2}); treatment time = 1–8 min	80–100	Khoobdel et al. (2010)
Treatment using O_3/H_2O_2	Atrazine	$[O_3] = 3.5–4.5$ g^{-3}; $H_2O_2/O_3 = 0.4$ gg^{-1}	>80	Galey and PaSlawski (1993)
	Simazine	time = 10 min; $[\text{Atrazine}]_0 = 1.0$ μg L^{-1};	>80	Galey and PaSlawski (1993)
	Terbutryn	$[\text{Simazine}]_0 = 1.0$ μg L^{-1}; $[\text{terbu.}]_0 = 1.0$ μg L^{-1}	>80	Galey and PaSlawski (1993)
	Atrazine	$[\text{Atrazine}]_0 = 0.46.10^{-5}$ M; $[O_3] = 4.2.10^{-5}$ M; $[H_2O_2] = 5.9.10^{-2}$ M	91	Nélieu et al. (2000)
Photo-fenton oxidation	Atrazine	$[\text{Atrazine}]_0 = 227$ μM; [Fe(III)] = 5.0×10^{-5} M; $[H_2O_2] = 1.0 \times 10^{-2}$ M; Irradiation = 300–400 nm; T = 25.0°C; pH = 2.8; Treatment time < 30 min	99	Huston and Pignatello (1999)
Fenton process	Atrazine	$[\text{Atrazine}]_0 = 4.6$ μM; [ferrihydrite] = 0.2 g L^{-1}; $[H_2O_2] = 10$ mM; pH = 3;	21	Barreiro et al. (2007)
Electro-oxidation	Atrazine	$[\text{Atrazine}]_0 = 10$ μg L^{-1}; I = 2.0 A; Ti/IrO_2 (anode)	95	Zaviska et al. (2011)

organic and inorganic pollutants associated with suspended solids. Thus, pesticides associated with suspended solids can be removed from water using the coagulation-flocculation process. According to Sheoran (2008), this process can only remove 4% of alachlor using 15 mg L^{-1} of aluminum sulphate, whereas 11% of metachlor can be removed by means of a concentration of 30 mg L^{-1} of alum. However, pesticides can be efficiently removed from water when coagulation is combined with powdered activated carbon. For instance, Kouras et al. (1995) recorded 98% of dodine removal from water by using ferric chloride combined with powdered activated carbon. By comparison, 15% of dodine removal was obtained when $FeCl_3$ was used alone.

5.1.2 Treatments Using Activated Carbon

Two types of activated carbon can be used to remove dissolved organic matter in water: powdered activated carbon and granular activated carbon. This material has good absorption properties towards a wide spectrum of micro-pollutants such as pesticides. It is often used in drinking water production units. For instance, powdered activated carbon injection upstream of the clarifier unit can be practiced to fight against temporary pollution of raw water by pesticides. Matsui et al. (2002) injected 84 mg L^{-1} of powdered activated carbon to remove pesticides, such as asulam and simazine from water. Then, 80% of asulam and 90% of simazine were removed. When treatment requires a concentration higher than 30 g powdered activated carbon per cubic meter over a long period, e.g. over 2–3 months/year, it is not economically viable; other treatment methods should be considered, notably granular activated carbon. Granular activated carbon is typically used for the design of columns in continuous flows or filter beds. It plays the role of an adsorbent and a filtering material involving physical and chemical phenomena (Svrcek and Smith 2004). Its efficiency can be attributed to a specific adsorption induced by chemical binding occurring between pesticides and the functional groups, e.g. R-COOH and R-OH, R- being an aromatic ring, on the surface of granular activated carbon. Pollutants can also be removed by surface electrostatic attraction between the sorbent, e.g. activated carbon and the sorbate. In electrostatic attraction, activated carbon can have both positive and negative charges capable of attracting the opposite charge of pesticide and remove them from the solution (Moreno-Castilla and Rivera-Utrilla 2001). This type of process has a high adsorption capacity and technical control is easier than with powdered activated carbon (Svrcek and Smith 2004). Adsorption on granular activated carbon offers an efficiency of 90–99% of removal for various pesticides, e.g. atrazine, lindane, malathion (Hameed et al. 2009; Sotelo et al. 2002a, b; Knappe et al. 1997). However, this process is often limited by the competition between different adsorbate organic molecules, and more importantly, by the organic matter naturally present in raw waters (Agbekodo et al. 1996).

5.1.3 Chlorination

Chlorine is a chemical commonly used as an oxidizing agent in water treatment plants. For pH close to neutral values, chlorine is mainly in the form of hypochlorous acid. Chlorine is capable of oxidizing a wide range of organic compounds with one or many double bonds, including aromatic groups, a major element of the molecular structure of pesticides. Westerhoff et al. (2005) showed that the chlorination of atrazine yielded 35% of removal at a chlorine dose of 3.8 mg L^{-1} for a pH of 5.5, whereas only 20% was recorded by Ormad et al. (2008) when injecting 18 mg L^{-1} of Cl_2 in water contaminated by 551 ng L^{-1} of atrazine. By comparison, Ormad et al. (2008) obtained 50% of simazine removal using the same dose of chlorine in water containing 554 ng L^{-1} of simazine. Atrazine was more difficult to oxidize than simazine due to the molecular structure of these compounds. Given the molecular structure of hypochlorous acid characterized by the polarization of the bond Cl–O in the direction $Cl^{\delta+} \rightarrow OH^{\delta-}$, its action is most often reduced to a modification of the structure of organic molecules in water: formation of more oxidized compounds and generation of organochlorinated compounds, such as trihalomethanes (Doré 1989). Trihalomethanes are proven to be carcinogenic compounds (Ormad et al. 2008).

5.2 Alternative Methods for Pesticide Removal

5.2.1 Ozonation

Nowadays, in water treatment plants, ozone is most often used in the oxidation step rather than chlorine or hypochlorous acid due to its numerous advantages, such as high oxidizing power, improvement of the taste of water and the fact that it does not generate hazardous organochlorinated by-products, despite its higher costs (Ormad et al. 2008; Von Gunten 2003). Ozone can react with organic matter, such as pesticides via two distinct mechanisms: direct and indirect oxidation. The indirect reaction takes place via radicals generated in the decomposition of ozone into hydroxyl, superoxide, ozonide and hydroperoxide radicals. Ormad et al. (2008) removed 65% of simazine, 50% of atrazine, and 85% of aldrin by injecting 4.3 mg O_3 L^{-1} in natural water (Table 3). Sometimes oxidation can be insufficient due to the presence of natural organic matter and bicarbonate ions that consume a portion of hydroxyl radical (Aguiar et al. 1993; Zaviska et al. 2009). Large amounts of organic molecules such as pesticides can be fragmented into small molecules by ozone (Siddiqui et al. 1997), and non-biodegradable compounds can be transformed into biodegradable compounds, e.g. acids and aldehydes. These by-products are a carbon source for bacteria, which promotes bacterial growth in the distribution network (Siddiqui et al. 1997).

5.2.2 Treatment by Advanced Oxidation Processes

Advanced oxidation processes have been designed to optimize the oxidative capacity of oxidizing agents such as ozone or UV by accelerating the kinetics of their redox reactions (Zhou and Smith 2001). These processes ensure the sufficient production of hydroxyl radicals to purify water under normal atmospheric pressure and room temperature (Zaviska et al. 2009; Drogui et al. 2007; Parsons 2004). Advanced oxidation processes can be divided into four groups: Homogenous chemical oxidation (H_2O_2/Fe^{2+} and H_2O_2/O_3), Homogenous/heterogeneous photo-catalytic (H_2O_2/UV, O_3/UV and Fe^{2+}/H_2O_2/UV; TiO_2/UV), sonification and electrochemical oxidation processes. In the treatment of drinking water, the system H_2O_2/O_3 is commonly used for pesticide removal. This method is more efficient than ozonation alone, because the hydrogen peroxide accelerates the decomposition of ozone in water and thus produces more hydroxyl radicals. Hydrogen peroxide combined with ozone is efficient for removing micropollutants or toxic compounds (pesticides, hydrocarbons, etc.) present in drinking water, industrial wastewater and groundwater (Chromostat et al. 1993; Paillard 1994). Oxidation by O_3/H_2O_2 is inserted between the steps of sand filtration and filtration on granular activated carbon. The objective of this treatment is to reduce the micropollutant content, namely pesticides, before granular activated carbon filtration in order to increase the lifetime of granular activated carbon. Hydrogen peroxide is injected in the ozonation tank having a hydraulic residence time of 10 min in order to maintain the ratio O_3/H_2O_2 at an optimal value. It should be noted that the residual concentration of H_2O_2 cannot exceed the maximum value of 0.5 mg L^{-1} recommended in treated waters. Galey and Paslawski (1993) described treatment conditions imposed to remove pesticides, e.g. atrazine, simazine, and terbutryn, and organochlorinated compounds, e.g. lindane and endosulfan, operated at a pre-industrial pilot scale in three drinking water treatment plants in France located in Choisy-le-Roi, Neuilly-sur-Marne and Mery-sur-Oise. Tests were carried out in pilot plants operating with ozonation tanks composed of three to four compartments with a flow rate of up to 10,000 m^3 day^{-1}. Ozone was injected at a concentration ranging between 3.5 and 4.5 g m^{-3}, whereas a ratio of 0.4 H_2O_2/O_3 (g g^{-1}) was imposed. The water was artificially contaminated with 1.0 μg L^{-1} of pesticides. In all cases, residual concentrations of pesticides were below the European permissive level recommended of 0.1 μg L^{-1} for each individual substance after treatment. Over 80% of triazines, e.g. atrazine, simazine and terbutryn were removed, atrazine being the most difficult to remove.

5.2.3 Treatment by Electro-Oxidation Processes

The degradation of the herbicide atrazine in aqueous medium having an initial concentration of 100 μg L^{-1} has been studied through electro-oxidation using titanium coated with iridium oxide (Ti/IrO_2) and titanium coated with tin oxide (Ti/SnO_2) circular anode electrodes (Zaviska et al. 2011). The performance of the electrolytic cell resulted from its capability to react with the pollutants by using the indirect effect of electrical current where active chlorine is electrochemically generated. A factorial experimental design was used at first to determine the parameters that influenced the herbicide atrazine degradation. Current intensity and treatment time were the main parameters affecting the degradation rate. Using a 2^4 factorial matrix, the best performance for atrazine degradation of 95% was obtained by selecting a Ti/IrO_2 anode operated at a current intensity of 2.0 A during 40 min of treatment time in the presence of 1.0 $g L^{-1}$ NaCl.

5.2.4 Treatment by Membrane Techniques

The molecular weight cut-off of microfiltration membranes from 0.1 to 10 μm and ultrafiltration membranes from 10 to 50 nm are too high to retain micro-pollutants such as pesticides. However, the addition of PAC in the recirculation loop of ultrafiltration processes can lead to a good removal of micro-pollutants like atrazine (Ivancev-Tumbas and Hobby 2010; Lee et al. 2000). Membrane technologies, especially nanofiltration and reverse osmosis have emerged as suitable separation processes for micro-pollutant and salt removal. In fact, molecular weight cut-off of nanofiltration membranes from 200 to 250 Da is much lower and can remove all or a portion of pesticides. The rates of pesticide removal, which varied from 50 to 100% were recorded by Agbekodo et al. (1996) after treating water contaminated by atrazine whose concentration varied from 100 to 400 ng L^{-1} and simazine whose concentration ranged from 500 to 1,000 ng L^{-1} using a semi-industrial pilot unit equipped with nanofiltration polyamide polysulfone membrane modules having a molecular weight cut-off of 200 Da. Over 99% of isoproturon was removed while Sarkar et al. 2007 treated river water contaminated by the same pollutant using a coagulation-adsorption-nanofiltration approach. Coagulation using polyaluminum chloride was carried out before adsorption using powdered activated carbon, followed by nanofiltration having a molecular weight cut-off of 200 Da. It is worth noting that in the presence of natural organic matter in water, a complexation can occur with pesticides, e.g. atrazine and simazine and induce an increase of their apparent molecular weight. Likewise, a negative charge can appear after the complexation of pesticides, which contributes to increase the level of rejection by negatively charged membranes (Agbekodo et al. 1996).

5.3 Quantitative and Qualitative Comparison of Methods Used to Remove Pesticides from Water

Among organic compounds tested, atrazine herbicide remains by far the compound which has been most studied by several authors to emphasize the performance of physicochemical and chemical oxidation processes for the removal of pesticides. The herbicide atrazine is most often encountered in surface waters at concentrations higher than 0.1 μg L^{-1} (Giroux 1998; Berryman and Giroux 1994). From Table 7, it

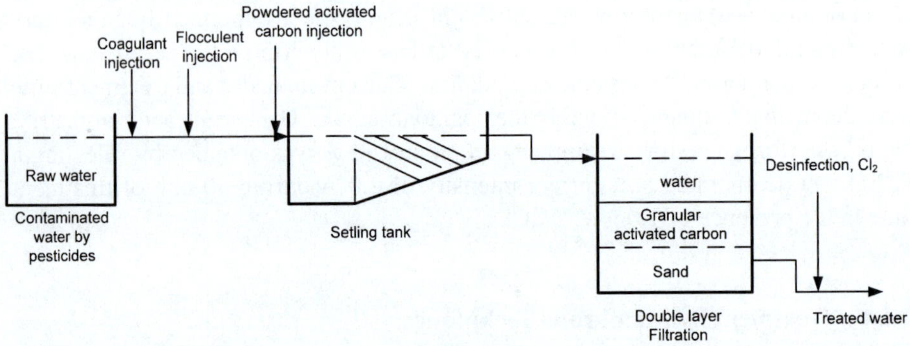

Fig. 3 Proposed flow sheet for drinking water treatment plants in order to remove micro-pollutant such as pesticides (Source: Personal diagram made by P. Drogui, INRS-ETE 2010)

can be seen that physicochemical treatments using granular activated carbon and membranes using nanofiltration are the most effective processes to remove atrazine. A removal rate up to 100% can be recorded using either granular activated carbon or membrane techniques, compared with the coagulation process using alum for which 10–25% of atrazine was removed. By comparison, 55% of atrazine was removed using powdered activated carbon (Ormad et al. 2008; Knappe et al. 1997; Agbekodo et al. 1996). The use of granular activated carbon can be particularly interesting when combined with sand filtration. A schematic flow sheet for drinking water treatment plants using activated carbon remove micro-pollutants such as pesticides is indicated in Fig. 3.

Actually, the integration of a double layer filtration unit using this alternative technique would allow increasing the capacity of drinking water treatment plants for water production without necessarily building new infrastructures (Zeng and Yang 2008). Among oxidation processes, photo-assisted Fenton oxidation, a treatment using O_3/H_2O_2 and electrochemical oxidations were found to be the most efficient processes to remove atrazine. Over 95% of atrazine can be removed, compared to ozonation alone for which 50% of atrazine was removed and chlorination for which only 20–35% of atrazine was removed depending on the initial concentration of atrazine (Zaviska et al. 2009; Nélieu et al. 2000; Ormad et al. 2008; Hua et al. 2006; Huston and Pignatello 1999) (Table 8). For a practical application, e.g. full scale application, the system H_2O_2/O_3 is commonly used for pesticide removal.

6 Future Trends and Perspectives

6.1 Some Trends in the Production of Pesticides and Management of Pest Control

Since the 1960s, integrated pest management approaches have been widely implemented and designed to use a variety of natural controls and cultural methods to suppress pest populations (BANR 2000). Conventional breeding will most probably continue to play an essential role in the improvement of agricultural crops. However, many believe that traditional breeding methods will not be sufficient to meet the increasing demand for staple crops in developing countries. Over the last two decades, scientists have focused on expanding genetically modified crops such as plant-pesticide and herbicide-tolerant crops. The next 20 years are expected to see a substantial increase in the use of genetically modified plants, with a substantial impact on crop protection. This should contribute to the implementation of integrated pest managements such as no-tillage practices in order to minimize the contamination of water by pesticides. Genetically modified plants can actually reduce some of the harmful side effects of insecticides. However, actual observations suggest that the rapid diffusion of genetically modified *Bacillus thuringiensis*-based crops will lead to pest resistance. A similar problem may occur in the future with some genetically modified herbicide-tolerant crops. The spreading of these crops could result in the transfer of their genetic qualities to weeds, creating new generations of weeds that could resist herbicides and thus reduce crop yields. These last two paradoxes are supported by actual observations made in 2010 under specific circumstances. It is then expected that new synthetic pesticides, e.g. molecular modeling and synthetic chimiomimetic compounds of bi-pesticides found in natural plant extracts will appear in future markets. Also, new generations of genetically modified crops would be needed to maintain the diversity and levels of crop production. The use of these new products will change the market of pesticides and thus the contamination of waters. There are also growing concerns about the long-term effects of the increased consumption of genetically altered materials on both humans and animals. More adapted and specific production approaches will be needed in order to preserve the soil quality, integrity of eco-agrosystems and human health.

6.2 *Trends and Perspectives for Water Treatment Contaminated by Pesticides*

The use of pesticides for agricultural purposes is responsible for numerous cases of groundwater and river contamination in the surroundings of treated areas. The discharge of such pollutants into the environment is one of the major and global public health issues that call for urgent action. Consequently, different strategies must be adopted by the municipalities and industries to deal with refractory pollutants like pesticides. Conventional methods such as chlorination can be inappropriate, notably because of their inefficiency to fully oxidize pesticides and the associated risk of forming toxic by-products, e.g. organochlorinated compounds. Alternative methods such as the combination of ozone and hydrogen peroxide are more reliable to eliminate pesticides, since there is a residual effect out of the reactor. In addition, this type of treatment can be easily implemented in water treatment plants that are already using ozonation system. The use of activated carbon, particularly in the form of powder, can be efficient in cases of punctual contamination of raw water by pesticides. On the other hand, the integration of a double layer filtration unit composed of granular activated carbon and sand can constitute an interesting alternative method for municipalities to modernize their drinking water treatment plants without necessarily building new infrastructures. Such a combination of granular activated carbon filter and sand filtration would allow increasing the capacity of drinking water treatment plants to produce potable water.

There are three major areas for future development and research: (i) Alternative oxidation methods and the use of activated carbon should be further adapted and integrated into existing water treatment plants to efficiently remove pesticides. The coupling O_3/H_2O_2 is a good example; it can be advantageously applied in water treatment plants that are already using ozonation; (ii) In addition to simple equipments, e.g. small installations and easy operation, e.g. reduction of operation costs, the research works have to focus on clearly identifying by-products and evaluating the toxicity of these intermediates after the oxidation of pesticides; and (iii) Using the best combination of biopesticides and chemical pesticides possible.

References

Agbekodo KM, Legube B, Dard S (1996) Atrazine and simazine removal mechanisms by nanofiltration: Influence of natural organic matter concentration. Water Res 30(11):2535–2542

Aguiar A, Carbonnière F, Paillard H, Legube B (1993) Oxydation des pesticides et coagulation des substances humiques par le peroxyde d'hydrogène à faibles doses et le fer ferreux. Water Supply 11:129–138

Aspelin AL, Grube AH (1999) Pesticide industry sales and usage. 733-R-99-001. US Environmental Protection Agency, Office of Pesticide Programs, Washington, DC

Barreiro JC, Capelato MD, Martin-Neto L, Hansen HCB (2007) Oxidative decomposition of atrazine by a Fenton-like reaction in a H_2O_2/ferrihydrite system. Water Res 41:55–62

Berryman D, Giroux I (1994) La contamination des cours d'eaux par les pesticides dans les régions de cultures intensives de maïs au Québec, Campagne d'échantillonnage de 1992 et 1993. Ministère de l'Environnement du Québec, Direction des Écosystèmes Aquatiques, Envirodoq EN940594, report no. PES-4, Québec, 134 pp

Board on Agriculture and Natural Resources (BANR) (2000) Genetically modified pest-protected plants: science and regulation. Committee on Genetically Modified Pest-Protected Plants, National Research Council, National Academy Press, Washington, DC, 292 p

Board on Agriculture and Natural Resources (BANR) (2010) Toward sustainable agricultural systems in the 21st century. Committee on Twenty-First Century Systems Agriculture, Division on Earth and Life Studies, National Research Council, National Academy Press, Washington, DC, 598 p

Caron E, Lafrance P, Duchemin M (2007) Quantification des pertes d'herbicides par ruissellement de surface et par infiltration dans des dispositifs tampon: enherbés et enherbés+arborés. In: Oturan M, Mouchel J-M (eds) Pesticides: impacts environnementaux, gestion et traitement. Presses de l'École Natioale de Ponts et Chaussées, Paris, pp 153–163

Caron E, Lafrance P, Auclair JC, Duchemin M (2010) Impact of grass and grass of polar buffer strips on atrazine and metolachlor losses in surface runoff and subsurface infiltration from agricultural plots. J Envion Qual 39(2):617–629

Chromostat N, De Laat J, Doré M, Suty H, Pouillot M (1993) Étude de la dégradation de triazine par O_3/H_2O_2 et O_3. Cinétique et sous-produits de dégradation. Water Supply 11:149–157

Commission of Life Sciences (2000) The future role of pesticides in US agriculture. Committee on the future role of pesticides in US agriculture, Board on agriculture and natural resources and

Board on environmental studies and toxicology, National Research Council, National Academy Press, Washington, DC, 332 p

Doré M (1989) Chimie des oxydants et traitement des eaux. Technique et documentation–Lavoisier, Paris, p 505. ISBN 2852065622

Drogui P, Blais JF, Mercier G (2007) Review of electrochemical technologies for environmental applications. Recent Patents Eng 1(3):257–272

Dymond M, Hurr K (2010) The global status of commercialised genetically modified plants – 1 July 2008 to 31 December 2009. Plant Imports, MAF and Biosecurity, New Zealand, p 35

Environment Canada (2001) Protocols for the derivation of water quality guidelines, Canadian environmental quality guidelines. Retrieved from http://ceqg-rcqe.ccme.ca

Environmental Protection agency (EPA) (2000) Drinking water standards and health advisories, Office of Water 4304. EPA 822-B-00-001, U.S. Environmental Protection Agency, Washington, DC

European Union (1997) Common position (EC) 12767/97. Adopted by the council on 16 December 1997 with a view to the adoption of council directive on the quality of water intended for human consumption. Interinstitutional file 95/0010 (SYN)

Fernandez-Cornejo J, Just RE (2007) Researchability of modern agricultural input markets and growing concentration. Am J Agric Econ 89(5):1269–1275, American Agricultural Economics Association

Flury M (1996) Experimental evidence of transport of pesticides through fields soils – a review. J Environ Qual 25:25–45

Galey C, PaSlawski D (1993) Élimination des micropolluants par l'ozone couplé avec le peroxyde d'hydrogène dans le traitement de potabilisation des eaux. Eau Ind Nuisances 161:46–49

Gaynor JD, McTavish DC, Findlay WI (1995) Atrazine and metolachlor loss in surface and subsurface runoff as affected by cultural practices. J Environ Qual 24:246–256

Giroux I (1998) Impact de l'utilisation des pesticides sur la qualité de l'eau des bassins versants des rivières Yamaska, L'Assomption, Chaudière et Boyer, Ministère de l'Environnement et de la Faune du Québec, Direction des Écosystèmes Aquatiques, Québec, 48 p

Gorse I, Dion S (2010) Bilan des ventes de pesticides au Québec pour l'année 2003. Ministère du Développement durable, de l'Environnement et des Parcs (MDDEP), Gouvernement du Québec, ISBN: 978-2-550-58570-1, 81 p. Available from http://www.mddep.gouv.qc.ca/pesticides/bilan/bilan2007.pdf

Hameed BH, Salman JM, Ahmad AL (2009) Adsorption isotherm and kinetic modeling of 2,4-D pesticide on activated carbon derived from date stones. J Hazard Mater 163:121–126

Hua W, Bennett ER, Letcher RJ (2006) Ozone treatment and the depletion of detectable pharmaceuticals and atrazine herbicide in drinking water sourced from the upper Detroit River, Ontario, Canada. Water Res 40:2259–2266

Huston PJ, Pignatello JJ (1999) Degradation of selected pesticide active ingredients and commercial formulations in water by the photo-assisted Fenton reaction. Water Res 33:1238–1246

Ivancev-Tumbas I, Hobby R (2010) Removal of organic xenobiotics by combined out/in ultrafiltration and powdered activated carbon adsorption. Desalination 255:124–128

Jusoh A, Hartini WJH, Nora'aini A, Endut A (2011) Study on the removal of pesticide in agricultural run off by granular activated carbon. Bioresour Technol. doi:10.1016/j.biortech.2010.12.074

Khoobdel M, Shayeghi M, Golsorkhi S, Abtahi M, Vatandoost H, Zeraatii H, Bazrafkan S (2010) Effectiveness of ultrasound and ultraviolet irradiation on degradation of carbaryl from aqueous solutions. Iran J Arthropod-Borne Dis 4(1):47–53

Kiely T, Donaldson D, Grube A (2004) Pesticides industry sales and usage: 2000 and 2001 market estimates. Biological and Economic Analysis Division, Office of Pesticide Programs, Office of Prevention, Pesticides, and Toxic Substances, U.S. Environmental Protection Agency (EPA), Washington, DC, p 33

Knappe DRU, Snoeyink VL, Roche P, Prados MJ, Bourbigot MM (1997) The effect of preloading on rapid small-scale column test predictions of atrazine removal by GAC adsorbers. Water Res 31(11):2899–2909

Kouras A, Zouboulis A, Samara C, Kouimtzis T (1995) Removal of pesticides from surface waters by combined physicochemical processes. Part I: Dodine. Chemosphere 30(12):2307–2315

Kouras A, Zouboulis A, Samara C, Kouimtzis T (1998) Removal of pesticides from aqueous solutions by combined physicochemical processes – The behavior of lindane. Environ Poll 103:193–202

Lafrance P, Banton O, Bernard F. inc. (1993–1996) Évaluation environnementale des pratiques culturales sur maïs pour la réduction des pertes d'herbicides, Rapport scientifique No. 390, Institut National de la Recherche Scientifique, INRS-Eau, Université du Québec, Québec, 919 p

Lafrance P, Banton O, Gagné P (1997) Exportation saisonnière d'herbicides vers les cours d'eau mesurée sur six champs agricoles sous quelques pratiques culturales du maïs (Basses-Terres du Saint-Laurent). Rev Sci Eau 4:439–459

Lafrance P, Guibaud G, Bernard C (2001) Rendement de zones tampon herbacées pour limiter les pertes d'herbicides en phase dissoute par ruisellement de surface. In: Proceedings of the 30e Congrès du Groupe Français des Pesticides, Presses de l'Université de Reims Champagne-Ardenne, Rheims, pp 408–417

Lebel J (2003) La Santé – Une approche écosystemique. Centre de recherche pour le développement international (CRDI), Ottawa, Canada, 84 p

Lee SJ, Choo KH, Lee CH (2000) Conjunctive use of ultrafiltration with powdered activated carbon adsorption for removal of synthetic and natural organic matter. J Ind Eng Chem 6(6):357–364

Matsui Y, Knappe DRU, Takagi R (2002) Pesticide adsorption by granular activated carbon adsorbers. 2. Effects of pesticide and natural organic matter characteristics on pesticide breakthrough curves. Environ Sci Technol 36:3432–3438

Ministry of Health NZ (2000) Drinking-water standards for New Zealand, Ministry of Health, New Zealand. Available from www.moh.govt.nz

Moreno-Castilla C, Rivera-Utrilla J (2001) Carbon materials as adsorbents for the removal of pollutants from the aqueous phase. MRS Bull 26(11):890–894

Nélieu S, Kerhoas L, Einhorn J (2000) Degradation of atrazine into ammeline by combined ozone/hydrogen peroxide treatment in water. Environ Sci Technol 34(3):430–437

Ormad MP, Miguel N, Claver A, Matesanz JM, Ovelleiro JL (2008) Pesticides removal in the process of drinking water production. Chemosphere 71(1):97–106

Paillard H (1994) Étude de la minéralisation de la matière organique dissoute en milieu aqueux dilué par ozonation, oxydation avancée O3/H2O2 et ozonation catalytique hétérogène. Thèse de Doctorat, de l'université de Poitiers, Poitiers, 224 p

Parsons S (2004) Advanced oxidation processes for water and wastewater treatment. IWA Publishing, Alliance House (Ed.), London, p 356

Reichenberger S, Bach M, Skitschak A, Frede HD (2007) Mitigation strategies to reduce pesticide inputs into ground- and surface water and their effectiveness: a review. Sci Total Environ 384:1–35

Rousseau AN, Lafrance P, Lavigne MP, Savary S, Konan B, Quilbé R, Amrani M, Jiapizan P (2011) A hydrological modelling framework for defining watershed-scale achievable performance standards of pesticides beneficial management practices. J Environ Qual. doi:10.2134/jeq2010.0281

Sarkar B, Venkateswralu N, Nageswara R, Bhattacharjee C, Kale V (2007) Treatment of pesticide contaminated surface water for production of potable water by a coagulation–adsorption–nanofiltration approach. Desalination 212(1–3):129–140

Sheoran M (2008) Advanced oxidation processes for the degradation of pesticides. M.Tech thesis, Environmental Science & Technology Department. Department of Biotechnology & Environmental Sciences, Thapar University, Patiala. Available from http://hdl.handle.net/10266/689

Siddiqui MS, Amy GL, Murphy BD (1997) Ozone enhanced removal of natural organic matter from drinking water sources. Wat Res 31:3098–3106

Sotelo JL, Ovejero G, Delgado JA, Martınez I (2002a) Adsorption of lindane from water onto GAC effect of carbon loading on kinetic behavior. Chem Eng J 87:111–120

Sotelo JL, Ovejero G, Delgado JA, Martınez I (2002b) Comparison of adsorption equilibrium and kinetics of four chlorinated organics from water onto GAC. Water Res 36:599–608

Spencer WF, Cliath MW, Blair JW, LeMert RA (1985) Transport of pesticides from irrigated fields in surface runoff and tile drain waters, vol 31, Conservation research report. Agricultural Research Service, USDA, Washington, DC, p 71

Statistics Canada (2001) Protecting crops from pests. Canadian agriculture at a glance, Catalogue no. 96-325-XPB

Svrcek C, Smith DW (2004) Cyanobacteria toxins and the current state of knowledge on water treatment options: a review. J Environ Eng Sci 3:155–185

Templeton SR, Zilberman D, Yoo SJ (1998) An economic perspective on outdoor residential pesticide use. Environ Sci Technol 21:416–423

Todd AK (2010) Changes in urban stream water pesticide concentrations one year after a cosmetic pesticides ban, Environmental Monitoring and Reporting Branch, Ontario Ministry of the Environment, Queen's Printer for Ontario, Ontario, 30 p

Von Gunten U (2003) Ozonation in drinking water. Part I: Oxidation kinetics and product formation. Water Res 37:1443–1467

Wauchope RD (1978) The pesticide content of surface water draining from agricultural fields. J Environ Qual 7:459–472

Westerhoff P, Yoon Y, Snyder S, Wert E (2005) Fate of endocrine-disruptor, pharmaceutical, and personal care product chemicals during simulated drinking water treatment processes. Environ Sci Technol 39(17):6649–6663

World Health Organization (WHO) (1988) Revision of the WHO guidelines for drinking-water quality. Document WHO/PEP/89.4, WHO, Geneva

Yudelman M, Ratta A, Nygaard D (1998) Food, agriculture, and the environment, discussion paper 25, Pest Management and Food Production – Looking to the Future, International Food Policy Research Institute (IFPRI), Washington, DC, 53 p

Zaviska F, Drogui P, Mercier G, Blais JF (2009) Procédé d'oxydation avancée dans le traitement des eaux et des effluents urbains et industriels: Application à la dégradation des polluants réfractaires. Rev Sci Eau 22(4):535–564

Zaviska F, Drogui P, Mercier G, Blais JF, Lafrance P (2011) Experiment design methodology applied to electrochemical oxidation of the herbicide atrazine using Ti/IrO_2 and Ti/SnO_2 circular anode electrodes. J Hazard Mater 185:1499–1507

Zeng Z, Yang C (2008) Contrast experiment of the treatment of stable micro-polluted surface water between GAC-sand filtration and sand filtration. Gongye Shui chuli 28(9):43–46

Zhou H, Smith DW (2001) Advanced technologies in water and wastewater treatment. Can J Civil Eng 28:49–66

Nitrogen Use Efficiency by Annual and Perennial Crops

Corina Carranca

Abstract The amount of nitrogen fertilizer applied to plants is usually large. Only part of this fertilizer, of about 6–50%, is taken up by plants, depending on fertilizer, plant and soil type, climatic conditions, and agricultural practices. The unaccounted nitrogen can be emitted from the ecosystem as trace gas and ammonia volatilization, or lost by leaching and runoff in the nitrate or ammonium form. The goal of reducing mineral nitrogen usage will be to this twenty-first century what the goal of reducing pesticides was to the last century. In the present study we reviewed the different concepts for nitrogen use efficiency by annual and woody plants. The major points were (i) understanding the terminology and the context in which each concept for nitrogen use efficiency has been used for annual and perennial woody plants, (ii) identifying the critical steps of controlling plant nitrogen use efficiency, and (iii) addressing new approaches to improve the efficiency for annual and perennial woody plants.

Some factors have been extensively studied for arable crops, but not for woody plants, and included the source, timing and rate of fertilizer nitrogen, plant growth curves, environmental factors, and cultural practices. Precision nitrogen management, and intercropping and crop rotation including legumes are recommended techniques to improve nutrient efficiency. Maximum crop yield has been achieved using high-yielding bred plants, with an optimization of photosynthetic capacity, but without maximizing the nitrogen use efficiency. At low nitrogen availability, C_3 plants have greater nitrogen use efficiency than C_4 plants, whereas at high nitrogen the opposite is true. Consequently, identifying the regulatory elements controlling the balance between nitrogen allocation to maintain photosynthesis and the reallocation of the remobilized nitrogen to sink organs in C_3 and C_4 species is vital for improving fertilizer use efficiency and reducing excessive input of fertilizer, while maintaining an acceptable yield.

C. Carranca (✉)
INIAV, Unidade de Ambiente e Recursos Naturais, Quinta do Marquês,
Av. República, Nova Oeiras, 2784-505 Oeiras, Portugal

CEER, Instituto Superior de Agronomia, UTL, 1349-017 Lisbon, Portugal
e-mail: c.carranca.ean@clix.pt

E. Lichtfouse (ed.), *Farming for Food and Water Security*, Sustainable Agriculture Reviews 10, DOI 10.1007/978-94-007-4500-1_3,

In the present review we reinforced the importance of selecting plant genotypes from the ancient and modern germoplasm in order to improve nitrogen use efficiency. Efforts should include plant selection under low nitrogen, which has not been a priority for plant breeders. Plant breeders may need to find newer, more appropriate plant cultivars which can maximize the role of mycorrhizal in agriculture. The contribution of mycorrhizal in annual and perennial crops, particularly with ancient plant germoplasm, and the influence of rootstocks and plant N reserves in woody plants, and ammonia emissions from annual and perennial crops (1–4% of fertilizer nitrogen) should be considered for improving fertilizer nitrogen use efficiency. Beneficial traits can probably include the ability to maintain the plant photosynthetic capacity and nitrogen uptake under reduced (or high) nitrogen level perhaps through beneficial mycorrhizal associations and old (or modern) crop genotypes. Nitrogen absorption depends on fungi strain and plant cultivar. Rootstocks influence roots number and class, the nutrient uptake and translocation, and the bud break in woody plants which mechanism is not fully understood. This factor is probably crucial for attaining optimum nitrogen use efficiency and careful attention should be given for the choice of appropriate rootstock. Finally, we found that most studies for crop nitrogen recovery are based on recovery efficiency at harvest or fruit ripening which may underestimate the potential for fertilizer N utilization. Greater efficiency data arise when the efficiency is calculated on basis of mean nitrogen during the major growth period.

Keywords Annual crops • Concepts • Factors • Genotype • Growth curves • Mycorrhizal association • Plant N reserves • Rootstock • Woody plants

1 Introduction

Among 50–70 million people will be added annually to the world population in the middle of 2030. Almost all of these increases are expected to take place in developing countries, especially the group of 50 least developed countries. More food and fiber will be required to feed and clothe these additional people and to increase the daily food uptake of the still 830 million undernourished worldwide. Diets have shifted away from staples such as cereals, roots, tubers, and pulses towards more livestock products, oil, fruits, and vegetables (FAO 2008). These support continuing and increasing demand of fertilizers to restore and enhance fertility of the world agricultural land for higher yields and improve product quality. The sub-Saharan Africa is the only region in the world where per capita food production has stagnated over the last century leading to acute food insecurity and poverty. Inherent low soil fertility and low use of fertilizers are among the main biophysical constraints to increasing agricultural productivity (Sanchez et al. 1996).

FAO (2008) estimated that world chemical fertilizers supply [nitrogen (N), phosphate (P_2O_5) and potash (K_2O)] will increase by 34×10^6 t, representing an annual growth rate of 3% between 2007/2008 and 2011/2012 (Table 1), sufficient to cover the annual demand growth of about 2%.

Table 1 World nitrogen supply and demand balance in the 2007/2008–2011/2012 period

	2007/2008	2009/2010	2010/2011	2011/2012
	($\times 10^6$ t)			
Total supply	131,106	140,732	147,748	154,199
Total demand	127,820	133,059	136,198	139,140
Surplus	3,286	7,673	11,550	15,059

Source: FAO (2008)

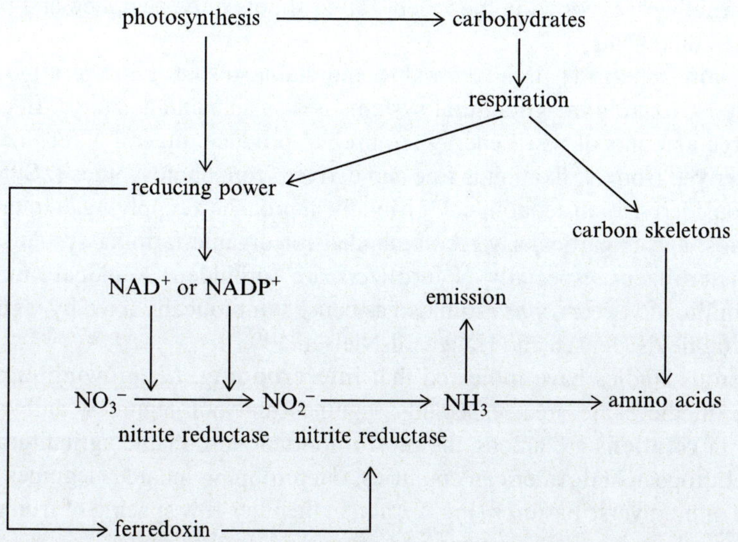

Fig. 1 Major pathways for ammonia (NH_3) assimilation in the plant and emissions in the reproductive phase (Source: Adapted from Mifflin and Lea 1976)

Nitrogen is the most abundant nutrient in nature and is often regarded as the single most important plant nutrient. In agro-ecosystems it undergoes several paths (chemical and biological transformations) and losses from the system, becoming the most limiting plant nutrient. It is mainly taken up by plants in inorganic forms [ammonium (NH_4^+) or nitrate (NO_3^-)]. In the foliar application, N is mostly supplied in the urea form with low biuret content. It is not for sure that crops can uptake N in the organic form (amino acids), but some authors support this hypothesis (Forsum et al. 2008; Krab et al. 2008; Gördes et al. 2011; Svennerstam et al. 2011).

In the plant, N is the main component of organic compounds such as nucleic and amino acids, proteins, enzymes, chlorophyll, nicotinamide adenine dinucleotide (NAD^+) and nicotinamide adenine dinucleotide phosphate ($NADP^+$), and several hormones (Fig. 1). Excess N promotes an excessive vegetative growth and plant development, leaves turn dark green, stems become more succulent and less fibrous, plants become more susceptible to pests and diseases, bloom is anticipated, fruit setting is reduced, maturity is delayed, peel thickness, fruit size and acidity are increased, and soluble solids content and maturity index of fruits are reduced. Low N reduces the plant growth, fruit yield and size, and causes leaf deficiency symptom with a light green color.

Because of the strong influence of N on plant growth, large amounts of fertilizer N are often applied to fields to maximize crop productivity.

Plants belonging to *Fabaceae* Family are able to use the atmospheric N_2 for its nutrition when symbiotically associated with the bacteria collectively called *Rhizobium.* These leguminous plants supply the symbiont bacteria with carbohydrates (malate, succinate) to synthesize large amounts of energy needed to capture the atmospheric N_2 and convert into ammonia (NH_3) within the infected root nodule tissue. Ammonia is then converted to amide, ureides, allantoin or allantoic acid in nodules which are exported in the xylem sap to shoots to be degraded and produce NH_4^+ and amino acids.

Since atmospheric N_2 is a renewable and stable resource in the atmosphere, symbiotic N_2 fixation in agricultural systems is a sustainable N source. In contrast to the large amounts of fossil energy required to produce mineral N fertilizers, the energy derived from N_2 fixation is free and derived from photosynthesis. Symbiosis is then considered as an economically friendly approach to supplying N to the agro-ecosystems, and is particularly recommended for organic farming systems where chemical fertilizers, especially N fertilizers are forbidden. A global amount of 40–120 million t N year^{-1} was estimated as being symbiotically fixed by legumes in agricultural fields (Jensen and Haugaard-Nielsen 2003).

Numerous studies have indicated that **intercropping**, i.e. growing more than one crop in the same area, especially legumes and non-legumes, and growing legumes in **rotations** are among the most important sustainable agricultural practices. In European and American countries, intercropping includes legumes mostly in mixed or biodiverse pastures (Fig. 2), and in the inter-row spacing of fruit (Fig. 3) and olive (*Olea europaea* L.) trees and vineyards. In the tropics, forest trees are intercropped with legumes. In China, wheat (*Triticum* sp.)/fababean (*Vicia faba* L.) and maize (*Zea mays* L.)/fababean are the main intercropping combinations adopted by farmers (Li et al. 2011). Interspecific competition and facilitation interactions are the main mechanisms involved in the advantages obtained from intercropping. Monocultures often cannot fully utilize available space and resources, such as soil resources and light, but two species intercropped together can use space and resources more efficiently than when they occupy different niches. When the niches are similar, species compete intensely for light, water and soil resources. However, environmental changes made by one species may provide benefits to the associated ones, for example by N_2 fixation. In legume/non-legume intercropping system, non-legumes such as cereals or grasses are more competitive than legumes for soil inorganic N and induce legumes to fix more atmospheric N_2 with a greater proportion of total N in legumes derived from N_2 fixation compared to monocropping systems. Intercropping combinations can significantly increase N use efficiency by both plant species and have a great potential for decreasing available soil NO_3^- since this N form is immediately used by non-legumes in the vicinity.

This is also true for cash crops preceding the legumes in rotations. The contribution of fixed N_2 by legumes to the soil following harvest will depend upon the N balance (credit N) at harvest which is determined by the difference between the amounts of N_2 fixed and removed N in the seed or feeding plant material (Carranca 2011).

Fig. 2 A Portuguese agro-forestry system consisting of a biodiverse pasture with about 20 intercropped plant species and cultivars (legumes and non-legumes) and cork trees (*Quercus suber* L.)

Fig. 3 A Portuguese apple (*Malus domestica* Borkh.) orchard intercropped with *Plantago lanceolata* L.

If N balance is positive, there is a positive addition of N to the soil (N benefit or N input), otherwise there is a negative credit N (N deficit). Final N balance can range from as little as −132 kg ha^{-1} in non-nodulated soybean (*Glycine max* L.) to as much +135 N kg ha^{-1} in lupine (*Lupinus* sp.) (Peoples et al. 1995; Carranca 2011).

While symbiotic fixation may be tailored to host-plant needs, inorganic fertilizers are usually applied in large doses and only up to 50% of which is used by crops, being lost from the soil. Improve nutrient use efficiency is thus the appropriate goal for sustainable agriculture, because of growing pressure for agriculture to minimize negative environmental and economical impacts. Sustainable efficiency is thus the nutrient input needed to sustain the system at the optimum productivity.

In the present study, (i) an overview on the existing concepts for the assessment of **nitrogen use efficiency** within different contexts (agronomic, physiological, and economical and environmental) is presented for annual and perennial woody plants, (ii) including factors affecting the N efficiency, and (iii) addressing new approaches to improve this efficiency.

Several definitions have been suggested for the N use efficiency in crop production. Nitrogen use efficiency is a term currently used to indicate the relative balance between the amount of fertilizer N taken up and used by the crop *versus* the amount of fertilizer N lost from the agricultural system. A common way to assess N use efficiency in crop production is based on the amount of yield produced per unit soil N, as proposed by Moll et al. (1982). In this approach, authors comprised N use efficiency into two components:

1. **Nitrogen uptake efficiency, crop N recovery efficiency** or **fertilizer N use efficiency**, defined as the N accumulated in the plant relative to its supply, i.e. total plant N per unit soil N, N_t/N_{soil}, important in low N systems:

$$\text{Nitrogen recovery efficiency (\%)} = \left(\text{plant N}_{\text{fertilized}},\ \text{kg ha}^{-1} - \text{plant N}_{\text{control}},\ \text{kg ha}^{-1}\right) \times 100 / \left(\text{fertilizer N applied, kg ha}^{-1}\right),$$

2. **Nitrogen utilization efficiency** or **physiological efficiency**, which represents grain yield or harvested material relative to crop N accumulation, i.e. crop yield per unit of plant N, G_w/N_t, important in high N conditions:

$$\text{Physiological efficiency}\left(\text{kg yield kg}^{-1}\text{N}\right) = \left(\text{crop yield}_{\text{fertilized}},\ \text{kg ha}^{-1} - \text{crop yield}_{\text{unfertilized}},\ \text{kg ha}^{-1}\right) / \left(\text{plant N}_{\text{fertilized}},\ \text{kg ha}^{-1} - \text{plant N}_{\text{unfertilized}},\ \text{kg ha}^{-1}\right)$$

The **agronomic efficiency, plant N use efficiency** or **partial factor productivity** was then defined by Moll et al. (1982) as the ratio of grain yield or harvested material to supplied N, and is composed of N uptake efficiency and physiological N use efficiency. Most fertilizer studies focus on this parameter. It is the product of recovery efficiency from added fertilizer N and physiological efficiency (Hardarson et al. 2008; Kara 2010):

$$\text{Agronomic efficiency}\left(\text{kg yield kg}^{-1}\text{N}\right)= \left(\text{crop yield}_{\text{fertilized}},\ \text{kg ha}^{-1} - \text{crop yield}_{\text{unfertilized}},\ \text{kg ha}^{-1}\right)/ \left(\text{fertilizer N applied, kg ha}^{-1}\right)$$

The crop N use efficiency is the basis for economic and environmental efficiency, which can be quite different. As agronomic efficiency improves, economic and environmental efficiency will also benefit (Roberts 2008). **Economic efficiency** is defined as the optimal farm income by the proper use of nutrient inputs, and the site-specific **environmental efficiency** is determined by studying nutrient not used by the crop which is at risk of loss in the ecosystem. The susceptibility of N loss varies with the crop, soil and climatic conditions, and management practices. In general, this loss to the environment is only a concern when fertilizers are applied at rates above the agronomic need.

Understanding the above terminology and the context in which each one is used is crucial to prevent misinterpretation and misunderstanding. In the agricultural context, crop N use efficiency needs to focus on the specific biomass fractions forming yield, which are defined by the harvested product and related to both the consumption of nutrient and product quality aspects, for instance the protein content (Weih et al. 2011). The agronomic N efficiency is apparently the meaningful concept that both allows the analysis of the functionally important N use efficiency components in an agronomic context (yield and quality), and takes advantage of the method in growth analysis and functional ecology.

Farmers have been requested to further improve crop N use efficiency and yield, simultaneously. Increasing crop N use efficiency and simultaneously decreasing N accumulation in the soil has become an urgent priority for agricultural production. Understanding the regulatory mechanisms of controlling plant N economy is vital for improving plant N efficiency and reducing excessive input, while maintaining an acceptable yield. It is therefore of major importance to identify the critical steps of controlling plant N use efficiency.

2 Factors Affecting Plant N Use Efficiency

Agronomic efficiency depends on several factors (Recous et al. 1988; Raun and Johnson 1999; Carranca 2010):

1. Crop, including the cultivar (germoplasm), expected yield, exported N, capacity for N accumulation and mobilization to the meristematic points, rootstock, NH_3 volatilization from leaves following anthesis, and root development, activity, and ability for an efficient fungi colonization;
2. Source, timing and rate of fertilizer N;

3. Environmental factors, e.g. soil type and climate;
4. Cultural practices, namely the intercropping and crop rotation, sowing date, planting density, soil mobilization, irrigation, and weeding.

Most factors for a proper agronomic efficiency, such as the source, timing and rate of fertilizer N, plant growth curves, environmental factors and cultural practices have been extensively studied for arable crops (Moll et al. 1982; Recous et al. 1988; Jensen et al. 1997; Carranca et al. 1999; Hardarson et al. 2008; Zhang et al. 2009; Carranca 2010; Weih et al. 2011), but not for permanent crops, especially the woody plants. The contribution of an efficient mycorrhizal system in annual and perennial woody plants (Cruz et al. 2007; Cruz and Carranca 2010), the plant N reserves in perennial crops (Menino et al. 2007; Neto et al. 2008), the rootstock influence in woody plants (Lea-Cox et al. 2001), and plant NH_3 losses (Stutte et al. 1979; Harper et al. 1987) have not been strongly discussed.

3 Crop

Crop productivity results from the integration of several environmental and genetic processes throughout the life of the crop that differs in intensity and timing of action. To achieve high yield, plants must establish optimal photosynthetic capacity and then maintain a high rate of photosynthesis during the reproductive phase. Nitrogen supply plays a major role in both processes by maximizing leaf area index and biochemical components of the photosynthetic apparatus. A positive response of photosynthetic rate to leaf N concentration has been widely reported (Correia 2001; Netto et al. 2005; Figueiredo 2011; Neto et al. 2011).

Photosynthetic N efficiency is determined in large part by specific leaf weight and by the allocation of leaf N to photosynthetic enzymes (Uribelarrea and Crafts-Brandner 2009). These authors (2009) observed that specific leaf weight was constant across N treatments, whereas leaf N accumulation increased with N supply, indicating that investment in photosynthetic machinery per unit leaf area increased with N supply.

3.1 Germoplasm

Both N recovery efficiency and agronomic efficiency are useful concepts. Nitrogen recovery (increase in N uptake per unit nutrient added) in annual and perennial crops rarely exceeds 50%, and is often much lower (Table 2). Under rainfed conditions, crop removal efficiency ranged 20–30%, and under irrigated conditions it can be improved up to 40% (Roberts 2008). The definition and components of N use efficiency in annual grains place an emphasis on grain yield and protein. The recovery N efficiency in the world cereal grain [wheat, maize, rice (*Oryza sativa* L.), barley (*Hordeum vulgare* L.), sorghum (*Sorghum bicolor* L.), oat (*Avena sativa* L.) and rye

Table 2 Some examples of nitrogen use efficiency estimated at harvest for several annual and perennial crops, in different environmental conditions

Crop	Location	Farming system	Soil/texture	N recovery efficiency (%)	References
Sugarbeet (*Beta vulgaris* L.)	Central Portugal	Surface irrigation	Eutric Fluvisol [a]	23–40	Oliveira et al. (1989)
Spinach (*Spinacia oleracea* L.)	West Portugal	Sprinkling irrigation	Antrosol [a]	18	Carranca (2005)
Wheat (*Triticum durum* L.)	Central Portugal	Rainfed	Ortic Luvisol [a,b]	20–40	Alves (1979); Carranca et al. (1999)
Wheat (*Triticum durum* L.)	India	Rainfed	–	18–49	Roberts (2008)
Sorghum	Australia	–	32–62		Hardarson et al. (2008)
Silage maize (*Zea mays* L.)	Central Portugal	Surface irrigation	Loamy sand	65–73	Carranca (1989)
Silage maize (*Zea mays* L.)	Central Portugal	Surface irrigation	Loam	76	Carranca (1989)
Grain maize (*Zea mays* L.)	North/South Portugal	Surface irrigation	Loamy sand	53–93	Carranca (1989)
Grain maize (*Zea mays* L.)	Central Portugal	Surface irrigation	Loam	52	Carranca (1989)
Grain maize (*Zea mays* L.)	North USA	–	–	37	Roberts (2008)
Rice (*Oryza sativa* L.)	Asia	Irrigated	–	31	Roberts (2008)
Rice (*Oryza sativa* L.)	Asia	Precision agriculture	–	40	Roberts (2008)
Rice (*Oryza sativa* L.)	Indonesia	Upland	–	9–18	Hardarson et al. (2008)
Rice (*Oryza sativa* L.)	Philippines	Flooded	–	44	Hardarson et al. (2008)
Pasture	Australia	–	–	38–55	Hardarson et al. (2008)
Orange trees (*Citrus sinensis* L.) (1–3 years)	South Portugal	Drip fertigation	Ortic Podzol [b]	6–30	Menino et al. (2007)
Pear trees (*Pyrus communis* L.) (1–3 years-old)	West Portugal	Drip fertigation	Antrosol [a]	6–33	Neto et al. (2008)
Pear trees (*Pyrus communis* L.) (5 years-old)	Italy	Rainfed	–	15–25	Righetti et al. (1994)
Apple trees (*Malus domestica* Borkh.) (3 years-old)	UK	Drip fertigation	–	22	Neilsen et al. (2001)
Eucalyptus trees (*Eucalyptus globulus* L.) (1 year-old)	West Portugal	Surface irrigation	Eutric Regosol [a]	4	Carranca et al. (2009)

nd not detected, – not available
[a] IUSS Working Group WRB (2006)
[b] FAO (1971–1981)

(*Secale cereale* L.)] would therefore be estimated at 20–50% (Cassman et al. 1993; Carranca et al. 1999; Raun and Johnson 1999; Li et al. 2011). Low recovery N efficiency is thought to be the major constraint to achieving higher yields.

Limited N losses may be achieved either with reduction in fertilizer application or with cultivars that better absorb and utilize N. Plant breeding programs must produce cultivars that absorb N and use it more efficiently to produce grain or harvested material rather than non-harvested material. In natural ecosystems, less is known about the genetic factors controlling crop N use efficiency and whether the genetic mechanisms differ significantly between high and low fertility environments (Dawson et al. 2008).

Wheat, maize and rice varieties with high harvest indices (yield produced divided by the total dry biomass) have high physiological N use efficiency. Plants with a low N accumulation in fallen leaves or non-harvested material have a high N harvest index and high N utilization efficiency. Increasing protein content in cereals by applying higher rates of fertilizer N is relatively inefficient, as physiological N use efficiency decreases with increasing N level. Cereal cultivars that produce more biomass with greater N accumulation are not necessarily the ones that use N more efficiently. Furthermore, high N assimilation after anthesis is needed to achieve high yields and high N recovery. Olea et al. (2004) and Zhang et al. (2009) observed that higher recovery efficiency was dependent on glutamine synthetase activity. In healthy plants, NH_4^+ is assimilated into amino acids by the action of these enzymes.

Considerable genetic variability exists among maize (C_4 plants, very efficient in utilizing soil N and water) hybrids for traits associated with the ability of kernel to utilize and assimilate N (Uribelarrea and Crafts-Brandner 2009). Differences among corn hybrids for N use efficiency are largely due to variation in the utilization of accumulated N before anthesis (physiological N efficiency), especially under low N supply (Moll et al. 1982; Raun and Johnson 1999). Bred varieties of irrigated maize in Portugal have an N recovery greater than 52% (Table 2). The inclusion of improved hybrid maize germplasm in East and Central Africa increased the agronomic efficiency from 17 to 26 kg grain/kg N, from 11 to 18 kg grain/kg N in Zambia, and from 38 to 52 kg grain/kg N in Malawi (Vanlauwe et al. 2011). In the last years, the agronomic efficiency of fertilizer N by maize in the United States of America has increased 39%. In Kenya, N recovery by maize varies from 10% to 55% (Kibunja et al. 2012).

The C_4 plants (e.g. maize, sorghum) are physiologically more efficient in utilizing soil N comparing with C_3 plants [e.g. wheat, sugarcane (*Saccharum* L.), some grasses, legumes] because the formers do not need much Rubisc molecules during the photosynthetic process as do the C_3 plants (De Varennes 2003). At low N availability, C_3 plants have greater N use efficiency than C_4 plants, whereas at high N, the opposite is true. Consequently, identifying the regulatory elements controlling the balance between N allocation to maintain photosynthesis and the reallocation of the remobilized N to sink organs such as young developing leaves and seeds in C_3 and C_4 species is of major importance (Hirel et al. 2007).

Using a simple selection criterion to improve the agronomic N recovery of linseed (*Linum usitatissimum* L.), an annual oil plant, may have negative implications on seed yield and quality. Dordas (2011) observed that agronomic efficiency was negatively correlated with relative seed yield, indicating that high yield was not associated with more efficient N uptake. A negative correlation between agronomic and recovery N efficiency at anthesis and at the green capsule stage was obtained. Therefore, evaluation and selection of different genotypes for agronomic N recovery should be based on multiple criteria rather than just one criterion and also should be accompanied by evaluation of seed yield.

Oilseed rape or canola (*Brassica napus* L.) is an important catch crop in China, receiving a high rate of N (200–330 kg N ha^{-1}) for the more common genotypes (Jensen et al. 1997; Zhang et al. 2009), with a low agronomic N recovery (<50%). Although this crop is very efficient in taking up N (root depth vary between 90 and 190 cm, and does not form mycorrhizal), with two-third of aboveground N located in the seeds (high N harvest index and high physiological N efficiency), the agronomic N recovery determined at harvest was low.

In certain plants, nutrients such as N can be accumulated (stored) during periods of external abundance and consumed in subsequent growth when they are externally limited [e.g. cabbage (*Brassica* sp.), spinach (*Spinacia oleracea* L.)]. In case of cauliflower (*Brassica oleracea* L.), Erley et al. (2010) concluded that selecting for cultivars without delayed head formation is more beneficial for N efficiency than for high yield under optimum N supply. Therefore, breeding for cultivars with a low susceptibility in leaf appearance to low temperature may be effective in enhancing fertilizer N use. A selection for cultivars with high harvest indices is advantageous without compromising storage ability of the heads by increasing the water content.

Concluding, genetic selection is often conducted with high fertilizer N input in order to eliminate N as a variable; however, this can mask efficiency differences among genotypes in accumulating and utilizing N to produce yield (Moll et al. 1982; Raun and Johnson 1999). Using plants grown under agronomic conditions at low and high N fertilization regimes is possible to develop whole-plant physiological studies combined with gene, protein, and metabolite profiling to build up a comprehensive picture depicting the different steps of N uptake, assimilation, and recycling to the final mobilization to the seed. Consequently, continued efforts are needed to include plant selection under low N, which has not often been considered a priority by plant breeders.

3.2 *Growth Curves*

Traditionally, crop growth functions that are used for decision support are derived from conventional experienced-based agronomic research, in which crop growth is related to some defined variables, including N rate (Jame and Cutforth 1996; Gwaze et al. 2002). Plant growth is normally expressed as polynomial or exponential functions of the defined variables, with a regression coefficient obtained through a nonlinear curve-fitting procedure based on observed values (Figs. 4 and 5). The growth and nutrient uptake processes during the period of major growth are best described by

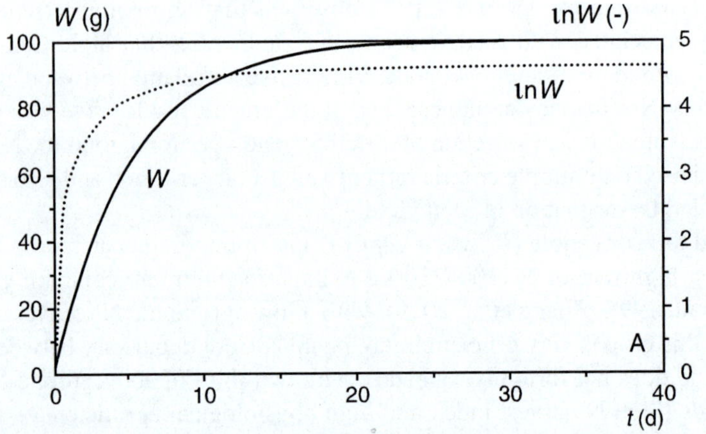

Fig. 4 Mitscherlich's growth curve for annual crops (*W* dry weight, *g*; *t* time, day) (Source: Adapted from Heinen 1999)

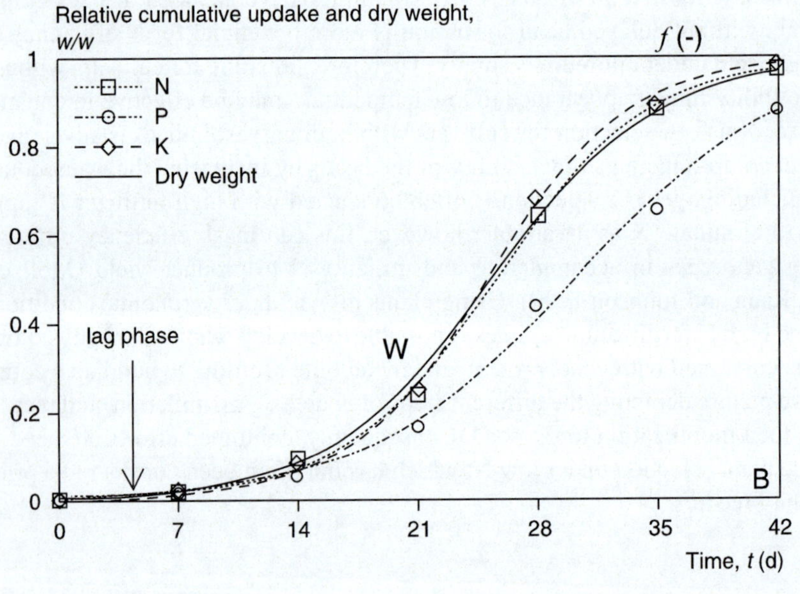

Fig. 5 Sigmoidal growth curve and nutrient uptake for woody perennial crops, but also for autumn-winter cereals (*W* dry weight, *g*; *t* time, day) (Source: Adapted from Kozlowski and Pallardy 1997)

exponential or linear functions; slower growth after flowering in annual crops and after ripening in fruit trees normally follows asymptotic curves.

The major growth period for winter cereals in Northern Europe starts with accelerated seedling growth in spring and ends with the beginning of grain-filling (Fig. 4). In South Europe, germination of autumn-winter cereals occurs slowly in autumn and the exponential vegetative growth takes place in spring (Fig. 5). In general, the major growth period is followed by a final period characterized by ceased growth and dominated by nutrient and carbohydrate relocation processes, which greatly affects the physiological N efficiency (Mikkelsen et al. 1995; Carranca 1996; Heinen 1999; Mengel and Kirkby 2001; Weih et al. 2011).

The general size/age relationship in perennial woody plants is represented by the cumulative growth curve which for biological organisms is sigmoidal (Fig. 5). A double sigmoid growth curve is also possible in several perennial plants like the cherry (*Prunus* L.), apricot (*P. armeniaca* L.), peach (*P. persica* L.), palm (*Palmae*), olive, grape (*Vitis vinifera* L.), and coffee (*Coffea* sp.). Neither of the two major types of curves is restricted to a particular type of fruit (Kozlowski and Pallardy 1997). For example, berries and apples may show both types of growth curves.

Plant N content varies with the plant growth stage, and higher N rates generally results in decreased recovery efficiency values. Since N is the main limiting nutrient for most plant species, it is not surprising that it is efficiently economized by plants in a tightly regulated process. This is the case of N mobilization during seed germination from seed storage proteins or in juvenile phase in woody plants to build up all components of photosynthetic apparatus. At this phase, and depending on size of seed and juvenile plant, the crop response to fertilizer N is at a slow rate (lag phase, Fig. 5). Moreover, during plant senescence, N is mobilized from senescent tissues and transferred to developing parts of the plant wherein it can be conveniently employed for the biosynthesis of a vast variety of plant molecules. Not all photosynthates required for grain-filling or fruit set are photosynthesized during the reproductive period (after flowering), depending on photosynthetic capacity after flowering and on the conditions for the long distance transport of photosynthate. Some of the carbohydrates are synthesized before flowering and stored in stems and leaves during the vegetative growth stage (Mengel and Kirkby 2001). The proportion of stored assimilates used in grain-filling can vary from very low rates up to 40% for rice, 5–20% for wheat, 12–15% for maize, and about 20% for barley.

Taking into account the growth curves, an optimal N use efficiency is obtained by applying N at phases of greater response (W phase and maturity phase in woody plants, Figs. 4 and 5), whereas lowest N use efficiency occurs at highest yield (asymptotic curve, lnW phase, Fig. 4) (Mikkelsen et al. 1995; Heinen 1999; De Varennes 2003; Roberts 2008).

Horticultural plants are generally high N input crops. The amount of nutrients they take up depends on the quantity of fruit and biomass produced, which in turn is influenced by a number of genetic (including root pattern) and environmental variables. In the absence of any other production constraints, nutrient uptake and yield are very closely related. In case of white cabbage, a high N availability leads to a high water accumulation in the leaf tissue because it increases the size of individual cells (Erley et al. 2010). The N effect on cell size seemed to be mediated by improved cell wall extensibility and not by an increased turgor pressure due to high amounts of osmotic like NO_3^-. Nitrate concentrations of the heads were not related to head fresh weight/dry weight ratio heads, not affecting the water accumulation of fully expanded cells, and increased with the increase on protein content. Relatively high protein may be stored instead of being used for current growth when N is mobilized to the head and at late head growth when growth rate is decreasing. Therefore, a high N supply to the heads during the late head growth may be less effective in enhancing yield (Erley et al. 2010). High N mobilization to the head improves head growth, and physiological and agronomic efficiency.

Splitting a total amount of 90 kg N ha^{-1} N, one-third applied 1 week after sowing and two-third applied 1 month latter during the short-period of spinach growth in Portugal (3 months), the vegetable did not accumulate excessive levels of NO_3^- in leaves, and the recovery efficiency estimated in a single cut was low (18%) (Table 2) due to the shallow root system (Carranca 2005), but also because estimate was done at final harvest.

Studies on N use efficiency are generally based on measurements obtained at harvest (Recous et al. 1988; Roberts 2008; Carranca 2010) which underestimate the potential fertilizer N uptake efficiency. Low values for crop N recovery were recorded at harvest (Table 2) for wheat grown under rainfed conditions in India (18–49%) and for *T. durum* L. grown in the Mediterranean area (20–40%), 37% N for maize grown in the North Central United States of America, and 31–40% for rice in Asia, and after fruit ripening in woody trees, ranging from <10% to 35% (Menino et al. 2007; Niederholzer 2007; Neto et al. 2008). Determination of recovery efficiency at harvest and fruit ripening is referred as a limitation for N use efficiency estimates for annual and perennial crops, respectively, where maximum N accumulation has been found to occur at or near flowering, and not at harvest or fruit ripening (Cassman et al. 1993). Average crop N recovery during the major growth period should be preferably used.

3.3 Plant N Reserve

The quality of plant material bought by farmers reflects the agricultural practices performed in the nursery and is crucial for a good establishment of the orchard, conditioning the future growth and development of newly planted trees. A large proportion of total tree N is used to support annual growth (Neilsen et al. 2001). In young deciduous trees, the importance of N reserves for sustaining growth and development (lag phase, Fig. 5) is greater compared to citrus trees. Three-years-old pear trees (*Pyrus communis* L. 'Rocha') only contained in the reserve organs 16% of the total N fertilizer applied in the 2 preceeding years (Neto et al. 2008), contrasting with the 50% recovered by Menino et al. (2007) in orange trees (*Citrus sinensis* L. 'Lane Late') of the same age and planted in the same Mediterranean area. In deciduous trees the importance of N reserves for sustaining growth and development seems greater compared with evergreen trees, where leaves are an important N reserve organ. Nitrogen mobilization efficiency, i.e. the net result of several processes such as the enzymatic breakdown of N containing products in leaves, the phloem loading and transport, and the formation of an abscission layer that cuts off the transport path and causes the leaf to fall is higher in deciduous trees than is in evergreen trees, although differences are small comparing with differences in the mean residence time (Lambers et al. 2008). Strong wind, water stress and frost can reduce mobilization efficiency, but leaves typically abscise only after resorption has ceased (Lambers et al. 2008).

Young non-bearing trees (juvenile tree phase) are only dependent on new inputs of N during the leaf flush and bloom, though total plant recovery of fertilizer N is relatively small: 6%, 14–20% and 15–33%, respectively in the 1st, 2nd, and 3rd years-old (Righetti et al. 1994; Neilsen et al. 2001; Menino et al. 2007; Neto et al. 2008).

From Table 2, the N recovery efficiency in young trees [pears, apples, oranges, eucalyptus (*Eucalyptus globulus* L.)] was rather low in the first year after transplanting (4–6%) since plants suffered from environmental stress. There is a need to produce more stress efficient plants that use less water and fertilizer and have greater resistance to environmental stress such as temperature and drought at the transplanting phase (Davies 2008). The best management practices occurring in the nursery industry must be highly conducive for mycorrhizal usage.

3.4 Rootstock

Nitrogen requirements of citrus and deciduous trees over the first years after planting were related with the increase in the trunk cross-sectional area which is the main N reserve organ for woody plants (Menino 2005; Neto 2008). Tree trunk is often also dependent on slow- or fast-growing rootstocks. Lower N rates are normally efficiently taken up by slow-growing rootstocks. However, Lea-Cox et al. (2001) observed that N use efficiency declined with increasing the N rate irrespective of rootstocks, showing the importance of tree N reserves for plant growth. The genetic variability of rootstocks was possibly the main cause for greater or smaller nitrogen use efficiency by coffee plants, probably as a consequence of the enzyme nitrate reductase activity which regulates the N metabolism (Tomaz et al. 2004).

Rootstocks influence roots number and class, but apparently they do not influence their dimension, distribution and function. In addition, translocation and nutrients uptake and distribution differ among rootstocks. They influence the percentage of N assimilation in roots, thus reducing the amount translocated to the aboveground plant material. In non-fruiting vines, more than 85% N was assimilated in roots, whereas a heavy crop load forced fruiting vines to accumulate less sugar in the roots (Morinaga et al. 2003).

Rootstocks also affect bud break, but the mechanism is not fully understood. This can be a crucial factor affecting the N use efficiency by woody plants and more attention should be given.

The influence of rootstocks for woody plants to use efficiently the fertilizer N should be considered when planning the fertilization.

3.5 Plant N Loss

Plant N recovery also depends on plant N losses (NH_3 volatilization) (Fig. 1). Ammonia is emitted from canopy (leaves), mostly through the stomata when the internal NH_3 concentration is higher than that in the ambient atmosphere, as may often be the case, particularly during periods with rapid N absorption by the roots, when dry matter production is restricted by poor growth conditions or diseases, during senescence induced N remobilization from leaves (Sommer and Hutchings 1995; Pearson et al. 1998; Sommer et al. 2004), or in windy and warm days. Gaseous plant N losses have been found to be significant from flowering to senescence (Harper et al. 1987).

Factors affecting this loss are not sufficiently well understood to justify suggestions of control measures, but surely are related to N nutrition (Sommer and Hutchings 1995). Ammonia emission is attributed to the decomposition of protein during translocation from the leaf to the seeds, and is temperature dependent (Mengel and Kirkby 2001). The translocation index reflects the ability of a plant genotype, or management practice, to incorporate the accumulated N at flowering into the grain. Since NH_3 emission depends on air temperature, photorespiration is a potential source of NH_3 (Mengel and Kirkby 2001). Ammonia losses are thus important in tropical and Mediterranean-type climates.

It can be speculated that plants which assimilate N primarily in their roots and tend to be slow-growing, or climax species of N poor habitats have a mechanism for preventing excessive NH_3 loss to the atmosphere, i.e. those plants having a low apoplastic pH (Pearson et al. 1998). Leaf apoplastic pH plays a significant role in mineral nutrition of higher plants, especially for mobile nutrients such as N (Zou and Zhang 2003).

Nitrogen assimilation after anthesis is needed to achieve high yields and high protein content. This shows the importance of high physiological efficiency, and the proper timing and rate of N application. Between 1% and 5% of shoot N and 1–4% of fertilizer N applied may be lost by gaseous emission (Hardarson et al. 2008). Wheat and rice cultivars with high N harvest indices and low forage yield have low plant N losses and a greater N recovery (Raun and Johnson 1999). Ammonia losses were also documented for N_2 fixing legumes such as soybeans (45 kg N ha^{-1} $year^{-1}$) (Stutte et al. 1979; Raun and Johnson 1999). In case of forages, potential gaseous N losses are avoided by harvesting plants before grain-fill, thus increasing the crop N use efficiency with an N recovery higher than 76% (Mengel and Kirkby 2001). Important plant gaseous N losses were also reported for evergreen citrus trees in a Mediterranean region (Menino 2005; Menino et al. 2007).

3.6 Root Development, Activity and Colonization

Roots are central for anchorage, water and nutrients acquisition and translocation, and storage of reserves, but there are few studies in which the importance of the root system was investigated in relation to N supply (Hirel et al. 2007; Menino et al.

2007; Neto et al. 2008; Sorgonà et al. 2011), and particularly the effect of root colonization on N uptake efficiency (Ames et al. 1983; Miller and Jackson 1998; Azcón et al. 2001; Jackson et al. 2002; Cruz et al. 2007; Cruz and Carranca 2010). The belowground functional trait which best describes the capacity of the root system to explore the soil and acquire soil resource is the root length. The root length is strictly associated with different ability of the plants for N acquisition.

Considering that the root system is constituted by different root types/orders which are distinct genetically, developmentally and functionally, it is still not clear which type of roots or root traits could be advantageous for the plant adaptation in a low N environment (Sorgonà et al. 2011). The first order roots can be mycorrhizal and have a primary development. Depending on plant species, second order roots, and to a lesser extent third and fourth order roots might be either primarily absorptive or primarily associated with transport and support (Venezuela-Estrada et al. 2008; Sorgonà et al. 2011). Moreover, the capacity to acquire N by the root is sustained by the plant aboveground activity, such as leaf area and/or stem height which could be involved in crop N use efficiency. Nitrogen deprivation can cause starch accumulation in leaves and an increase of the photosynthate allocation towards the roots resulting in a decline in the shoot:root ratio (Sorgonà et al. 2011). Results suggested that the investment in the second order lateral root represented the strategy utilized by (citrus) rootstocks for adapting to low N availability.

Symbiotic associations with ectomycorrhizal and arbuscular mycorrhizal (AM) fungi enhance the acquisition of water and several mineral nutrients (N, P, micronutrients) by plants. Nutrient acquisition *via* the AM fungal partner involves transfer across two interfaces: one between the soil and the extra-radical mycelium of the fungus and one between the intra-radical mycelium of the fungus and the root cortex cells. Soil-to-plant N transport by the extra-radical mycelium of endomycorrhizal (AM) fungi was first demonstrated using pots where ^{15}N fertilizer was applied to soil containing the extra-radical mycelium, but no roots (Cruz et al. 2007). Extra-radical mycelium took 30–80% of total plant N.

Very few papers have addressed the topic of N-AM-host plant relationships, and no data are available on the efficiency of mycorrhizal for crop N use. Microbial inoculants are promising components for integrated solutions to agro-environmental problems because inoculants possess the capacity to promote plant growth, enhance nutrient availability, uptake and efficiency, and support the health of plants. The goal of reducing fertilizer usage will be to this century what the goal of reducing pesticides was to the last century. Appropriate management of mycorrhizal in agriculture should allow a substantial reduction in chemical use (e.g. fertilizers, pesticides) and production costs (Beauregard et al. 2008).

Agricultural crops under intensive management are normally highly fertilized. Mycorrhizal colonization generally decreases as nutrient concentration in the rhizosphere increases and it is assumed that mycorrhiza would not contribute significantly to an efficient nutrition of intensively managed crops. But beneficial effects of endomycorrhizal fungi have been reported in almost all the vegetables worldwide. Their use has shown to minimize chemical fertilizer use for onion (*Allium cepa* L.) and garlic (*Allium sativum* L.) by 30–50% (Chadha and Choudhari 2007).

The benefits of reducing fertilizer input to optimize conditions for the endomycorrhizal has also been noted for strawberry (*Fragaria* × *ananassa*) where different endomycorrhizal stimulated either increased stolon production or earlier flowering. Bred cultivars for high N input can show early growth depression as the fungus drains extra carbon (C) from the plant to establish its mycelium rather than to form compatible relationships with AM (Dodd 2000). The effective colonization however, depends not so much on the ultimate level of colonization or the overall effect on nutrient absorption, as on the timing of the colonization and nutrient uptake efficiency (Cruz et al. 2007).

Nitrogen absorption depends on fungi strain and plant cultivar. Lettuce (*Lactuca sativa* L. 'Romana') and tomato (*Lycopersicon esculentum* Miller 'Roma') were tested and significant increases on plant biomass and plant N were observed in symbiosis with the fungi *Glomus intraradices* (Cruz et al. 2007; Cruz and Carranca 2010). A depressing effect was noticed when both crops were inoculated with *G. clarum*. Tomato still responded positively to inoculation with *G. etunicatum*. In lettuce, increasing N supply in relation to phosphorus (P) tended to increase root colonization (Miller and Jackson 1998), but results on crop N use efficiency were not reported. Jackson et al. (2002) explained the plant N acquisition by lettuce (*L. taxa* inoculated with *G. intraradices*) by the increased exploration of the soil volume. Ames et al. (1983) also reported a correlation between mycorrhizal hyphal length and total N derived from applied mineral-^{15}N. According to these authors (1983), celery plants (*Apium graveoleus* L.) inoculated with *G. mossae* derived significantly more ^{15}N from two different sources of N (mineral-^{15}N and organic-^{15}N) than did non-mycorrhizal plants. Azcón et al. (2001) observed that the proportion of N derived from ^{15}N fertilizer decreased in lettuce plants colonized with endomycorrhizal in high N concentrations. This implies that mycorrhizal colonization may be beneficial for plants when fertilizer N is not readily available (organic fertilizer or in relatively low amount).

Most horticultural crops (e.g. lettuce) and cereals [e.g. oat, tef (*Eragrostis tef* Zucc.)] have shallow rooting systems and N recovery efficiency is reduced. If roots are inoculated with endomycorrhizal, an increase on N use efficiency can be expected. It is widely accepted that plants such as *Graminae* with highly branched root systems are less mycotrophic, i.e. less dependent on the fungus for normal growth than those with coarser roots like onion and cassava (*Manihot esculenta* Crantz), and this also determines the symbiotic dependence of the plant (Dodd 2000).

When planting a bare root nursery plant, the amount, timing, and method of N application has a pronounced effect on plant development and allocation pattern that influences growth in the following season. This material coming from the nursery is frequently highly fertilized and not colonized, and accumulates large amount of nutrient reserves, particularly N. Root pruning at planting delays N uptake in the first year, making the newly planted trees even more dependent on the N reserves accumulated in the nursery for the initial vegetative growth.

During the remobilization of N in woody plants which occurs early in spring (temperate countries), root uptake of N is inhibited. A rapid N uptake is associated with the initiation of shoot growth and bloom. If roots of woody plants coming from the nursery are inoculated with fungi, they will adapt more easily to the stress conditions in the field, reducing the importance of a starter fertilizer N in the first year and using more efficiently water and nutrients.

The increasing consumer demands for organic or sustainable production will require changes to incorporate cultural practices which increase arbuscular fungi diversity. What is important is that the maximum pool of AM diversity is maintained at as a high level as possible for subsequent crops, if dependent on endomycorrhizal for growth, particularly under reduced inputs. What is often forgotten is that large areas are sown to non-mycorrhizal hosts such as oilseed rape, cabbage, cauliflower, sugarbeet (*Beta vulgaris* L.), and grasses, and their effects on endomycorrhizal can only be one of reducing infective propagules and hence diversity for subsequent mycorrhizal crops.

Plant breeders may need to find newer more appropriate plant cultivars which can maximize the role of AM in agriculture. For sustainability, the aim is for practices that can give high enough yields with fewer environmental costs such as the optimization of the soil microbiota, and specifically the biotechnological AM approach for maximum N use efficiency (Dodd 2000).

4 Source, Timing and Rate of Fertilizer N

The growth, production, and N use efficiency of most crops are greatly affected by environmental conditions, soil water content in particular. Water stress in arid and semi-arid regions in the tropical and Mediterranean environments may be constrained. Efficiently mycorrhized roots help the plants for a higher water and nutrients uptake under these situations.

Over N application will result in reduced crop N use efficiency and losses in crop quality and fertilizer N. Greater synchrony between crop demand and nutrient supply (Figs. 4 and 5) is necessary to improve N uptake efficiency. Split applications of N during the growing season, rather than a single large application prior to planting are known to be effective in increasing N use efficiency for most crops, including the horticultural crops which have a short growth period, and woody plants with high internal N reserves (Carranca et al. 1999; Cassman et al. 2002; Carranca 2005; Menino 2005; Neto 2008; Roberts 2008). Topdress or sidedress N applications in the middle of the season can also result in greater crop N recovery. Because the risk of N loss is greater with fall N application, N should be applied in the spring and summer in temperate countries.

Soil testing remains one of the most powerful tools available for determining the nutrient supplying capacity of the soil, but to be useful for making appropriate fertilizer recommendations a good calibration data is necessary (Roberts 2008).

Tissue testing (foliar, floral and phloem analyses) is a well known method used to assess N status of growing crops, but other diagnostic tools are also available. Chlorophyll meters have proven useful and non-destructive in season N management for citrus, pears, peaches, cherry, papaya (*Carica papaya* L.), grapevine, coffee and rice (Correia 2001; Porro et al. 2001; Netto et al. 2002, 2005; Figueiredo 2011; Neto et al. 2011), and leaf color charts have been highly successful in guiding farmers, especially for vegetables, golf courses and lawn (Carranca 2005; Hardarson et al. 2008; Roberts 2008).

Determining the right placement of fertilizer N is as important as determining the right application of N rate. Prior to planting, nutrient is broadcast, i.e. incorporated in the soil or applied uniformly on surface, usually as ternary or binary composts. Applied at planting, fertilizer N can be banded with the seed, below the seed, and to the side of the seed. After planting, placement can be as a topdress or a subsurface sidedress.

When fertilizer N is applied at rates in excess of that needed for maximum crop yield, NO_3^- leaching can be significant. In cooler temperate climates, N losses through drainage have approached 23% for corn, corresponding to 26 kg N ha^{-1} $year^{-1}$ under conventional tillage (Raun and Johnson 1999). Because NH_4^+ is normally less subject to leaching or denitrification losses, N maintained as NH_4^+ in the soil should be advisable for late-season uptake for most crops. Losses of N in surface runoff also contribute to a lower crop N use efficiency and may range 1–13% of the total N applied, depending on slope, soil cover, rainfall intensity and distribution, irrigation type and frequency, soil tillage, and fertilizer N rate, source and timing (Carranca 1996; Raun and Johnson 1999).

Gaseous N losses from the soil due to denitrification from applied fertilizer N also reduce N use efficiency by the crops and include 9.5–35% for winter wheat, 10% for lowland rice, and 10–22% for maize (Moll et al. 1982; Carranca et al. 1999). The addition of straw to the soil can double these denitrification losses (IPCC 2007), particularly from paddy rice fields, as also observed by increased dehydrogenase and β-glucosidase activities in flooded soils (Figueiredo et al. 2011).

Low crop N recovery in flooded rice and other clay soils can be due to a greater NH_4^+ fixation in clay minerals 2:1 (Carranca 1989, 2010; Recous et al. 1988; Carranca et al. 1999).

Urea and NH_4^+ fertilizers have generally a lower crop N use efficiency compared to NO_3^- fertilizers when applied to the surface soil, especially at high temperature, windy days, and high soil pH. Ammonia losses from soil can exceed 40% (Raun and Johnson 1999; Ahmed et al. 2009). Proper cultural practices are recommended to avoid this loss, including the use of NO_3^- fertilizers in alkaline soils, avoiding the application of NH_4^+ or urea fertilizers in windy and hot days, and by incorporating sludge, manure and other organic residues in the soil. Ahmed et al. (2009) reported that the use of triple superphosphate can reduce NH_3 losses from alkaline soils because it makes the microsite immediately around the fertilizer N acidic. The addition of zeolite to soils also helps the temporary reduction of NH_3 volatilization. The small internal tunnels of clinoptilolite zeolite have been found to physically protect NH_4^+ from too much nitrification by microorganisms. This process does not only reduce NH_3 loss but also helps in releasing NH_4^+ slowly into the soil increasing NH_4^+ fertilizers or urea-N efficiency by the crop.

Plant residues and other organic material for soil amendment may be considered as slow-release fertilizers, since residues have first to mineralize before N is available to the crop, depending on their amount and composition, especially the long C chains. Controlled-release fertilizers, i.e. compounds of low solubility or coated fertilizers have been successfully used to add small amounts of available N during the season, although this is an expensive technology (Hardarson et al. 2008). Most slow-release fertilizers are more expensive than coated water-soluble N fertilizers and have traditionally been used for high-value horticulture crops (Roberts 2008).

It is important to know how much fertilizer N is used by the plant and how much is lost out of the soil-plant ecosystem. Precision agriculture allows timely and precise application of fertilizer N to meet plant needs as they vary across the landscape, but is a high cost technology. It uses N sensors to automatically correct crop N deficiency on a site-by-site basis. In a certain area of land, soil characteristics and climate may vary and a site-by-site N fertilization is greatly recommended.

Actually, only few studies have investigated foliar N dynamics and efficiency (Kara 2010). Absorption of foliar N is a diffusion process that is governed by time and various other factors. To increase efficiency, urea application should avoid the windy and warm days. Foliar application of urea with a low biuret content at flowering increased wheat grain protein by as much as 4.4% (Raun and Johnson 1999). Smith et al. (1991) also observed positive increases in grain protein content of wheat with foliar N application close to flowering. Woolfolk et al. (2002) concluded that late-season foliar N applications before or immediately after flowering may significantly enhance grain protein content in wheat. Foliar urea applied at barley tillering in three sprayings to minimize leaf damage also increased grain protein more effectively than broadcast ammonium nitrate fertilizer. Unlike, in Turkey, N recovery by wheat was higher in conventional N application than late season foliar N application (Kara 2010).

Foliar urea application to perennial plants in the autumn is an effective way to increase reserve N and consequently improve plant performance the following year, but there is little information regarding the woody plant N use efficiency.

These factors are the most studied and controlled in farming systems, but still N recovery by most plants has not been greater than 50%.

5 Conclusion

The best hope for reducing fertilizer N requirements lies in finding more efficient ways to fertilize crops or using more efficient crops to absorb and use N. Understanding the regulatory mechanisms of controlling plant N economy is vital for improving crop N use efficiency and for reducing excessive input of fertilizers, while maintaining an acceptable yield of main crop species cultivated in the world.

Maximum crop yields have been achieved using new high-yielding breeding plants, with an optimization of photosynthetic capacity, but no maximization on N recovery has been achieved. Breeding for crops that have high yields when there is plenty of available N probably creates less efficient plants for N use. There is a significant margin to improve N use efficiency by selecting genotypes from the available ancient (native) and modern germplasm collections for performance in sustainable agricultural systems. Efforts are needed to include plant selection under low N, which has not been a priority for plant breeders, but when an excess soil N cannot be totally avoided, it is also important to search for genotypes or plant species that are able to absorb and accumulate high N concentrations. The complexity of the role of the enzyme Rubisco in primary CO_2 assimilation in the photorespiratory process and as a storage N pool needs further investigation to optimize crop N use efficiency under low N input for both C_3 and C_4 plant species.

Nitrogen use efficiency can apparently be optimized by reducing crop NH_3 losses, by controlling the photorespiration and improving the N harvest index, maximizing the physiological N use, and applying the proper N rate at the right time.

Beneficial genetic traits also include the ability to maintain photosynthesis and N uptake under reduced N to optimize N use efficiency perhaps through beneficial soil microorganism associations such as mycorrhiza, as was the well succeeded case with strawberries. The optimization of this symbiosis with available germoplasm, probably ancient, can apparently contribute to improve the agronomic efficiency. The effective colonization depends mostly on the timing of the colonization and N use efficiency. Plant N acquisition apparently depends on increased soil volume exploration mainly by the first and second order lateral roots, which in case of woody plants depends perhaps on rootstock type. The slow-growing rootstocks can be possibly more efficient for absorbing and using available N.

Finally, depending on environmental conditions, a rigorous application of fertilizer N at right form, rate, time and place (if possible, specific site-by site) to improved cultivars (not exactly bred cultivars), targeting both sustainable yields and high N use efficiency (by increasing plant mechanisms to accumulate and utilize N) will benefit farmers, society, and the environment alike.

Precision N management, and intercropping or crop rotation systems including legumes are sustainable agricultural techniques that improve efficiency of crop N use.

Most studies for crop N use efficiency are based on N recovery values at harvest (annual crops) or fruit ripening (woody plants) which may underestimate the potential for fertilizer N utilization. Greater efficiency arises when it is calculated on basis of the mean N during major growth period.

Studies are missing for natural ecosystems to evaluate the contribution of genetic factors to control the N use efficiency, and the genetic mechanisms involved in environments of high and low fertility.

References

Ahmed OH, Hussin A, Ahmad HMH, Jalloh MB, Rahim AA, Majid NMA (2009) Enhancing the urea-N use efficiency in maize (*Zea mays*) cultivation on acid soils using urea amended with zeolite and TSP. Am J Appl Sci 6:829–833

Alves JA (1979) Efeito dos adubos azotados sobre o peso e o teor de azoto do grão, a exportação e a recuperação do azoto aplicado na cultura do trigo. INIA, Oeiras

Ames RN, Reid CP, Porterf PLK, Cambardella C (1983) Hyphal uptake and transport of nitrogen from two ^{15}N-labelled sources by *Glomus mossae*, a vesicular-arbuscular mycorrhizal fungus. New Phytol 95:381–396

Azcón R, Ruiz-Lozano JM, Rodriguez R (2001) Differential contribution of arbuscular mycorrhizal fungi to plant nitrate uptake (^{15}N) under increasing n supply to soil. Can J Bot 79:1175–1180

Beauregard MS, Hamel C, Arnaud MS (2008) Arbuscular mycorrhizal fungi communities in major intensive North American grain production. In: Siddiqui ZA (ed) Mycorrhizae: sustainable agriculture and forestry. Springer-Science, Dordrecht, pp 135–157

Carranca C (1989) Nitrogen availability in some maize cultivated soils of Portugal. In: Germon JC (ed) Management systems to reduce impact of nitrates. Elsevier Applied, London, pp 197–209

Carranca C (1996) Nitrogen cycling in Portuguese soils and its assessment by ^{15}N. Ph.D. thesis (Agronomy), University of Lisbon, Portugal

Carranca C (2005) Crop management and postharvest quality of spinach for fresh consumption and processing. In: Dris R (ed) Vegetables: growing, environment and mineral nutrition. WFL Publ, Helsinki, pp 260–270

Carranca C (2010) Gestão eficiente do azoto e da matéria orgânica na produção agrícola e na sustentabilidade dos ecossistemas agrários. Programas de investigação e de pós-graduação para habilitação para o exercício de funções de coordenação científica da carreira de investigação do INIA, Oeiras

Carranca C (2011) The role of legumes for sustainability: research needs for future prospects. In: Symbiosis: evolution, biology and ecological effects. NOVA publishers (in press)

Carranca C, De Varennes A, Rolston DE (1999) Variation in N-recovery to winter wheat under Mediterranean conditions, studied with ^{15}N-labelled fertilizers. Eur J Agron 11:145–155

Cassman KG, Kroff MJ, Gaunt J, Peng S (1993) Nitrogen use efficiency of rice reconsidered: what are the key constraints? Plant Soil 155(156):359–362

Cassman KG, Dobermann A, Walters DT (2002) Agroecosystems, nitrogen use efficiency, and nitrogen management. Ambio 31:132–140

Chadha KL, Choudhari ML (2007) Report of the working group on horticulture, plantation crops and organic farming for the XI five year plan (2007–12). Government for India planning commission, New Delhi

Correia MFOP (2001) Physiological and nutritional characterization of iron chlorosis in citrus: assessment of the mechanisms underlying the tolerance to HCO_3 effects. Ph.D. thesis, University of Algarve, Faro

Cruz C, Carranca C (2010) A colonização micorrízica no uso eficiente do azoto pelas culturas hortícolas. Rev APH (Associação Portuguesa de Horticultura) (102), Julho-Agosto 102:9–12

Cruz C, Egsgaard H, Trujillo C, Per A, Requena N, Martins-Loução MA, Jakobsen I (2007) Enzymatic evidence for the key role of arginine in nitrogen translocation by arbuscular mycorrhizal fungi. Plant Physiol 144:782–792

Davies FT Jr (2008) Opportunities from down under: how mycorrhizal fungi can benefit nursery propagation. Comb Proc Int Plant Propag Soc 58:539–548

Dawson JC, Huggins DR, Jones SS (2008) Characterizing nitrogen use efficiency in natural and agricultural ecosystems to improve the performance of cereal crops in low-input and organic agricultural systems. Field Crops Res 107:89–101

De Varennes A (2003) Produtividade dos solos e ambiente. Escolar Editora, Lisboa

Dodd JC (2000) The role of arbuscular mycorrhizal fungi in agro- and natural ecosystems. Outlook Agric 29:63–70

Dordas CA (2011) Nitrogen nutrition index and its relationship to N use efficiency in linseed. Eur J Agron 34:124–132

Erley GS, Dewi ER, Nikus O, Horst WJ (2010) Genotypic differences in nitrogen efficiency of white cabbage (*Brassica oleracea* L.). Plant Soil 328:313–325

FAO (Food and Agriculture Organization of the United Nations) (1971–1981) FAO/UNESCO soil map of the world (1:5 000 000)

FAO (Food and Agriculture Organization of the United Nations) (2008) Current world fertilizer trends and outlook to 2011/12. Food and Agriculture Organization of the United Nations, Rome

Figueiredo N (2011) Dinâmica do azoto em campos alagados para produção de arroz, em Salvaterra de Magos. Eficiência do uso do nutriente pela cultura. Thesis for master degree in agricultural engineering, ISA/UTL, Lisbon
Figueiredo N, Fareleira P, Menino R, Marques P, Vargues A, Carranca C (2011) A produção de arroz em Portugal. Livro das comemorações dos 75 anos da Estação Agronómica Nacional, Oeiras
Forsum O, Svennerstam H, Ganeteg U, Na¨sholm T (2008) Capacities and constraints of amino acid utilization in Arabidopsis. New Phytol 179:1058–1069
Gördes D, Kolukisaoglu Ü, Thurow K (2011) Uptake and conversion of D-amino acids in *Arabidopsis thaliana*. Amino Acids 40:553–563
Gwaze DP, Bridgwater FE, Williams CG (2002) Genetic analysis of growth curves for a woody perennial species, *Pinus taeda* L. Theor Appl Genet 105:526–531
Hardarson G, Hood-Nowotny R, Jensen ES, Peoples MB, Van Cleemput O, Van Kessel C, Vanlauwe B, Zapata F (2008) In: IAEA (ed) Guidelines on nitrogen management in agricultural systems, vol 29, Training course series. International Atomic Energy Agency, Vienna
Harper LA, Sharpe RR, Langdale GW, Evans JE (1987) Nitrogen cycling in a wheat crop: soil, plant, and aerial nitrogen transport. Agron J 79:967–973
Heinen M (1999) Analytical growth equations and their Genstat 5 equivalents. Neth J Agric Sci 47:67–89
Hirel B, Le Gouis J, Ney B, Gallais A (2007) The challenge of improving nitrogen use efficiency in crop plants: towards a more central role for genetic variability and quantitative genetics within integrated approaches. J Exp Bot 58:2369–2387
IPCC (2007) Contribution of Working Group II (WGII): Climate change 2007: Impacts, vulnerabilities and adaptation in vulnerable countries. United Nations Framework Convention on Climate Change
IUSS (International Union of Soil Science), Working Group WRB (World Reference Base) (2006) World reference base for soil resources 2006, vol 103, 2nd edn, World soil resources reports. FAO, Rome
Jackson LE, Miller D, Smith SE (2002) Arbuscular mycorrhizal colonization and growth of wild and cultivated lettuce in response to nitrogen and phosphorus. Sci Hortic 94:205–218
Jame YW, Cutforth HW (1996) Crop growth models for decision support systems. Can J Plant Sci 76:9–19
Jensen ES, Haugaard-Nielsen H (2003) How can increased use of biological N_2 fixation in agriculture benefit the environment? Plant Soil 252:177–186
Jensen LS, Christensen L, Mueller T, Nielsen NE (1997) Turnover of residual ^{15}N-labelled fertilizer N in soil following harvest of oilseed rape (*Brassica napus* L.). Plant Soil 190:193–202
Kara B (2010) Influence of late-season nitrogen application on grain yield, nitrogen use efficiency and protein content of wheat under Isparta ecological conditions. Turk J Field Crop 15:1–6
Kibunja CN, Mwaura FB, Mugendi DN, Kitonyo EM, Salema MP (2012) Crop N uptake and fertilizer N use efficiency under a continuous maize-bean farming system in the semi-humid highlands of Kenya. Afr J Agric Res (in press)
Kozlowski TT, Pallardy SG (1997) Physiology of woody plants. Academic, New York
Krab EJ, Cornelissen JHC, Lang SI, van Logtestijn RSP (2008) Amino acid uptake among wide-ranging moss species may contribute to their strong position in higher-latitude ecosystems. Plant Soil 304:199–208
Lambers H, Chapin FS, Pons JL (2008) Plant physiological ecology, 2nd edn. Springer, New York
Lea-Cox JD, Syvertsen JP, Graetz DA (2001) Springtime ^{15}Nitrogen uptake, partitioning, and leaching losses from young bearing *Citrus* trees of differing nitrogen status. J Am Soc Hortic Sci 126:242–251
Li Q-Z, Sun J-H, Wei X-J, Christie P, Zhang F-S, Li L (2011) Overyielding and interspecific interactions mediated by nitrogen fertilization in strip intercropping of maize with faba bean, wheat and barley. Plant Soil 339:147–161
Mengel K, Kirkby EA (2001) Principles of plant nutrition, 5th edn. Kluwer Academic, London, p 849
Menino MR (2005) Eficiência de utilização do N em jovens laranjeiras 'Lane Late', na Campina de Faro (ZVN). Tese de Doutoramento em Engenharia Agronómica, ISA, Lisbon

Menino MR, Carranca C, De Varennes A (2007) Distribution and remobilization of nitrogen in young non-bearing orange trees grown under Mediterranean conditions. J Plant Nutr 30:1083–1096

Mifflin BJ, Lea PJ (1976) The pathway of nitrogen assimilation in plants. Phytochemistry 15:873–885

Mikkelsen DS, Jayaweera GR, Rolston D (1995) Nitrogen fertilization practices of lowland rice. In: Bacon P (ed) Nitrogen fertilization in the environment. Marcel Dekker Inc, New York, pp 171–223

Miller RL, Jackson LE (1998) Survey of vesicular-arbuscular mycorrhizae in lettuce production in relation to management and soil factors. J Agric Sci, Cambridge 130:173–182

Moll RH, Kamprath EJ, Jackson WA (1982) Analysis and interpretation of factors which contribute to efficiency to nitrogen utilization. Agron J 74:562–564

Morinaga K, Imai S, Yakushiji H, Koshita Y (2003) Effects of fruit load on partitioning of ^{15}N and ^{13}C, respiration and growth of grapevine roots at different fruit stages. Sci Hortic 97:239–253

Neilsen D, Millard P, Neilsen GH, Hogue EJ (2001) Nitrogen uptake, efficiency of use, and partitioning for growth in young apple trees. J Am Soc Hortic Sci 126:144–150

Neto CS (2008) Estudo da dinâmica do azoto em pereira (*Pyrus communis* L.). Tese de Doutoramento em Engenharia Agronómica, ISA, Lisboa

Neto CS, Carranca C, Clemente J, De Varennes A (2008) Nitrogen distribution, remobilization and re-cycling in young non-bearing 'Rocha' pear trees. Sci Hortic 118:299–307

Neto C, Carranca C, Clemente J, de Varennes A (2011) Assessing the nitrogen status of young non-bearing 'Rocha' pear trees grown in a Mediterranean region by using chlorophyll meter. J Plant Nutr 34:627–639

Netto AT, Campostrini E, de Oliveira JG, Yamanishi OK (2002) Portable chlorophyll meter for the quantification of photosynthetic pigments, nitrogen and the possible use for assessment of the photochemical process in *Carica papaya* L. Braz J Plant Physiol 14:203–210

Netto AT, Campostrini E, Bressan-Smith SE (2005) Photosynthetic pigments, nitrogen, chlorophyll fluorescence and SPAD-502 readings in coffee leaves. Sci Hortic 104:199–209

Niederholzer F (2007) Improving the nutrient efficiency of tree crops. Plant and Soil Conference, UCCE Farm Advisor, Sutter/Yuba Counties, Yuba City, pp 63–70

Olea F, Pérez-Garcia A, Cantón FR, Rivera ME, Cañas R, Ávila C, Cazorla FM, Cánovas FM, de Vicente A (2004) Up-regulation and localization of asparagine synthetase in tomato leaves infected by the bacterial pathogen *Pseudomonas syringae*. Plant Cell Physiol 45:770–780

Oliveira MD, Gusmão MR, Carranca C, Gonçalves MG (1989) Some forms of nitrogen in a sugarbeet cultivated soil of Portugal. In: Germon JC (ed) Management systems to reduce impact of nitrates. Elsevier Applied, London, pp 210–223

Pearson J, Clough ECM, Woodall J, Havill DC, Zhang X-H (1998) Ammonia emissions to the atmosphere from leaves of wild plants and *Hordeum vulgare* treated with methionine sulphoximine. New Phytol 138:37–48

Peoples MB, Herridge DF, Ladha JK (1995) Biological nitrogen fixation: an efficient source of nitrogen for sustainable agricultural production? Plant Soil 174:3–28

Porro D, Bertamini M, Dorigatti C, Stafanini M, Ceschini A (2001) Lo SPAD nella diagnosi dello stato nutrizionale della vite. L'Informatore Agrario 26:49–55

Raun WR, Johnson GV (1999) Improving nitrogen use efficiency for cereal production. Agron J 91:357–363

Recous S, Machet JM, Mary B (1988) The fate of labeled ^{15}N urea and ammonium nitrate applied to a winter wheat crop. II- Plant uptake and N efficiency. Plant Soil 112:215–224

Righetti TL, Khemira H, Sugar D, Sanchez EE (1994) Nitrogen requirements and nitrogen uptake efficiency for *Comice* pears. Acta Hortic 367:304

Roberts TL (2008) Improving nutrient use efficiency. Turk J Agric Forest 32:177–182

Sanchez PA, Izac A-MN, Valencia I, Pieri C (1996) Soil fertility replenishment in Africa: a concept: 200–207. In: Breth SA (ed) Achieving greater impact from research investments in Africa. Sasakawa Africa Association, Mexico City

Smith CJ, Frency JR, Sherlock RR, Galbally IE (1991) The fate of urea nitrogen applied in a foliar spray to wheat at heading. Fert Res 28:129–138

Sommer SG, Hutchings N (1995) Techniques and strategies for the reduction of ammonia emission from agriculture. Water Air Soil Pollut 85:237–248

Sommer SG, Schjoerring JK, Denmead OT (2004) Ammonia emission from mineral fertilizers and fertilized crops. Adv Agron 82:557–622

Sorgonà A, Lupini A, Abenavoli MR (2011) Nitrate use-efficiency: a morphological analysis of the above- and below-ground. Functional traits in two citrus rootstocks. Glob J Plant Ecophysiol 1:26–37

Stutte CA, Weiland RT, Blem AR (1979) Gaseous nitrogen loss from soybean foliage. Agron J 71:95–97

Svennerstam H, Jämtgård S, Ahmad I, Huss-Danell K, Näsholm T, Ganeteg U (2011) Transporters in Arabidopsis roots mediating uptake of amino acids at naturally occurring concentrations. New Phytol. doi:10.1111/j.1469-8137.2011.03699.x

Tomaz MA, Sakiarna NS, Martinez HEP, Cruz CD, Zambolim L, Pereira AA (2004) Comparison of nutritional efficiency among hydroponic grafted young coffee trees for N, P, and K. Crop Breed Appl Biotechnol 4:92–99

Uribelarrea M, Crafts-Brandner SJ (2009) Physiological N response of field-grown maize hybrids (*Zea mays* L.) with divergent yield potential and grain protein concentration. Plant Soil 316:151–160

Valenzuela-Estrada LR, Vera-Caraballo V, Ruth LE, Eissenstat DM (2008) Root anatomy, morphology, and longevity among root orders in *Viccinium corymbosum* (Ericaceae). Am J Bot 95:1506–1514

Vanlauwe B, Kihara J, Chivenge P, Pypers P, Coe R, Six J (2011) Agronomic use efficiency of N fertilizer in maize-based systems in sub-Saharan Africa within the context of integrated soil fertility management. Plant Soil 339:35–50

Weih M, Asplund L, Bergkvist G (2011) Assessment of nutrient use in annual and perennial crops: a functional concept for analyzing nitrogen use efficiency. Plant Soil 339:513–520

Woolfolk CW, Raun WR, Johnson GV, Thompson WE, Mullen RW, Wyn KJ, Freeman KW (2002) Influence of late-season foliar nitrogen applications on yield and grain nitrogen in wheat. Agron J 94:429–434

Zhang Z-H, Song H-X, Liu Q, Rong X-M, Peng J-W, Xie G-X, Zhang Y-P (2009) Study on differences of nitrogen efficiency and nitrogen response in different oilseed rape (*Brassica napus* L.) varieties. Asian J Plant Sci 1:105–112

Zou C, Zhang F (2003) Ammonium improves iron nutrition by decreasing leaf apoplastic pH of sunflower plants (*Helianthus annuus* L. cv. Frankasol). Chin Sci Bull 48:2215–2220

Microalgae for Bioremediation of Distillery Effluent

Nirbhay K. Singh and D.B. Patel

Abstract Microalgae are single-celled autotrophic photosynthetic microorganisms, and constitute a heterogeneous group of unicellular microorganisms. Microalgae are valuable sources of food and feed products, high-value oils, biofuels, chemicals, medicinal products and pigments. Microalgae are potent candidates for bioremediation of a large number of pollutants. Distillery effluents, also referred to as spentwash/stillage/slop/vinasses, are one of the most environmentally aggressive industrial effluents. Distillery effluents often have low pH, strong odor, dark brown color, and extremely high nutrients content. They often have chemical oxygen demand (COD) ranging from 80,000 to 100,000 mg L^{-1}; biological oxygen demand (BOD) of 40,000–50,000 mg L^{-1}; and nutrients like nitrogen of 1,660–4,200 mg L^{-1}; phosphorus of 225–3,038 mg L^{-1}; and potassium of 9,600–17,475 mg L^{-1}. With the development of economies and resultant growth of distillery industries, large volume of spentwash is produced which is likely to cause extensive soil and water pollution due to the presence of high amount of organic matter and dark brown colored recalcitrant compounds. There have been many isolated studies for treatment of distillery effluents and related compounds using microalgae. This review tries to weave these isolated studies in a string to reflect a clear picture of the utility of microalgae in bioremediation of distillery effluents. In view of the wider applicability of microalgal strains in remediation of domestic wastewaters, inorganic nutrients like nitrogen and phosphorus, heavy metals, pesticides, phenols, aromatic hydrocarbons, textile dyes, and detergents; we reviewed the potentiality of these unicellular microorganisms for bioremediation of distillery effluents.

For treatment of wastewaters native microalgal strains are a favorable alternative to the traditional wastewaters treatment systems. The traditional physico-chemical methods of waste water treatment are costly, energy expensive, environmentally unfriendly,

N.K. Singh (✉) • D.B. Patel
Department of Microbiology, C.P. College of Agriculture, S.D.A.U., S.K. Nagar, Palanpur, Gujarat 385506, India
e-mail: nirbhaysingh78@gmail.com

E. Lichtfouse (ed.), *Farming for Food and Water Security*, Sustainable Agriculture Reviews 10, DOI 10.1007/978-94-007-4500-1_4,

unsustainable, and have tendency to form toxic intermediates. Moreover, the anaerobic degradation of aromatic amino acids by many heterotrophic bacteria and fungi may further aggravate the pollution problem due to further liberation of phenol and cresol. However, biological processes using microorganisms are comparatively less expensive and can lead to almost complete mineralization of the compounds. The environment friendly approach of activated sludge process using consortium of microorganisms is regarded more efficient for mineralization of toxic organic compounds. A dark brown recalcitrant pigment called melanoidin, formed in the molasses as a result of Maillard reaction, can form stable complexes with metal cations. *Oscillatoria boryana* can utilize melanoidins as a carbon and nitrogen source; and decolorize pure melanoidins by about 75% and crude pigment by 60% in 30 days. A consortium of *Oscillatoria, Lyngbya* and *Synechocystis* decolorized melanoidin by 98% by absorption followed subsequently by degradation of the organic compounds. The microalgal strains like *Anabaena cylindrica*, *Phormidium foveolarum*, *P. valderianum*, *Synechococcus*, *Ankistrodesmus braunii* and *Scenedesmus quadricauda* have been reported instrumental in degradation of phenol and its derivatives, whereas the performances of *Phormidium ambiguum*, *Chroococcus minutus*, *Oscillatoria*, and *Anabaena azollae* were found satisfactory for degradation of lignin. *Phormidium ambiguum* and *Chroococcus minutus* were found to reduce lignin by over 73.0% from the pulp and paper mill wastes in 5 days; whereas *Phormidium*, *Oscillatoria*, and *Anabaena azollae* were able to degrade lignin by 89% and hemicellulose by 92% from coir waste.

The ability of microalgae to grow under mixotrophic growth conditions enable them to survive under low light or carbon dioxide shortage and represents an alternative to other biological treatments for remediation of phenol-containing wastewaters. The microalgae carry out *ortho*-fission of the phenolic substances extracellularly in the dark. This reaction is catalyzed by a protein; however, such transformation is inhibited by heat, proteases and metal ions. *Ochromonas danica* can enzymatically carry out *meta*-cleavage of phenol and its methylated homologues and the compounds produced are metabolized as intermediates of the Krebs cycle. Several species of cyanobacteria are also known to possess phenol-degrading enzymes like lignin peroxidase, laccase, polyphenol oxidase, superoxide dismutase, catalase, peroxidase and ascorbate peroxidise. The lignolytic and anti-oxidative enzymatic activities increases in the presence of phenol, if microalgae are subjected to nitrogen limiting condition. Moreover the photosynthetic nature of the microalgae enable them to produce toxic active oxygen species like O^{2-}, OH^-, and H_2O_2 which have strong oxidising agent and are involved in degradation of melanoidin. Presence of molecular oxygen is indispensable for enzymatic breakdown of aromatic ring of phenols by microalgae; which involves hydroxylation of the aromatic ring and formation of catechol followed subsequently by oxidation. Moreover, the microalgal growth and the rate of biodegradation can be enhanced under increased light intensities and by addition of carbon dioxide and sodium bicarbonate. The microalgae offers spectacular prospects of their use in bioremediation because of their ubiquitous distribution, cost efficient, central role in nitrogen fixation, turnover of carbon and other nutrient elements, almost complete mineralization of compounds, and ability to scavenge nutrients.

Keywords Bioremediation • Distillery wastewaters • Melanoidins • Microalgae • Pollution

Abbreviations

COD Chemical oxygen demand
BOD Biochemical oxygen demand
NAD Nicotinamide adenine dinucleotide
$NADH_2$ Reduced nicotinamide adenine dinucleotide
LC Lethal concentration

1 Introduction

Microalgae are microscopic photosynthetic cell factories, have high surface area-to-volume ratio, and are primary synthesizers of organic matter in aquatic habitats. They occur as discrete individuals alone, in pairs, in clusters, or in sheet of individuals all looking alike; but cannot form roots, stems, or leaves. The most abundant microalgae are single-cell drifters, generally called phytoplankton. They are capable of rapid uptake of nutrients and CO_2, possess faster cell growth and have much higher photosynthetic efficiency (Singh and Dhar 2011). They constitute a heterogeneous group of photosynthetic and/or heterotrophic unicellular microorganisms containing chlorophylls and other photosynthetic pigments and are used in human health food products, feeds for fish and livestock, high-value oils, chemicals, and pharmaceutical products (Hallmann 2007; Pulz and Gross 2004; Spolaore et al. 2006).

Production of ethanol from molasses generates large volumes of wastewater, which is often characterized by low pH, strong odor, dark brown color, and extremely high chemical oxygen demand – 80,000–100,000 mg L^{-1} and biochemical oxygen demand – 40,000–50,000 mg L^{-1} (CPCB 2003). Distilleries are rated as one of the most polluting industries because the effluent emitted from it contains high amount of organic matter and nutrients like nitrogen – 1,660–4,200 mg L^{-1}, phosphorus – 225–3,038 mg L^{-1}, and potassium – 9,600–17,475 mg L^{-1} (Mahimairaja and Bolan 2004); which can lead to eutrophication of water bodies. The dark color of such effluents blocks penetration of sunlight deep into the water and hinders photosynthesis by aquatic flora. Therefore, it deteriorates the water quality of aquatic bodies and is deleterious to the existence of aquatic life (FitzGibbon et al. 1998). Effluents from distilleries contain a large number of pollutants in varying concentrations. Wide variation in composition of industrial effluents is because of the presence of various recalcitrant compounds that complicate the treatment of wastewater streams. In distilleries, the main source of wastewater generation is the distillation step wherein large volumes of dark brown effluent, termed as spentwash, stillage, slop or vinasses, are generated (Nandy et al. 2002; Patil et al. 2003).

Table 1 Characteristics of vinasses/spentwash generated from cane and beet molasses

Characteristics	Cane molasses [a]	Beet molasses [b]
Chemical oxygen demand, mg L^{-1}	104,000–134,400	91,100
Biological oxygen demand, mg L^{-1}	46,100–96,000	44,900
Total dissolved solids, mg L^{-1}	79,000–87,990	–
Total nitrogen, mg L^{-1}	1,660–4,200	3,569
Total phosphorus, mg L^{-1}	225–3,038	163
Potassium, mg L^{-1}	9,600–17,475	10,030
Sulfur as SO4, mg L^{-1}	3,240–3,425	3,716
pH	3.9–4.3	5.35

mg L^{-1} represent the amount of various ingredients in milligrams per litre of the effluents

"Note the high level of chemical oxygen demand, biological oxygen demand, nitrogen, phosphorous, potassium and sulfur in the cane molasses spentwash in comparison to the beet molasses spentwash. However, the pH in case of cane molasses spentwash is much less than the beet molasses spentwash"

[a] Mahimairaja and Bolan (2004)

[b] Wilkie et al. (2000)

The characteristics of the vinasses depend on the raw material used. It is estimated that about 88% of the molasses constituents end up as wastes (Jain et al. 2002). Molasses spentwash has very high levels of BOD, COD, potassium, phosphorus, and sulfate content (Table 1). In addition, cane molasses spentwash contains many low molecular weight compounds such as lactic acid, glycerol, ethanol and acetic acid (Wilkie et al. 2000). Cane molasses also contains around 2% of a dark brown pigment called melanoidins that impart dark color to the spentwash (Kalavathi et al. 2001). Melanoidins are low and high molecular weight complex biopolymers that are largely recalcitrant to microbial decomposition and are formed as one of the final products of Maillard reaction (Martins and van Boekel 2004). Apart from melanoidins, spentwash contains other colorants such as phenolics, caramel and melanin. Phenolics are more pronounced in cane molasses wastewater whereas melanin is significant in beet molasses (Godshall 1999). These chemical compounds are discharged in huge quantities by molasses based distilleries and fermentation industries (Kumar and Chandra 2006).

Removal of the recalcitrant compounds from effluents can be attained by physico-chemical processes *e.g.* solvent extraction, activated carbon adsorption, advanced oxidation (Ugurlu and Kula 2007) as well as biological processes (Sreekanth and Sivaramakrishna 2009). However, the physico-chemical removal or treatment technologies have been found to have inherent drawback owing to their high cost and tendency to form secondary toxic intermediates (Talley and Sleeper 1997). On the other hand, biological processes using microorganisms are comparatively less expensive and can lead to complete mineralization of the compounds, with practically no toxic end-products. Therefore, biological removal especially activated sludge process has turned out to be a favorable alternative in comparison to the traditional physical and chemical processes (Thavasi and Jayalakshmi 2003). An environment friendly approach of activated sludge process using consortium of microorganisms was employed to completely mineralize toxic organic compounds (Rajani Rani et al. 2011).

Globally, wastewater treatment is an area of fundamental importance in environment based research due to the ongoing pollution problems posed by various sources. Microalgae-based processes have been designed earlier for the biotransformation of environmental pollutants (Olaizola 2003). The use of several microalgae including cyanobacteria in wastewater treatment is known for many years. The strains of *Chlorella vulgaris*, *Scenedesmus dimorphus*, *Nostoc muscorum*, *Anabaena variabilis*, *Plectonema*, *Oscillatoria*, *Phormidium* and *Spirulina* present a good option for biological treatment of agro-industrial wastewater (Gonzalez et al. 1997; Singh and Dhar 2006). Most of these cyanobacterial species perform more efficiently in suspension cultures. Their application toward the removal of BOD, COD, fats, oils, grease, Zn and Cu; as well as physical contaminants like suspended and dissolved solids from mixed domestic–industrial wastewater in a relatively short duration is well documented. Besides removing pollutants, the cyanobacteria have the ability to easily acclimatize to extreme environments and produce many valuable compounds (Pulz and Gross 2004). Certain microalgal species can be used for energy like biohydrogen production through the process of bio-photolysis, while few others may naturally synthesize hydrocarbons suitable for direct use as high-energy liquid biofuels (Singh and Dhar 2011).

Thus it is imperative to say that the distillery effluents have the capacity to be potentially problematic with respect to environmental pollution; and therefore, treatment of these effluents is of serious environmental concern. This study encourages the feasibility of using microalgae as a supplementary tool for bioremediation of distillery effluent owing to their wider adaptability and unique physiological and biochemical attributes. Increasingly large number of researchers and entrepreneurs have realized about the broad potential of microalgae. In this background this review analyzes the potential of microalgae for the treatment and decontamination of distillery wastewaters and their capacity to decontaminate aromatic pollutants present in such effluents.

2 Microalgal Treatment of Wastewaters

The utilization of microalgal for treatment of wastewaters has an edge over the conventional systems and is getting more attention than other microorganisms isolated from the soil, especially in the tropical and subtropical regions (Abdel Hameed and Hammouda 2007). The photoautotrophic nature of microalgae and ability of some species to fix atmospheric nitrogen enable them to be used as producers, as opposed to consumers, and thus make their growth and maintenance inexpensive (Hensyl 1989). The idea of using microalgae in decontamination of wastewater was initially proposed about six decades before; since then a number of reports have been published regarding their use in bioremediation of distillery wastewaters and related xenobiotic compounds (Table 2). Wastewater treatments by microalgal cultures do not generate additional pollution when the biomass is harvested; at the same time they offer efficient recycling of nutrients (Perez- Garcia et al. 2011). The major problem in using microalgae for wastewater treatment is their recovery from the

Table 2 Bioremediation of distillery spentwash and related xenobiotic compounds by microalgae

Substrate	Organism	Observations	References
Distillery vinasses	*Oscillatoria, Lyngbya*, and *Synechocystis*	Treatment of anaerobically treated distillery vinasses with *Oscillatoria boryana* attained 54% decolorisation of spentwash (final concentration- 5%, v/v) after 30 days of treatment in open field conditions without nutrient amendment. Strains of *Oscillatoria, Lyngbya*, and *Synechocystis* reduced color of spentwash respectively by 96%, 81% and 26% through bio-flocculation. Consortium of these three cyanobacterial strains showed maximum decolorisation of spentwash by 98%	Kalavathi et al. (2001), Patel et al. (2001)
	Chlorella vulgaris	Use of *Chlorella vulgaris* for 4 days followed by macrophyte treatment for 6 additional days reduced color of the effluent up to 52%. Whereas, macrophyte treatment alone reduced the original color only up to 10%	Valderrama et al. (2002)
		Laboratory-scale microalgae pond permitted 83.2% and 88.0% reduction of Chemical oxygen demand and Biological oxygen demand respectively, of distillery effluent pre-treated in an anaerobic fixed bed reactor	Travieso et al. (2008)
Pure melanoidins	*Oscillatoria boryana* BDV 92181	This cyanobacterium used melanoidin as nitrogen and carbon source and decolorized the pure melanoidin (0.1% w/v) by about 75%. However, no growth was observed at melanoidin concentrations above 0.5%	Kalavathi et al. (2001)
Phenol and methylated homologues	*Ochromonas danica*	This microalga is a nutritionally versatile and mixotrophic chrysophyte that could grow on phenol as the sole carbon source, removed phenol carbon from the growth medium, and incorporated this phenol into the protein, nucleic acid, and lipid fractions. This study provided the Enzymology of phenol degradation and evidence that the enzymes involved in phenol catabolism are inducible in nature. It was the first definitive identification of the *meta*-cleavage pathway for aromatic ring degradation in a eukaryotic alga	Semple and Cain (1996)
	Chlorella vulgaris, Chlorella VT-1	Both microalgae could tolerate phenol. However, *Chlorella* VT-1 was comparatively more tolerant and could grow in the presence of phenol (at 400 mg L^{-1}). Phenol was not removed from the medium in the presence of light by Chlorella VT-1 but it was capable of growth on phenol in the dark where phenol was removed from the medium	Scragg (2006)
	Phormidium valderianum BDU 30501	This organism was able to tolerate phenol, grow at a phenol concentration of 50 mg L^{-1} and removed 38 mg L^{-1} of phenol within a retention period of 7 days. It possessed inducible polyphenol oxidase and laccase enzymes	Shashirekha et al. (1997)
	Autotrophic and heterotrophic microalgae	A dynamic energy budget model for describing aerobic biodegradation of phenolic compounds by autotrophic and heterotrophic microalgae was proposed. This model can be applied to the biodegradation of a wide variety of aromatic compounds	Lika and Papadakis (2009)

Low molecular weight phenols (tyrosol, hydroxytyrosol, xytyrosol, catechol, sinapic acid, ferulic acid, 4-OH-benzoate, *p*-coumaric acid, vanillic acid)	*Scenedesmus quadricauda*, *Ankistrodesmus braunii*	These phenol resistant strains of green microalgae degraded phenols (at 400 mg L^{-1}) to more than 70% within 5 days in effluents from olive-oil mill wastewaters. These strains were less efficient in degradation of tannins and lignin, but removed over 50% of low molecular weight phenols in dark. These phenols were not completely removed, but were biotransformed into non-identified aromatic compounds	Pinto et al. (2002, 2003)
Nitrophenols and halophenols	*Chlorella fusca*, *Anabaena variabilis*	*Chlorella fusca* and *Anabaena variabilis* in photoautotrophic conditions degraded 2-nitrophenol and 3-nitrophenol. *C. fusca* also revealed capacity to degrade 2,4-dichlorophenol	Hirooka et al. (2003), Tsuji et al. (2003)
	Coenochloris pyrenoidosa, Chlorella vulgaris	*Coenochloris pyrenoidosa* and *Chlorella vulgaris* degraded 4-nitrophenol during 3 days However, p-chlorophenol was degraded by these microalgae at a concentration of 150 mg L^{-1} in 5 days	Lima et al. (2003, 2004)
	Tetraselmis marina	This microalga removed mono-chlorophenols from the growth medium with higher efficiency as compared to *p*-chlorophenol. Under a 16 h light:8 h dark photoregime, 8.5 mg L^{-1} *p*-chlorophenol, 7.5 mg L^{-1} *m*-chlorophenol and 4.0 mg L^{-1} *o*-chlorophenol were removed. However, *p*-chlorophenol removal efficacy (13 mg L^{-1}) under continuous illumination was 3.6 times higher when compared to that achieved under 8 h light:16 h dark photoregime. *p*-chlorophenol removal also got enhanced with increase in concentration of $NaHCO_3$	Petroutsos et al. (2007)
	Scenedesmus obliquus	Halophenols (chlorophenols, bromophenols, iodophenols) regarded as toxic to microalgae. Degradation of halophenols required exogenous supply of carbon and energy sources as organic carbon (glucose), inorganic carbon (CO_2) or both	Papazi and Kotzabasis (2007)

(continued)

Table 2 (contiued)

Substrate	Organism	Observations	References
Pulp and paper industry wastewater	Mixed culture of green algae (*Chlorella, Chlorococcum Chlamydomonoas, Pandorina, Eudorina*), diatoms, flagellates, and cyanobacteria (*Microcystis, Anabaena*)	Removed 55–60% Chemical oxygen demand, 80–85% color, and 50–80% absorbable xenobiotic compounds after a treatment period of 42 days in batch reactors	Tarlan et al. (2002)
Lignin	*Anabaena azollae*	This report showed oxidative degradation of coir waste and usefulness of coir waste as carrier for cyanobacterial biofertilizer with supporting enzymatic study	Malliga et al. (1996)
Naphthalene	*Agmenellum quadruplicatum, Oscillatoria* sp, other cyanobacteria and eukaryotic microalgae	Microalgal cells were able to remove naphthalene from the growth medium by accumulation within the cells, but were unable to metabolize it. Cyanobacteria and eukaryotic microalgae were capable of biotransforming naphthalene to four major metabolites, 1-naphthol, 4-hydrox-4-tetralone, *cis*-naphthalene dihydrodiol and *trans*-naphthalene dihydrodiol at non-toxic concentrations	Cerniglia et al. (1979, 1980), Cerniglia (1992)
	Scenedesmus obliquus	*Scenedesmus obliquus* utilized naphthalene sulfonic acids as sulphur source for the production of biomass and released desulfonated carbon ring into the medium. It also utilized nitro and amino substituents from aminonaphthalenes and amino and nitrobenzoates as nitrogen sources. *S. obliquus* desulfonated these compounds in a matter of hours in sulphur-deficient medium	Luther (1990); Luther and Soeder (1987)
Textile dye, Azo-aniline dyes	*Oscillatoria formosa* NTDM02, *Phormidium valderianum, Oscillatoria* sp., *Chlorella* sp.	*Oscillatoria formosa* decolorized the textile dye and the dye Amido Black. *Phormidium valderianum,* marine cyanobacterium, produced hydrogen and removed dyes from the solution. *Oscillatoria* and *Chlorella* used aniline dye as carbon and energy source and completely mineralized aniline to CO_2 on extended incubation	Ali et al. (2011), Jingi and Houtian (1992), Shah et al. (2000)
	Phormidium ceylanicum, Gloeocapsa pleurocapsoides, and *Chroococcus minutes*	These cyanobacteria were isolated from sites polluted by industrial textile effluents. *Gloeocapsa pleurocapsoides* and *Phormidium ceylanicum* decolorized Acid Red 97 and Sky Blue dyes by more than 80% after 26 days. *Chroococcus minutus* was also reported to decolorize Amido Black 10B (55%). Decolorisation had an inverse relation with the molecular weight, chemical structure and presence of inhibitory functional groups such as $-NO_2$ and $–SO_3Na$	Parikh and Madamwar (2005)

treated effluent. However, this problem is better handled when algae are harvested in a two-step process involving flocculation followed by centrifugation, filtration, or micro-straining to get a solid concentration (Singh and Dhar 2011). Moreover, immobilization technology is well known to entrap the microalgal cells into a matrix and thereby, solves the harvest problems to a considerable limit. This technology offers a greater degree of operational flexibility, easy separation and higher nutrient removal efficiency in the immobilized algal biomass than the freely suspended cells of the same species (Abdel Hameed 2002).

Microalgae are regarded as effective accumulators and degraders of different kinds of environmental pollutants, used in low-cost methods for remediating domestic wastewater, and removal of heavy metals and inorganic nutrients including nitrogen and phosphorus (Lefebvre et al. 2007; Mansy and El-Bestawy 2002; Podda et al. 2000; Singh and Dhar 2007). They are also efficient agents for the assimilation of organic matter from various contaminated media (Michelou et al. 2007; Zubkov et al. 2007). Some cyanobacteria have also been successfully used in the remediation of solid-waste and wastewaters containing pesticides (Kuritz 1999), phenols (Pinto et al. 2002, 2003), aromatic hydrocarbons (Ibraheem 2010), textile dyes (Parikh and Madamwar 2005) and detergents (Cain 1994); primarily due to their capacity to metabolize these compounds as nitrogen, phosphorus, carbon and sulfur sources. Efforts are under progress for sequencing and characterization of a large number of cyanobacterial members in order to optimize their beneficial use for reclamation of wastewaters (Herrero and Flores 2008; Rocap et al. 2003; Tabei et al. 2007). *Scenedesmus*, *Chlorella*, *Phormidium* and *Oscillatoria* are examples of few frequently used microalgal genera employed in wastewater treatment systems leading to a progressive reduction in chemical oxygen demand (COD), biological oxygen demand (BOD), and heavy metal content in industrial wastewaters, both in batch or in continuous systems (Mc Hugh 2003). Mohana et al. (2009) considered cyanobacteria to be ideal for treatment of distillery effluent as they, apart from degrading the polymers, also oxygenated water bodies and reduced the BOD and COD levels.

Traditional wastewater treatment processes require high amount of energy for mechanical aeration to provide oxygen to aerobic bacteria to consume the organic compounds present in the wastewater whereas; microalgae provide an efficient way to consume nutrients and provide the aerobic bacteria with the needed oxygen through photosynthesis (Oswald et al. 1953). Approximately 1 kg of BOD removal in an activated sludge process requires 1 kWh of electricity for aeration, which produces 1 kg of fossil CO_2 from power generation. By contrast, 1 kg of BOD removed by photosynthetic oxygenation requires no energy inputs and produces enough algal biomass to generate methane that can produce 1 kWh of electric power (Oswald 2003). Some hydrogen producing cyanobacteria are also being considered as an alternative energy source (Singh and Dhar 2011).

The efficiency of microalgal wastewater treatment largely depends upon the capacity of the naturally occurring strains to degrade pollutants present in the aquatic habitats, necessities of the communities, and cost competitiveness of the systems. The adaptation of microalgae to various growth regimes often requires varying acclimatisation periods depending upon the strength of the pollutant and the genetic and physiological abilities of the strains (de-Bashan et al. 2008). Therefore, it is necessary that the microalgal strains destined for remediation of distillery wastewater must pass the scrutiny of adaptability to the prevailing environmental conditions and the local wastewater scheme (de-Bashan and Bashan 2004). Microalgae are frequently known to inhabit heavily polluted environments, are widely distributed, and dominate the micro floral populations in such environments (Sorkhoh et al. 1992). These may acquire natural resistance and selectivity against environmental pollutants due to their presence in such polluted systems (Al-Hasan et al. 2001). Therefore, native strains of microalgae from a particular area are considered more suitable for elimination of phenol and phenolic compounds as compared to the non-native strains used by other researchers in other parts of the world (Kafilzadeh et al. 2010).

Microalgae are quite different from other microbial systems in many of the physiological, morphological, biochemical, and molecular characteristics. They are quite diverse and flexible in meeting several requirements for adaptations which other systems cannot achieve; and therefore, they are gaining increased prominence in the field of environmental microbiology and biotechnology. Specialised attention has also been given on the selection and cultivation of appropriate microalgal strains and on their subsequent processing for specific bioremoval/bioremediation applications. Additionally, increased recovery of biomass from the treated effluent with involvement of immobilization technology, together with the use of metabolically active microalgae to optimize the scale-up of these new bioremediation processes will offer a greater degree of operational flexibility.

Therefore, we can say that the native microalgal strains are a favorable alternative to the traditional costly wastewaters treatment systems having tendency to form toxic intermediates. However, biological processes using microorganisms are comparatively less expensive and can lead to almost complete mineralization of the compounds. Microalgae are known to have excellent bioremediation potential against diverse environmental pollutants like domestic wastewater, heavy metals, pesticides, phenols, aromatic hydrocarbons, textile dyes, detergents and inorganic nutrients like nitrogen and phosphorus, primarily due to their capacity to metabolize these compounds as nitrogen, phosphorus, carbon and sulfur sources.

3 Environmental Impact of Distillery Effluent

The amount of waste produced during distillation process varies from 88% to 95% by volume of the alcohol distilled. An average molasses based distillery generates about 15 L of spentwash L^{-1} of alcohol produced (Beltran et al. 2001). The amount and characteristics of spentwash produced during distillation are highly variable and depend on the nature of feedstocks, besides other aspects of the processes of ethanol production. The sources of wastewater in a distillery plant are stillage, fermenter and condenser cooling water, and fermenter wastewater. A large amount of water is used in distillation and this lead to production of high level of liquid residues including liquor, sugarcane washing water, water from the condensers, cleaning of the equipments, and other residual water. These liquid residues produced during the industrial phase of the production of alcohol is extremely polluting because it contains approximately 5% organic material and fertilizers such as potassium, phosphorus and nitrogen besides many recalcitrant or slowly biodegradable compounds (Borrero et al. 2003). The effluents from molasses based distilleries contain large amounts of dark brown colored molasses spentwash, which is one of the most hazardous waste products to dispose because of low pH, high temperature, dark brown color, high ash content, and high BOD and COD (Agarwal et al. 2010; Beltran et al. 1999; Nandy et al. 2002).

The dark color, high pollution load, and putriciable organics make the spentwash a potential water pollutant. The dark nature of it can block out sunlight from rivers and streams thus may reduce the oxygenation of water by photosynthesis; and hence becomes detrimental to aquatic life (Pant and Adholeya 2007). The high pollution load would result in enrichment of nitrogen and phosphorus in the streams besides many harmful metals, which ultimately leads to eutrophication of the contaminated water courses. Putriciable organics like skatole, indole, and other sulphur compounds present in the spentwash produces obnoxious smell, if it is disposed in canals or rivers (Mahimairaja and Bolan 2004). Such effluents, if highly concentrated may have toxic effects on a large number of flora and fauna present in such ecosystems. The LC_{50} value for distillery spentwash has been observed to be 0.5% using a biotoxicity study on the fresh water fish *Cyprinus carpio* (Mahimairaja and Bolan 2004). The stress caused by distillery effluent resulted in a shift towards anaerobiosis at organ level during sub-lethal intoxication (Ramakritinan et al. 2005).

Phenolic compounds and their derivatives e.g. hydroxyl, methyl, amide, sulphonic are one of the most potential pollutants present in the distillery effluent. They may be colorless solids or thick liquids, often containing a pink tint owing to the presence of oxidation products. The concentration of phenolic compounds in the industrial effluents range from 6 to 2,000 mg L^{-1} whereas; it is approximately 50 mg L^{-1} in the vinasses. However, the admissible limit of phenolic compounds in the water bodies is only 3 mg L^{-1} (Shashirekha et al. 1997). These compounds are considered to be a serious threat to the environment because of their toxicity at low concentration and are recalcitrant to degradation.

Acute exposure of phenolic compounds may cause central nervous system disorders, muscular convulsions and comma. It damages the mucus membrane and causes a burning effect on the skin, ultimately leading to its whitening and erosion. Phenol exposure may also lead to hypothermia, muscle weakness, myocardial depression, and tremors. Renal damage and salivation may even be induced by continuous exposure to phenol. It may further result in irritation of the eye, conjunctional swelling, corneal whitening and finally blindness. Other effects include hepatic damage and frothing from the nose and mouth followed by headache. However, chronic exposure may lead to anorexia, dermal rash, dysphasia, gastrointestinal disturbance, vomiting, weakness, weightlessness, muscle pain, hepatic tenderness, nervous disorder, paralysis, cancer and genetofibre striation (EPA 1979).

The crude spentwash contains high concentrations of K, P, S, Fe, Mn, Zn, Cu and heavy metals. Excessive and continuous discharge of spentwash may lead to significant levels of soil pollution and acidification. It is reported to inhibit seed germination, reduce soil alkalinity, cause soil manganese deficiency and damage agricultural crops (Agrawal and Pandey 1994; Kannabiran and Pragasam 1993). The root meristem assay of settlings grown on crude showed a detrimental effect on mitotic efficiency and also induction of *de novo* chromosomal aberrations like clump formation, chromosome stickiness, laggards and micronuclei formation. Inhibitory effects of spentwash were also visible on bud sprouting and settling height (Srivastava and Radha 2010). However, effect of distillery effluent on seed germination is governed by its concentration and is crop-specific (Ramana et al. 2002).

On the other hand, organic wastes contained in distillery effluent are valuable source of plant nutrients, especially N, P, K and organic substances; if properly utilized (Pathak et al. 1999). Distillery effluent in combination with bio-amendments such as farm yard manure, rice husk and *Brassica* residues have been reported to improve the properties of sodic soil (Kaushik et al. 2005). The use of fungi for bioconversion of distillery waste into microbial biomass and some useful metabolites has been reviewed by Friedrich (2004). Enhanced production of oyster mushrooms, *Pleurotus* sp. using distillery effluent as a substrate amendment has also been reported (Pant et al. 2006).

One of the most important components of vinasses is melanoidins, which represent nearly 2% of the vinasses. These are polymers synthesized by Maillard reactions and have complex chemical structures that are yet not fully understood (Kalavathi et al. 2001; Morales and Jimenez-Perez 2001). The color of vinasses is caused by the presence of melanoidins, together with phenolics from the feedstock such as tannic, gallic and humic acids, caramels from overheated sugars, and furfurals from acid hydrolysis (Kort 1979). Although, melanoidins are known to be degraded by certain bacteria and fungi, they are quite recalcitrant to biodegradation. However, heterotrophic organisms are not particularly suited for use in the treatment of vinasses since they tend to deplete the available dissolved oxygen present within the system. On the other hand, autotrophic oxygen-evolving cyanobacteria have been successfully employed for the decolorisation of vinasses (Chandra et al. 2008). The ability of these photoautotrophs to detoxify these wastewaters is largely based upon their capacity to degrade diverse phenolic compounds as well as melanoidin pigments.

The large amount of liquid residues produced by distilleries contain high load of salts of potassium, phosphorus, nitrogen and heavy metals, putriciable organic material, phenolic and sulphonic derivatives which produces obnoxious smell, bad odor, and lead to anaerobiosis of the aquatic habitats. Low pH, high temperature, dark brown color, high ash content, and high BOD and COD of the distillery effluents may cause eutrophication of the contaminated water courses, result in to death of many organisms, reduce the aesthetic value of the ecosystems and are even toxic to mammals.

4 Degradation of Phenolic Compounds

Till date, research on phenol biodegradation has almost exclusively concentrated on the use of bacteria and fungi. However, these bacteria and fungi may further aggravate the pollution problem due to anaerobic degradation of the aromatic amino acids present in the wastewater that may further lead to liberation of phenol and cresol (Evans and Fuchs 1988). Although little is known about the ability of microalgae to metabolize aromatic substances, their remarkable ability to adapt to unfavorable environmental conditions makes them a potential candidate for the bioremediation of polluted areas (Radwan and Al-Hasan 2000). The microalgae offers spectacular prospects of their use in bioremediation because of their ubiquitous distribution, central role in fixation, turnover of carbon and other nutrient elements, ability to scavenge nutrients, and adaptation to heterotrophic condition.

Even at low concentrations phenolic compounds are very toxic to most of the organisms (Koch 1989). However, some microorganisms are able to utilize them as carbon and energy source (Wurster et al. 2003). While many phenols show acute toxicity to algae (Shigeoka et al. 1988), both cyanobacteria and eukaryotic microalgae are capable of bio-transforming aromatic compounds, including phenols (Semple and Cain 1996). The first report on phenol degradation by the freshwater cyanobacteria, namely, *Anabaena cylindrica* and *Phormidium foveolarum*, was given by Ellis (1977). Subsequently, the metabolism of naphthalenes, anilines, biphenyl and phenanthrene and the formation of different hydroxylated metabolites by one unicellular and two filamentous cyanobacterial strains have been reported (Cerniglia et al. 1984; Narro et al. 1992). Investigations of the unicellular marine cyanobacterium *Synechococcus* PCC 7002 revealed its ability to metabolize phenol under non-photosynthetic conditions up to 100 mg L^{-1} (Wurster et al. 2003). However, Shashirekha et al. (1997) reported inhibitory effects on the growth of *Phormidium valderianum* at this phenol concentration. Moreover, Ishihara and Nakajima (2003) reported removal of bisphenol from a medium spiked with 40 μM of bisphenol by two marine microalgal strains of *Nannochloropsis* sp. and *Chlorella gracilis* upto a level of 34% and 53% respectively over a period of 6 days under light.

Pinto et al. (2002) reported a removal of more than 70% of phenol by two green microalgae, *Ankistrodesmus braunii* and *Scenedesmus quadricauda* from olive-oil mill wastewaters within 5 days of treatment; which was comparable to those obtained with fungi and bacteria. However, a limited reduction of phenol content was observed after 5 days of treatment, irrespective of algal concentration (Pinto et al. 2003). Cyanobacterial sensitivity to phenolics is related to the number and polarity of substituents in the aromatic ring (Della Greca et al. 1992). Wurster et al. (2003) provided the first record of an *ortho*-fission of a phenolic substance by cyanobacteria, and muconic acid as the major product observed during the degradation of phenol in the dark. They further concluded that a light-controlled metabolism was responsible for phenol degradation by cyanobacteria, because under photosynthetic conditions no phenol was degraded and that no metabolites were produced. This breakdown was probably catalyzed by a protein and the transformation was an extracellular process inhibited by heat, proteases and metal ions.

Biodegradation of phenolic compounds like lignin and tannins is also of interest due to their greater structural complexity and abundance in the distillery wastewater. These substances are structurally similar to humic acid which is present in vinasses as well as in many other waste streams (Murugan and Al-Sohaibani 2010; Robinson et al. 2001). Lignin is a three-dimensional amorphous aromatic hetero-polymer located in the middle lamella of the plant cell wall and supplies structural support to

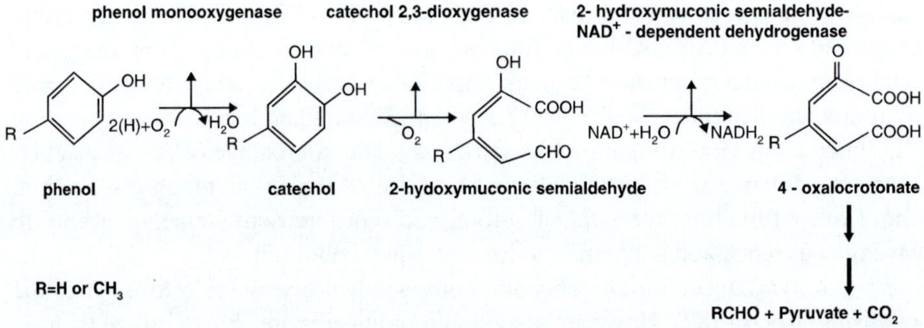

Fig. 1 *Meta-cleavage* pathway of phenol biodegradation in *Ochromonas danica* (Adapted from Semple and Cain 1996). Note the catabolic pathway for degradation of phenol and its assimilation and mineralization as proved by the whole-cell and enzymatic studies. This is the first definitive demonstration of this mechanism for aromatic ring degradation in a eukaryotic alga

wood in the trees. However, tannins are water-soluble phenolic polymers that precipitate with proteins and their structure is very similar to that of lignin. Both lignin and tannin are very much stable and are in fact toxic to many species of microorganisms. While anaerobic catabolic reactions are not efficient enough to cleave the aromatic rings in both types of polymers, aerobic metabolism can result in degradation, but the rates are slow and the degradation is only partial (Heider and Fuchs 1996; Kuhad et al. 1997).

There are few reports on the biodegradation of lignin and tannin by microalgae. It was Bharati et al. (1992), who for the first time reported a decline of over 73.0% of lignin from the pulp and paper mill wastes by two species of cyanobacteria *Phormidium ambiguum* and *Chroococcus minutus* over a period of 5 days. The extracellular enzymes like phenol oxidases like peroxidases and laccase produced by the growing microalgae were held responsible for such degradation. Malliga et al. (1996) have reported that *Anabaena azollae* while being used as a biofertilizer exhibited lignolysis of the coir waste. In another study, three different cyanobacterial strains of *Phormidium*, *Oscillatoria*, and *Anabaena azollae* found to degrade coir waste showing 89% degradation of lignin and 92% degradation of hemicellulose (Anbuselvi and Jeyanthi 2009).

It was observed that at low concentrations phenols have no toxic effects and can be used as alternative carbon source for microalgae. The biodegradation of phenols by microalgae occurs only under aerobic conditions. The first step in biodegradation of halophenols is dehalogenation i.e. split of halogen substituent which is determined by the bond dissociation energy of the corresponding substituent. The energetic requirements for the biodegradation of different halophenols were highest for chlorophenol, followed by bromophenol and iodophenol. Additionally, the meta-position of the halogen on the phenol ring needs more energy than the ortho- and the para-one (Papazi and Kotzabasis 2007). A typical pathway for degradation of phenol involving hydroxylation of the ring and formation of catechol followed subsequently by oxidation has been explained in the Fig. 1 (Semple and Cain 1996). Although both aerobic and anaerobic based activated sludge processes are known to facilitate microorganism mediated degradation of phenol, aerobic based processes are preferred (Contreras et al. 2008). Biodegradation of phenols by microalgae however occurs only under aerobic condition, and therefore presence of molecular oxygen is indispensable in order to initiate enzymatic attack on the aromatic ring of phenols (Semple and Cain 1996; Semple et al. 1999).

The microalgal degradation of phenolic compounds is a bioenergetic process depending on the growth conditions, especially on the exogenously supplied carbon and light sources (Papazi and Kotzabasis 2007, 2008). At high light intensities, the degradation of some phenolic compounds reduced significantly, which may be due an increased toxicity resulting from autoxidation processes enhanced by light (Nakai et al. 2001). This is an important aspect in the view of photosynthetic nature of microalgae however; further research is required in this regard. Nevertheless, most microalgae species are mixotrophs, they are capable of employing both photoautotrophic and heterotrophic metabolism. Microalgae, sensitive to phenolic pollutants when grown under either phototrophic or heterotrophic conditions, have an improved ability to mineralize phenolic compounds when grown under mixotrophic growth conditions (Tikoo et al. 1997). This characteristic allows such organisms to survive under conditions of low light or carbon dioxide shortage and represents an alternative to other biological treatments used for the biodegradation of phenol-containing wastewaters (Pinto et al. 2002).

The aerobic degradation of phenolic compounds by microalgae under photoautotrophic, heterotrophic, or mixotrophic conditions have been described by a mathematical model by Lika and Papadakis (2009). This model demonstrates that degradation of phenolic compounds is enhanced by increasing the light intensity in a wide range, which is mainly attributed to increase in photosynthetic oxygen production. It further illustrates that; both the microalgal growth and the rate of biodegradation can be enhanced under increased light intensities, by addition of inorganic carbon sources such as carbon dioxide and sodium bicarbonate. Lack of oxygen may however be a limiting factor during the peak phase of phenol biodegradation. Use of carbon source such as glucose increases the specific growth rate whereas; it reduces the biodegradation rate of phenolic compounds. This reduction in degradation of phenolic compounds may be due to competition for oxygen between glucose utilization and phenol assimilation. Briefly speaking, degradation of phenolic compounds mediated by microalgae is a dynamic process and is controlled by several parameters. It therefore requires structural analysis of the phenolic compounds, its influence on microalgal growth, as well as involvement of the photosynthetic machinery in order to have a better understanding of phenol biodegradation at both the cellular and cultural level.

Anaerobic degradation of the aromatic amino acids by many bacteria and fungi may further aggravate the pollution problem due to further liberation of phenol and cresol. However, aerobic microalgae have capacity to metabolize aromatic substances due to their ability to adapt to unfavorable environmental conditions. The microalgae carry out *ortho*-fission of the phenolic substances extracellularly in the dark. This reaction is catalyzed by a protein; however, such transformation is inhibited by heat, proteases and metal ions. The microalgal strains like *Anabaena cylindrica*,

Phormidium foveolarum, *P. valderianum*, *Synechococcus*, *Ankistrodesmus braunii* and *Scenedesmus quadricauda* has been reported instrumental in degradation of phenol and its derivatives. Whereas, the performances of *Phormidium ambiguum*, *Chroococcus minutus*, *Oscillatoria*, and *Anabaena azollae* has been found satisfactory in degradation of lignin. Microalgal biodegradation of phenols occurs in presence of molecular oxygen, which is indispensable for enzymatic breakdown of aromatic ring of phenols. It involves hydroxylation of the aromatic ring and formation of catechol followed subsequently by oxidation. It has also been proved that both, the microalgal growth and the rate of biodegradation can be enhanced under increased light intensities and by addition of inorganic carbon sources such as carbon dioxide and sodium bicarbonate.

5 Melanoidin Degradation

Melanoidin, a dark brown recalcitrant pigment present in the molasses spentwash, is an adduct formed as a result of the reaction between an amino acid and a carbohydrate during Maillard reaction (Akarsubasi et al. 2006; Cammerer et al. 2002). The empirical formula of melanoidin has been worked out to be $C_{17-18}H_{26-27}O_{10}N$ and the molecular weight is between 5,000 and 40,000. It consists of acidic, polymeric and highly dispersed colloids, which are negatively charged due to the dissociation of carboxylic acids and phenolic groups (Manisankar et al. 2004). These recalcitrant anionic hydrophilic polymers can form stable complexes with metal cations (Plavsic et al. 2006). Conventional biological treatments are reported to accomplish the degradation of melanoidin to about 6% (Agrawal and Pandey 1994). Presently, only certain heterotrophic bacteria and fungi are mainly reported to degrade melanoidin. But heterotrophs are not adequately suitable for treating spentwash since they deplete oxygen in the effluent and cannot be easily adopted for aquatic habitats (Kalavathi et al. 2001). However, oxygen-evolving photoautotrophs have also been reported to be useful for the treatment of wastewaters, have ability to remove nutrients like nitrogen and phosphorus from sewage waters, and can accomplish the improvement of physico-chemical properties of sewage effluents (Singh and Dhar 2007).

Research regarding cyanobacterial degradation of melanoidin started in the present millennium with the work of Kalavathi et al. (2001), who screened 13 marine cyanobacterial strains of *Oscillatoria*, *Phormidium*, *Spirulina* and *Synechococcus* to study their capacity to decolorize anaerobically treated spentwash. The decolorisation efficiency of the strains decreased in the order *Oscillatoria* > *Phormidium* > *Spirulina* > *Synechococcus*. Among these, *Oscillatoria boryana* BDV 92181, an obligate aerobic marine cyanobacterium, utilized melanoidins both as a carbon and nitrogen source, and decolorized pure melanoidins pigment (0.1% w/v) by about 75% and crude pigment (5% v/v) by about 60% in 30 days. Melanoidin degradation and decolorisation capacity of *Oscillatoria boryana* was due to the production of H_2O_2, hydroxyl, per-hydroxyl, and active oxygen radicals. However, decolorisation by this cyanobacterium slowed down at melanoidin concentrations above 0.5%, due to the inhibition of growth by blockade of penetration of light by dark color of the medium. *Oscillatoria*, *Lyngbya*, and *Synechocystis* species have been reported to decolorize distillery effluents through bio-flocculation by 96%, 81% and 26%, respectively. However, a consortium of these three strains achieved a maximum decolorisation of 98% (Patel et al. 2001). This decolorisation was initially attributed to absorption processes, followed subsequently by the degradation of the organic compounds.

A consortium of microalgae *Chlorella vulgaris* and macrophyte *Lemna minuscule* was also reported to decolorize wastewater and removed organic matter from ethanol producing units. The macrophyte treatment without microalgal pre-treatment reduced the dark coloration by only up to 10%. However, a combined process using *Chlorella* for 4 days followed by treatment with the *Lemna minuscula* for 6 additional days reduced the color to about 52%, which highlight the importance of microalgae in decolorisation of spentwash (Valderrama et al. 2002). Microalgae permit the removal of nutrients and organic matter from the wastewater and produce oxygen for the other organisms; while macrophyte growth is useful because it permits precipitation of the microalgal population and subsequent reuse of the treated wastewater. Moreover, certain *Chlorella vulgaris* strains were reported to be highly adaptable to load changes and resistant to colored media and performed efficiently in removing the pollutant load and improving the oxygen status of the wastewaters (Travieso et al. 2008).

A dark brown recalcitrant pigment called melanoidin, formed in the molasses as a result of Maillard reaction, can form stable complexes with metal cations. *Oscillatoria boryana* can utilize melanoidins as a carbon and nitrogen source and decolorized pure melanoidins by about 75% and crude pigment by 60% in 30 days. A consortium of *Oscillatoria*, *Lyngbya* and *Synechocystis* decolorized melanoidin by 98% by absorption followed subsequently by degradation of the organic compounds. Melanoidin degradation and decolorisation capacity of *Oscillatoria boryana* was due to the production of H_2O_2, hydroxyl, per-hydroxyl, and active oxygen radicals.

6 Mechanisms for Microalgal Degradation of Spentwash

Microalgae are capable of rapid uptake of nutrients and CO_2 and possess faster cell growth. They have been studied for their ability to degrade pollutants such as phenols, aromatic compounds, synthetic dyes and recalcitrant biopolymers such as melanoidin, but very few of them have focussed on the enzymatic aspects of such processes. A conceptual model for integrated microalgal biomass production, their utilization in biofuel production, bioremediation, and fertigation is represented in Fig. 2. The experiment with *Ochromonas danica* has revealed the involvement of enzymes of *meta* cleavage pathway which is commonly found in bacteria in the degradation of phenol and its methylated homologues (Semple and Cain 1996). This pathway involves the use of molecular oxygen by the enzyme 'phenol monooxygenase' to add a second hydroxyl group in the *ortho* position; leading to the formation

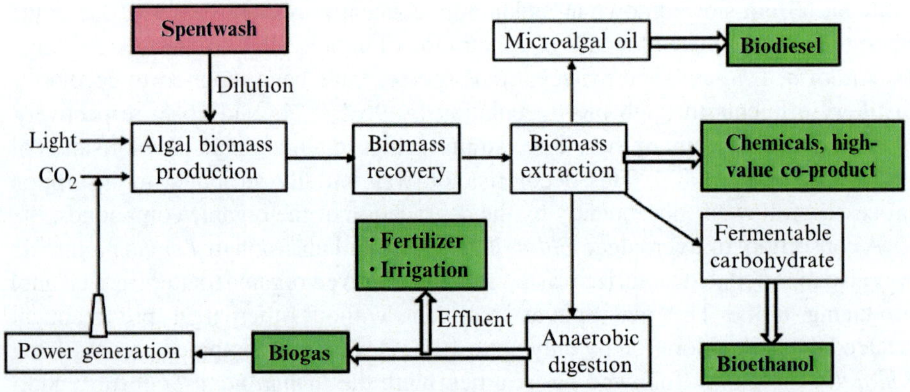

Fig. 2 A conceptual integrated model for treatment of distillery effluent including microalgal biomass production and utilization. Note the utilization of diluted spentwash for algal biomass production and its conversion to different types of biofuels e.g. biodiesel, bioethanol and biogas. They have excellent capacity of producing a wide range of high-value chemicals and bioactive compounds. The energy produced in such integrated plants may be used for power generation. Moreover, the CO_2 from the exhaust of such power plants may be routed back into the tank for algal biomass production because CO_2 more often becomes the limiting factor in intensive biomass production. Even the effluent after treatment may be used for fertigation

of catechol (Fig. 1). The resulting catechol molecule is degraded by the enzyme catechol 2,3-dioxygenase and lead to ring fission adjacent to the two hydroxyl groups of the catechol. Subsequently, the enzyme 2-hydroxymuconic semialdehyde NAD^+-dependent dehydrogenase gets involved, leading to the production of reduced pyridine nucleotide ($NADH_2$). The compounds obtained are metabolized further to the intermediates of the Krebs cycle (Bayly and Barbour 1984). The presence of a ring hydroxylating dioxygenase system has also been described in the chlorophyte alga *Selenastrum capricornutum* (Schoeny et al. 1988).

Several species of cyanobacteria are also known to possess polyphenol oxidases and laccase however; research in this area is still in its infancy. Cultures of *Phormidium valderianum* containing phenols at a concentration of 50 mg L^{-1} showed fivefold and twofold higher activity of polyphenol oxidase and laccase, respectively; than in the presence of 25 mg L^{-1} of phenol (Shashirekha et al. 1997). They further reported an increase of protein concentration in the medium which was attributed to the *de-novo* synthesis of phenol-degrading enzymes. Laccases are regarded as key enzymes for decolorisation of spentwash by Basidiomycetes fungi. Moreover, melanoidins isolated from distillery effluents results in selective induction of laccase gene expression in microalgae (Gonzalez et al. 2008).

Recently, constitutive expression of laccase and polyphenol oxidase for the decolorisation of the lignin model polymeric dye Poly R-478 has been observed in strains of *Oscillatoria* and *Phormidium* (Palanisami et al. 2010). The use of such model substrates are regarded as useful tool in identification of cyanobacterial bioremediation potential. In this study, laccase production was also induced by the known fungal laccase elicitors such as veratryl aldehyde, caffeic acid, guaiacol and tannic acid. In another study, *Oscillatoria willei*, when grown under conditions of nitrogen limitation but supplemented with phenolic compound, resulted in enhanced oxidative stress with a concomitant increase in lignolytic and antioxidative enzyme activities; such as lignin peroxidase, laccase, polyphenol oxidase, superoxide dismutase, catalase, peroxidase and ascorbate peroxidase. It was realised that the cyanobacterium *Oscillatoria willei* decolorized 52% of the substrate following 7 days of incubation (Saha et al. 2010). Moreover, cyanobacteria growing in media containing both pigment and effluent have been demonstrated to possess higher glutamine synthetase activity (Kalavathi et al. 2001). Such enzymes help the cyanobacteria in the scavenging of ammonia from amino acid component of the pigment (Stewart et al. 1975).

While oxygen is essential for the existence of aerobic life, it is well established that toxic active oxygen species namely, superoxide anion- O^{-2}, hydroxyl radical- OH^-, and hydrogen peroxide- H_2O_2, are generated in all photosynthetic organisms through electron transport systems (Asada 1999). Hydrogen peroxide (H_2O_2), a strong oxidising agent produced by cyanobacteria during the photosynthetic process, is regarded to be fundamentally involved in melanoidin degradation. Evidence for this comes from the fact that *in vitro* addition of H_2O_2 to melanoidin containing effluent has shown a color reduction of 97% (Patil and Kapadnis 1995). Another experiment demonstrated the decolorisation of synthetic melanoidins after addition of hydrogen peroxide (Hiramoto et al. 1997). The cyanobacterium *Phormidium laminosum* is reported to produce hydrogen peroxide, responsible for decolorisation of synthetic melanoidins (Ikan et al. 1992). In another study, *Oscillatoria boryana* BDU 92181 was found to produce hydrogen peroxide at the rate of 0.538 μmole μg^{-1} Chl^{-1} h^{-1} and was predicted to be a major cause of melanoidin degradation (Kalavathi et al. 2001).

Cyanobacteria has also been reported to produce the enzyme 'superoxide dismutase' that lead to the generation of hydroxyl radical and the resultant shift in pH to alkalinity (Campbell and Laudenbach 1995; Priya et al. 2010), which in turn influence the decolorisation and degradation of synthetic melanoidins. Thus decolorisation and degradation of synthetic melanoidin is known to be influenced by pH of the reaction mixture in the alkaline range due to hydroxyl ions (Hayase et al. 1984). It is advocated that the melanoidin pigment may be degraded into utilizable forms of nitrogen and carbon by the generation of hydroxyl ions (Kalavathi et al. 2001). Further, any hydrogen peroxide generated in cyanobacteria during photosynthesis can react with hydroxyl anion to give mainly per-hydroxyl anion [HOO-] which has a strong nucleophilic activity and can also help in decolorisation of melanoidin (Jones et al. 1998). Superoxide dismutase, present in aerobic organisms, catalyzes the conversion of superoxide anion radical into molecular oxygen and hydrogen peroxide (H_2O_2); thereby protecting the organism from its toxic effects. The superoxide radical is highly reactive and toxic, and commonly appears in the laccase producing organisms during the redox cycling of quinone. Therefore, this enzyme not only protects the cell, but may also facilitate the action of peroxidase enzymes through the production of the enzyme's main cofactor H_2O_2 (Leonowicz et al. 2001). A simplified mechanism for decolorisation of spentwash has been represented in the Fig. 3 (modified after Kalavathi et al. 2001 and Saha et al. 2010). It was also proved that the degradation of melanoidin pigment only occurs under light conditions and not in the darkness (Kalavathi et al. 2001). This finding suggests

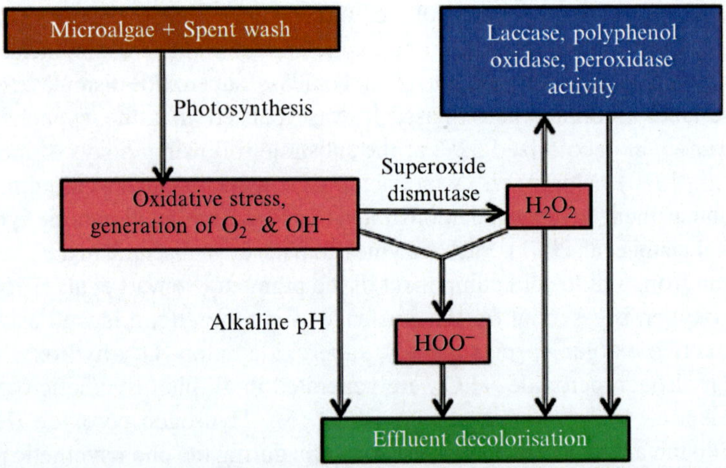

Fig. 3 Microalgal metabolic mechanisms involved in the degradation of distillery effluent. Note the production of toxic active oxygen species like superoxide anion, O_2^-; hydroxyl radical, OH^-; and hydrogen peroxide, H_2O_2 during photosynthesis. During such an oxidative stress, particularly in nitrogen-limiting condition, there is an over expression of the antioxidative enzymes like laccase, polyphenol oxidase, superoxide dismutase, peroxidase, catalase; and low molecular weight antioxidants like ascorbic acid and glutathione. H_2O_2 and HOO^- generated in cyanobacteria during photosynthesis also have a strong nucleophilic activity and can also help in decolorisation of the spentwash

a possible involvement of active oxygen released during the photolysis of water by the cyanobacterium.

Ochromonas danica is reported to carry out *meta*-cleavage of phenol and its methylated homologues involving the enzymes phenol monooxygenase and 2-hydroxymuconic semialdehyde NAD^+-dependent dehydrogenase and the compounds produced are metabolized further as intermediates of the Krebs cycle. Several species of cyanobacteria are also known to possess phenol-degrading enzymes like lignin peroxidase, laccase, polyphenol oxidase, superoxide dismutase, catalase, peroxidase and ascorbate peroxidise. The lignolytic and antioxidative enzymatic activities increases in the presence of phenol, if microalgae are subjected to nitrogen limiting condition. Moreover the photosynthetic nature of the microalgae enable them to produce toxic active oxygen species like O^{-2}, OH^-, and H_2O_2 through electron transport chain which have strong oxidising agent and are involved in degradation of melanoidin.

7 Conclusion

Distillery effluents are one of the most environmentally aggressive industrial effluents. These industries generally produce a large volume of spentwash which may lead to extensive soil and water pollution due to presence of high amount of organic matter and dark brown colored compounds, recalcitrant to biodegradation. Most studies regarding alternative biological treatments of spentwash have till date focused on the use of fungal and bacterial strains; but efficient decontamination of vinasses is yet a largely unsolved problem. More recently, microalgae have emerged as a new, low cost and attractive agent for the detoxification of spentwash. A low-cost and environment friendly approach for wastewater treatment is the use of activated sludge process having consortium of microorganisms which could nearly mineralize the toxic organic compounds (Rajani Rani et al. 2011). Microalgae are known to dominate the micro floral populations in many polluted environments and may acquire natural resistance and selectivity against environmental pollutants due to their presence in such polluted systems (Al-Hasan et al. 2001; Sorkhoh et al. 1992). Native strains of microalgae are considered more suitable for elimination of phenol and phenolic compounds as compared to the non-native strains (Kafilzadeh et al. 2010). However, a greater insight is needed into their adaptation mechanisms in such polluted environments, the extracellular mechanisms for transformation and degradation of phenolic compounds, and the exact effect of increasing light intensities on the biodegradation of organic compounds. Microalgae have capacity to treat a number of hazardous contaminants like aromatic hydrocarbons, sewage water, dyes, heavy metals, and pesticides and are known to produce high-value metabolites for food, feed, biofuels, chemicals, and pharmaceutical products. The highly coloured nature of distillery effluent leads to light blockade and reduce the photosynthetic ability of microalgae. This necessitates the dilution of such highly colored effluents to avoid light-blockade into water. Moreover, many microalgae have ability to grow under mixotrophic growth conditions and possess the advantage of trophic independence for nitrogen and carbon. Another requirement is that the effluent flow of the distillery-wastewaters must be very slow or the effluents should be retained in a tank to allow the microalgae to multiply and degrade the harmful compounds present in it. Further studies are necessary to analyze the degradation of the colored compounds within spentwash under different light intensities, physiological and environmental conditions. In addition, further work needs to be undertaken to study the metabolism and co-metabolism of various components of distillery wastewater, microalgal biomass production, kinetic parameters of biodegradation, costs of industrial scale-up, and characterization of by-products formed and their applications. Hence, it is our view that a comprehensive study focusing the expression of various enzyme complexes in different microalgae strains as well as their response to different components and physiological parameters of distillery effluent is needed to gain a complete understanding of their role in the degradation/assimilation of various toxic compounds; thereby facilitating the development of appropriate microalgae based processes which can be employed for efficient biological treatment of colored effluents such as spentwash.

Acknowledgements We acknowledge the facilities provided by the Department of Microbiology, C.P. College of Agriculture (S.D.A.U.) and financial support provided by the Institute (S.D.A.U.) under the scheme - Strengthening of the existing Department of Microbiology (B.H.11867) for preparation of this manuscript.

References

Abdel Hameed MS (2002) Effect of Immobilization on growth and photosynthesis of the green alga *Chlorella vulgaris* and its efficiency in heavy metals removal. Bull Fac Sci Assiut Univ 31(1-D):233–240

Abdel Hameed MS, Hammouda O (2007) Review: biotechnological potential uses of immobilized algae. Int J Agric Biol 9(1):183–192

Agarwal R, Lata S, Gupta M, Singh P (2010) Removal of melanoidin present in distillery effluent as a major colorant: a review. J Environ Biol 31:521–528

Agrawal CS, Pandey GS (1994) Soil pollution by spentwash discharge: depletion of manganese (II) and impairment of its oxidation. J Environ Biol 15:49–53

Akarsubasi AT, Ince O, Oz NA, Kirdar B, Ince BK (2006) Evaluation of performance, acetoclastic methanogenic activity and archaeal composition of full-scale UASB reactors treating alcohol distillery wastewaters. Process Biochem 41:28–35

Al-Hasan RH, Khanafer M, Eliyas M, Radwan SS (2001) Hydrocarbon accumulation by picocyanobacteria from the Arabian Gulf. J Appl Microbiol 91(3):533–540

Ali DM, Suresh A, Praveen Kumar R, Gunasekaran M, Thajuddin N (2011) Efficiency of textile dye decolorization by marine cyanobacterium, *Oscillatoria formosa* NTDM02. Afr J Basic Appl Sci 3(1):9–13

Anbuselvi S, Jeyanthi R (2009) A comparative study on the biodegradation of coir waste by three different species of marine cyanobacteria. J Appl Sci Res 5:2369–2374

Asada K (1999) The water-water cycle in chloroplasts: scavenging of active oxygen and dissipation of excess photons. Annu Rev Plant Physiol Plant Mol Biol 50:601–639

Bayly RC, Barbour MG (1984) The degradation of aromatic compounds by the *meta* and gentisate pathways. In: Gibson DT (ed) Microbial degradation of organic compounds. Marcel Dekker Incorporation, New York, pp 253–294

Beltran FJ, Garcia-Araya JF, Alvarez PM (1999) Wine distillery wastewater degradation 2. Improvement of aerobic biodegradation by means of an integrated chemical (ozone)-biological treatment. J Agric Food Chem 47:3919–3924

Beltran FJ, Alvarez PM, Rodriguez EM, Garcia-Araya JF, Rivas J (2001) Treatment of high strength distillery wastewater (cherry stillage) by integrated aerobic biological oxidation and ozonation. Biotechnol Prog 17:462–467

Bharati SG, Salanki AS, Taranath TC, Acharyulu MVRN (1992) Role of cyanobacteria in the removal of lignin from the paper mill wastewaters. Bull Environ Contam Toxicol 49:738–742

Borrero MAV, Pereira JTV, Miranda EE (2003) An environmental management method for sugar cane alcohol production in Brazil. Biomass Bioenerg 25:287–299

Cain RB (1994) Biodegradation of detergent. Curr Opin Biotechnol 5:266–274

Cammerer B, Jalyschkov V, Kroh LW (2002) Carbohydrate structures as part of the melanoidins skeleton. Int Congr Ser 1245:269–273

Campbell WS, Laudenbach DE (1995) Characterization of four superoxide dismutase genes from a filamentous cyanobacterium. J Bacteriol 177:964–972

Cerniglia CE (1992) Biodegradation of polycyclic aromatic hydrocarbons. Biodegradation 3:351–368

Cerniglia CE, Morgan JC, Gibson DT (1979) Bacterial and fungal oxidation of dibenzofuran. Biochem J 180:175–185

Cerniglia CE, Gibson DT, Van Baalen C (1980) Oxidation of naphthalene by cyanobacteria and microalgae. J Gen Microbiol 116:495–500

Cerniglia CE, Freeman JP, Althaus JR (1984) Biotransformation and toxicity of 1-and 2-methylnaphthalene and their derivatives in cyanobacteria. In: Liu D, Dutka BJ (eds) Toxicity screening procedures using bacterial systems. Marcel Dekker Incorporation, New York, pp 381–394

Chandra R, Bharagava N, Rai V (2008) Melanoidins as major colorant in sugarcane molasses based distillery effluent and its degradation. Bioresour Technol 99(11):4648–4660

Contreras EM, Albertario ME, Bertola NC, Zaritzky NE (2008) Modelling phenol biodegradation by activated sludges evaluated through respirometric techniques. J Hazard Mat 158:366–374

CPCB (2003) Charter on corporate responsibility for environmental protection. Workshop organized at Mumbai by MPCB on 01 Mar 2003. http://cpcb.nic.in//Charter/charter5.htm

de-Bashan LE, Bashan Y (2004) Recent advances in removing phosphorus from wastewater and its future use as fertilizer (1997–2003). Water Res 38:4222–4246

de-Bashan LE, Trejo A, Huss VAR, Hernandez JP, Bashan Y (2008) *Chlorella sorokiniana* UTEX 2805, a heat and intense, sunlight-tolerant microalga with potential for removing ammonium from wastewater. Bioresour Technol 99:4980–4989

Della Greca M, Monaco P, Pollio A, Previtera L (1992) Structure activity relationships of phenylpropanoids as growth inhibitors of the green alga *Selenastrum capricornutum*. Phytochemistry 31:4119–4123

Ellis BE (1977) Degradation of phenolic compounds by fresh water algae. Plant Sci Lett 8:213–216

EPA (1979) The Carcinogen Assessment Group's preliminary risk assessment on cresols: type 1 – air program. Prepared by the Office of Health and Environment Assessment for the Office of Air Quality Planning and Standards, Washington, DC

Evans WC, Fuchs G (1988) Anaerobic degradation of aromatic compounds. Annu Rev Microbiol 42:289–317

FitzGibbon FJ, Singh D, Mc Mullan G, Marchant R (1998) The effect of phenolics acids and molasses spentwash concentration on distillery wastewater remediation by fungi. Process Biochem 33:799–803

Friedrich J (2004) Bioconversion of distillery waste. In: Arora DK (ed) Fungal biotechnology in agriculture, food and environmental applications. Marcel Dekker Incorporation, New York, pp 431–442

Godshall MA (1999) Removal of colorants and polysaccharides and the quality of white sugar. In: Proceedings of sixth international symposium organized by association Andrew van Hook (AvH), Reims pp 28–35

Gonzalez LE, Canizares RO, Baena S (1997) Efficiency of ammonia and phosphorus removal from a Columbian agro-industrial wastewater by the microalgae *Chlorella vulgaris* and *Scenedesmus dimorphus*. Bioresour Technol 60:259–262

Gonzalez T, Terron MC, Yague S, Junca H, Carbajo JM, Zapico EJ, Silva R, Arana A, Tellez A, Gonzalez AE (2008) Melanoidin-containing wastewaters induce selective laccase gene expression in the white-rot fungus *Trametes* sp. I-62. Res Microbiol 159:103–109

Hallmann A (2007) Algal transgenics and biotechnology.Global Science Books. Transgenic Plant J 1:81–98

Hayase F, Kim SB, Kato H (1984) Decolorisation and degradation products of the melanoidins by hydrogen peroxide. Agric Biol Chem 48:2711–2717

Heider J, Fuchs G (1996) Anaerobic metabolism of aromatic compounds. Eur J Biochem 243:577–596

Hensyl WR (1989) Bergey's manual of systematic bacteriology. Williams & Wilkins, Baltimore

Herrero A, Flores E (2008) The cyanobacteria: molecular biology, genomics and evolution, 1st edn. Caister Academic Press, Norfolk. ISBN 978-1-190445515-8

Hiramoto K, Nasuhara A, Michikoshi K, Kato T, Kikugawa K (1997) DNA strand-breaking activity of 2,3-dihydro-3,5-dihydroxy-6-methyl-4H-pyran-4-one (DDMP), a Maillard reaction product of glucose and glycine. Mutat Res 395:47–56

Hirooka T, Akiyama Y, Tsuji N, Nakamura T, Nagase H, Hirata K, Miyamoto K (2003) Removal of hazardous phenols by microalgae under photoautotrophic conditions. J Biosci Bioeng 95:200–203

Ibraheem IBM (2010) Biodegradability of hydrocarbons by cyanobacteria. J Phycol 46:818–824

Ikan R, Ioselis P, Rubinsztain Y, Aizenshtat Z, Miloslavsky I, Yariv S, Pugmire R, Anderson LL, WoolfendenWR KIR, DorseyT PKE, Boon JJ, de Leeuw JW, Ishiwatari R, Morinaga S, Yamamoto S, Macihara T, Muller-Vonmoos M, Rub A (1992) Chemical, isotopic, spectroscopic and geochemical aspects of natural and synthetic humic substances. Sci Total Environ 117–118:1–12

Ishihara K, Nakajima N (2003) Improvement of marine environmental pollution using eco-system: decomposition and recovery of endocrine disrupting chemicals by marine phyto- and zooplanktons. J Mol Catal B 23:419–424

Jain N, Minocha AK, Verma CL (2002) Degradation of pre-digested distillery effluent by isolated bacterial strains. Indian J Exp Bot 40:101–105

Jingi L, Houtian L (1992) Degradation of azo dyes by algae. Environ Pollut 75:273–278

Jones PD, Briffa KR, Barnett TP, Tett SFB (1998) High-resolution palaeoclimatic records for the last millennium; interpretation, integration and comparison with general circulation model control-run temperatures. The Holocene 8:455–471

Kafilzadeh F, Farhangdoost MS, Tahery Y (2010) Isolation and identification of phenol degrading bacteria from Lake Parishan and their growth kinetic assay. Afr J Biotechnol 9(40):6721–6726

Kalavathi DF, Uma L, Subramanian G (2001) Degradation and metabolization of the pigment-melanoidin in distillery effluent by the marine cyanobacterium *Oscillatoria boryana* BDU 92181. Enzyme Microb Technol 29:246–250. doi:10.1016/S0141-0229(01)00383-0

Kannabiran B, Pragasam A (1993) Effect of distillery effluent on seed germination, seedling growth and pigment content of *Vigna mungo* (L.) Hepper (CVT9). Geobios 20:108–112

Kaushik A, Nisha R, Jagjeeta K, Kaushik CP (2005) Impact of long and short term irrigation of a sodic soil with distillery effluent in combination with bio amendments. Bioresour Technol 96:1860–1866

Koch R (1989) Umweltchemikalien. VCH, Weinheim

Kort MJ (1979) Color in the sugar industry. In: de Birch GG, Paker KJ (eds) Sugar: science and technology. Applied Science, London, pp 97–130

Kuhad RC, Singh A, Eriksson KEL (1997) Microorganisms and enzymes involved in the degradation of plant fibre cell walls. Adv Biochem Eng Biotechnol 57:45–125

Kumar P, Chandra R (2006) Decolorisation and detoxification of synthetic molasses melanoidins by individual and mixed cultures of *Bacillus* spp. Bioresour Technol 97:2096–2102

Kuritz T (1999) Cyanobacteria as agents for the control of pollution by pesticides and chlorinated organic compounds. J Appl Microbiol Symp Suppl 85:186–195

Lefebvre DD, Kelly D, Budd K (2007) Biotransformation of Hg(II) by cyanobacteria. Appl Environ Microbiol 73(1):243–249

Leonowicz A, Cho N, Luterek J, Wilkolazka A, Wotjas M, Matus A, Hofritchter M, Wesenberg D, Rogalski J (2001) Fungal laccase: properties and activity on lignin. J Basic Microbiol 41:185–227

Lika K, Papadakis IA (2009) Modeling the biodegradation of phenolic compounds by microalgae. J Sea Res 62(2–3):135–146. doi:10.1016/j.physletb.2003.10.071

Lima S, Castro P, Morais R (2003) Biodegradation of *p*-nitrophenol by microalgae. J Appl Physiol 15:137–142

Lima SAC, Filomena M, Raposo J, Castro PML, Morais RM (2004) Biodegradation of *p*-chlorophenol by a microalgae consortium. Water Res 38:97–102

Luther M (1990) Degradation of different substituted aromatic compounds as nutrient sources by the green alga *Scenedesmus obliquus*. Dechema Biotechnol Conferences 4:613–615

Luther M, Soeder CJ (1987) Some naphthalene sulphonic acids as sulphur sources for the green microalga, *Scenedesmus obliquus*. Chemosphere 16:1565–1578

Mahimairaja S, Bolan NS (2004) Problems and prospects of agricultural use of distillery spentwash in India. In: SuperSoil, 3rd Australian New Zealand soils conference. 5–9 Dec 2004, University of Sydney, Sydney

Malliga P, Uma L, Subramanian G (1996) Lignolytic activity of the cyanobacterium *Anabaena azollae* ML2 and the value of coir waste as a carrier for BGA biofertilizer. Microbios 86:175–183

Manisankar P, Rani C, Vishwanathan S (2004) Effect of halides in the electrochemical treatment of distillery effluent. Chemosphere 57:961–966

Mansy AE, El-Bestawy E (2002) Toxicity and biodegradation of fluometuron by selected cyanobacterial species. World J Microbiol Biotechnol 18:125–131

Martins SIFS, van Boekel MAJS (2004) A kinetic model for the glucose/glycine Maillard reaction pathways. Food Chem 90(1–2):257–269

Mc Hugh DJ (2003) A guide to the seaweed industry, vol 441, FAO fisheries technical papers. FAO, Rome, pp 101–115

Michelou VK, Cottrell MT, Kirchman DL (2007) Light-stimulated bacterial production and amino acid assimilation by cyanobacteria and other microbes in the North Atlantic Ocean. Appl Environ Microbiol 73(17):5539–5546

Mohana S, Acharya KB, Madamwar D (2009) Treatment technologies and potential applications. J Hazard Mater 163:12–25

Morales FJ, Jimenez-Perez S (2001) Free radical scavenging capacity of Maillard reaction products as related to color and fluorescence. Food Chem 72:119–125

Murugan K, Al-Sohaibani SA (2010) Biocompatible removal of tannin and associated color from tannery effluent using the biomass and tannin Acylhydrolase (E.C.31120) enzyme of mango industry solid waste isolate *Aspergillus candidus* MTTC 9628. Res J Microbiol 5:262–271

Nakai S, Yutaca I, Masaaki H (2001) Algal growth inhibition effects and inducement modes by plant-production phenols. Water Res 35:1855–1859

Nandy T, Shastry S, Kaul SN (2002) Wastewater management in cane molasses distillery involving bioresource recovery. J Environ Manage 65(1):25–38

Narro ML, Cerniglia CE, Van Baalen C, Gibson DT (1992) Evidence for an NIH shift in oxidation of naphthalene by the marine cyanobacterium *Oscillatoria* sp. strain JCM. Appl Environ Microbiol 58:1360–1363

Olaizola M (2003) Commercial development of microalgal biotechnology: from the test tube to the marketplace. Biomol Eng 20:459–466

Oswald WJ (2003) My sixty years in applied algology. J Appl Phycol 15:99–106

Oswald WJ, Gotaas HB, Ludwig HF, Lynch V (1953) Algal symbiosis in oxidation ponds. III. Photosynthetic oxygenation. Sew Ind Wastes 25:692–705

Palanisami S, Saha SK, Uma L (2010) Laccase and polyphenol oxidase activities of marine cyanobacteria: a study with Poly R-478 decolorization. World J Microbiol Biotechnol 26:63–69

Pant D, Adholeya A (2007) Biological approaches for treatment of distillery wastewater: a review. Bioresour Technol 96:2321–2334. doi:10.1016/j.biortech.2006.09.027

Pant D, Reddy UG, Adholeya A (2006) Cultivation of oyster mushrooms on wheat straw and bagasse substrate amended with distillery effluent. World J Microbiol Biotechnol 22:267–275

Papazi A, Kotzabasis K (2007) Bioenergetic strategy of microalgae for the biodegradation of phenolic compounds - exogenously supplied energy and carbon sources adjust the level of biodegradation. J Biotechnol 129:706–716

Papazi A, Kotzabasis K (2008) Inductive and resonance effects of substituents adjust the microalgal biodegradation of toxic phenolic compounds. J Biotechnol 135:366–373. doi:10.1016/j.jbiotec.2008.05.009

Parikh A, Madamwar D (2005) Textile dye decolorization using cyanobacteria. Biotechnol Lett 27:323–326

Patel A, Pawar R, Mishra S, Tewari A (2001) Exploitation of marine cyanobacteria for removal of color from distillery effluent. Indian J Environ Prot 21(12):1118–1121

Pathak H, Joshi HC, Chaudhary A, Kalra N, Dwivedi MK (1999) Soil amendment with distillery effluent for wheat and rice cultivation. Water Air Soil Pollut 113:133–140

Patil NB, Kapadnis BP (1995) Decolorisation of melanoidin pigment from distillery spentwash. Indian J Environ Health 37:84–87

Patil PU, Kapadnis BP, Dhamankar VS (2003) Decolorisation of synthetic melanoidin and biogas effluent by immobilized fungal isolate of *Aspergillus niger* UM2. All India Distiller's Association (AIDA) newsletter, pp 53–56

Perez-Garcia O, Escalante FME, de-Bashan LE, Bashan Y (2011) Heterotrophic cultures of microalgae: metabolism and potential products. Water Res 45:11–36

Petroutsos D, Katapodis P, Christakopoulos P, Kekos D (2007) Removal of *p* -chlorophenol by the marine microalga *Tetraselmis marina*. J Appl Phycol 19:485–490. doi:10.1007/s10811-007-9160-0

Pinto G, Pollio A, Previtera L, Temussi F (2002) Biodegradation of phenols by microalgae. Biotechnol Lett 24:2047–2051. doi:10.1023/A:1021367304315

Pinto G, Pollio A, Previtera L, Stanzione M, Temussi F (2003) Removal of low molecular weight phenols from olive oil mill wastewater using microalgae. Biotechnol Lett 25:1657–1659. doi:10.1023/A:1025667429222

Plavsic M, Cosovic B, Lee C (2006) Copper complexing properties of melanoidins and marine humic material. Sci Total Environ 366:310–319

Podda F, Zuddas P, Minacci A, Pepi M, Baldi F (2000) Heavy metal co-precipitation with hydrozincite $[Zn_5(CO_3)_2(OH)_6]$ from mine waters caused by photosynthetic microorganisms. Appl Environ Microbiol 66(11):5092–5098

Priya B, Sivaprasanth RK, Jensi VD, Uma L, Subramanian G, Prabaharan D (2010) Characterization of manganese superoxide dismutase from a marine cyanobacterium *Leptolyngbya valderiana* BDU20041. Saline Syst 6:6. http://www.salinesystems.org/content/6/1/6

Pulz O, Gross W (2004) Valuable products from biotechnology of microalgae. Appl Microbiol Biotechnol 65:635–648

Radwan SS, Al-Hasan H (2000) Oil pollution and cyanobacteria. In: Potts M, Whitton BA (eds) The ecology of cyanobacteria. Kluwer Academic, Dordrecht/London/Boston, pp 307–316

Rajani Rani M, Sreekanth D, Himabindu V (2011) Degradation of mixture of phenolic compounds by activated sludge processes using mixed consortia. Int J Energy Environ 2(1):151–160

Ramakritinan CM, Kumaraguru AK, Balasubramanian MP (2005) Impact of distillery effluent on carbohydrate metabolism of freshwater fish, *Cyprinus carpio*. Ecotoxicology 14:693–707

Ramana S, Biswas AK, Singh AB (2002) Effect of distillery effluent on some physiological aspects of maize. Bioresour Technol 84:295–297

Robinson T, Mc Mullan G, Marchant R, Nigan P (2001) Remediation of dyes in textile effluent: a critical review on current treatment technologies with a proposed alternative. Bioresour Technol 77:247–255

Rocap G, Larimer FW, Lamerdin J, Malfatti S, Chain P, Ahlgren NA, Arellano A, Coleman M, Hauser L, Hess WR, Johnson ZI, Land M, Lindell D, Post AF, Regala W, Shah M, Shaw SL, Steglich C, Sullivan MB, Ting CS, Tolonen A, Webb ES, Zinser ER, Chisholm SW (2003) Genome divergence in two Prochlorococcus ecotypes reflects oceanic niche differentiation. Nature 424:1042–1047

Saha SK, Swaminathan P, Raghavan C, Uma L, Subramanian G (2010) Ligninolytic and antioxidative enzymes of a marine cyanobacterium *Oscillatoria willei* BDU 130511 during Poly R-478 decolorization. Bioresour Technol 101:3076–3084. doi:10.1016/j.jbiortech.2009.12.075

Schoeny R, Cody T, Warshawsky D, Radike M (1988) Metabolism of mutagenic polycyclic aromatic hydrocarbons by photosynthetic algal species. Mutat Res 197:289–302

Scragg AH (2006) The effect of phenol on the growth *Chlorella vulgaris* and *Chlorella* VT-1. Enzyme Microb Technol 39:796–799. doi:10.1016/j.enzmictec.2005. 12.018 DOI:dx.doi.org

Semple KT, Cain RB (1996) Biodegradation of phenolic by *Ochromonas danica*. Appl Environ Microbiol 62:1265–1273

Semple KT, Cain RB, Schmidt S (1999) Biodegradation of aromatic compounds by microalgae. FEMS Microbiol Lett 170:291–300

Shah V, Garg N, Madamwar D (2000) Record of the cyanobacteria present in the Hamisar pond of Bhuj, India. Acta Bot Mala 25:175–180

Shashirekha S, Uma L, Subramanian G (1997) Phenol degradation by the marine cyanobacterium *Phormidium valderianum* BDU 30501. J Ind Microbiol Biotechnol 19:130–133

Shigeoka T, Sato Y, Tadeka Y (1988) Acute toxicity of chlorophenols to green algae, *Selenastrum capricornutum* and *Chlorella vulgaris*, and quantitative structure-activity relationships. Environ Toxicol Chem 7:847–854

Singh NK, Dhar DW (2006) Sewage effluent: a potential nutrient source for microalgae. Proc Indian Natl Sci Acad 72:113–120

Singh NK, Dhar DW (2007) Nitrogen and phosphorous scavenging potential in microalgae. Indian J Biotechnol 6:52–56

Singh NK, Dhar DW (2011) Microalgae as second generation biofuel. A review. Agron Sustain Dev 31:605–629. doi:10.1007/s13593-011-0018-0

Sorkhoh N, Al-Hasan R, Radwan S, Hopner T (1992) Self-cleaning of the Gulf. Nature 359:109–109

Spolaore P, Joannis-Cassan C, Duran E, Isambert A (2006) Commercial applications of microalgae. J Biosci Bioeng 101:87–96

Sreekanth D, Sivaramakrishna D (2009) Thermophilic degradation of phenolic compounds in lab scale hybrid up flow anaerobic sludge blanket reactors. J Hazard Mater 164:1532–1539

Srivastava S, Radha J (2010) Effect of distillery spentwash on cytomorphological behaviour of sugarcane settlings. J Environ Biol 31(5):809–812

Stewart WDP, Haystead A, Dharmawardene MWN (1975) Nitrogen assimilation and metabolism in blue-green algae. In: Stewart WDP (ed) Nitrogen fixation by free-living microorganisms. Cambridge University Press, Cambridge, pp 129–158

Tabei Y, Okada K, Tsuzuki M (2007) SII1330 controls the expression of glycolytic genes in *Synechocystis* sp. PCC 6803. Biochem Biophys Res Commun 355(4):1045–1050

Talley JW, Sleeper PM (1997) Roadblock to the implementation of bio-treatment strategies. Ann NY Acad Sci 829:16–29

Tarlan E, Dilek FB, Yetis U (2002) Effectiveness of algae in the treatment of a wood-based pulp and paper industry wastewater. Bioresour Technol 84(1):1–5. doi:10.1016/S0960-8524(02)00029-9

Thavasi R, Jayalakshmi S (2003) Bioremediation potential of hydrocarbonoclastic bacteria in Cuddalore harbour waters (India). Res J Chem Environ 7:17–22

Tikoo V, Scragg AH, Shales SW (1997) Degradation of pentachlorophenol by microalgae. J Chem Technol Biotechnol 68:425–431

Travieso L, Benitez F, Sanchez E, Borja R, Leon M, Raposo F, Rincon B (2008) Performance of laboratory-scale microalgae pond for secondary treatment of distillery wastewater. Chem Biochem Eng Q 22(4):467–473

Tsuji N, Hirooka T, Nagase H, Hirata K, Miyamoto K (2003) Photosynthesis-dependent removal of 2,4-dichlorophenol by *Chlorella fusca* var. Vacuolata. Biotechnol Lett 25:241–244

Ugurlu M, Kula I (2007) Decolorization and removal of some organic compounds from olive mill wastewater by advanced oxidation processes and lime treatment. Environ Sci Pollut 5:319–325

Valderrama LT, Del campo CM, Rodriguez CM, de-Bashan LE, Bashan Y (2002) Treatment of recalcitrant wastewater from ethanol and citric acid production using the microalga *Chlorella vulgaris* and the macrophyte *Lemna minuscula*. Water Res 36:4185–4192

Wilkie AC, Riedesel KJ, Ownes JM (2000) Stillage characterization and anaerobic treatment of ethanol stillage from conventional and cellulosic feedstocks. Biomass Bioenerg 19:63–102

Wurster M, Mundt S, Hammer E, Schauer F, Lindequist U (2003) Extracellular degradation of phenol by the cyanobacterium *Synechococcus* PCC 7002. J Appl Phycol 15:171–176

Zubkov MV, Mary I, Woodward EMS, Warwick PE, Fuchs BM, Scanlan DJ, Burkill PH (2007) Microbial control of phosphate in the nutrient-depleted North Atlantic subtropical gyre. Environ Microbiol 9:2079–2089

No-Till Direct Seeding for Energy-Saving Rice Production in China

Min Huang, Bing Xia, Yingbin Zou, Peng Jiang, Yuehua Feng, Zhaowei Cheng, and Yali Mo

Abstract This report shows that no–till direct seeding of super hybrid rice reduced energy use by 21%. Modern agriculture depends heavily on the use of fossil energy. The energy use in agriculture has attracted more and more attention because the increased emissions of greenhouse gases has risen the global mean temperature over the past century, and future impacts on the climate are uncertain. Adoption of no–till farming for crop production, such as corn, wheat or maize, is considered as a possibility to save energy. However, limited information is currently available on the no–till effect on energy use in rice production. In China, significant progress has been made in super hybrid rice breeding, and simplified rice establishment methods have become increasingly attractive because of their labor–saving benefits. No–till direct seeding is a relatively new simplified rice establishment method. We compared yield performance and energy input between transplanted (traditional) and no–till direct seeded super hybrid rice production systems by using data from field experiments and production survey. The results showed that no–till direct seeded system for super hybrid rice production did not reduce grain yield but reduced total energy use by 21% compared to the transplanted system. The reduced total energy use in no–till direct seeded system was attributed to lower energy use in both field operations and producing external inputs. For field operations, no–till direct seeded system had lower energy use by labor and machinery, which contributed about 6%

Min Huang and Bing Xia are contributed equally to this work.

M. Huang • B. Xia • Y. Zou (✉) • P. Jiang • Y. Feng • Z. Cheng • Y. Mo
College of Agronomy, Hunan Agricultural University,
Changsha 410128, China
e-mail: jxhuangmin@163.com; ybzou123@126.com

E. Lichtfouse (ed.), *Farming for Food and Water Security*, Sustainable Agriculture Reviews 10, DOI 10.1007/978-94-007-4500-1_5,

and 38% of the reduced total energy use, respectively. For producing external inputs, energy use in N fertilizer remained the highest contributor to total energy use in both no–till direct seeded and transplanted systems, and the lower energy use in no–till direct seeded system was mainly due to reduced N fertilizer application. Our study suggests that great attention needs to be paid to establish knowledge–based N management practices for no–till direct seeded super hybrid rice in future studies to achieve further energy saving.

Keywords No-till farming • Rice production • Simplified rice establishment method • Energy analysis

1 Introduction

Modern agriculture depends heavily on the use of fossil energy in machines and chemicals. For example, gasoline and diesel fuel are used in tractors; natural gas is used to synthesize ammonia, the basic building block in nitrogen fertilizers (Triplett and Dick 2008). In recent years, the energy use in agriculture has received more and more attention (Dalgaard et al. 2001), mainly because the increased emissions of greenhouse gases has risen the global mean temperature over the past century and future impacts on the climate are uncertain (Flavin and Dunn 1998). It was estimated that agricultural activities were responsible for ≈50% of global atmospheric inputs of methane and agricultural soils were responsible for 75% of global nitrous oxide emissions (Li et al. 2006). Fortunately it has been shown that the development of agricultural systems with low input of energy compared to the output of food can help to reduce emissions of agricultural greenhouse gases (Dalgaard 2000). In this context, the adoption of no–till farming might be one possibility to save energy (Allmaras and Dowdy 1985). In the United States, for which reliable historical data are available, the combined use of gasoline and diesel fuel in agriculture has fallen from its historical high of 29 billion liters in 1973 to 17 billion in 2002, a decline of ≈40%. For a broad sense of the fuel efficiency trend in U.S. agriculture, the liters of fuel used per megagram of grain produced dropped from 138 in 1973 to 54 in 2002, an impressive decrease of ≈60%. One reason for this was a shift to no–till practices on roughly two fifths of U.S. cropland (Brown 2006). There have been reports showing that no–till system reduced energy use in corn, wheat and soybean production by 7%, 13% and 18%, respectively, when compared to conventional–till system (Phillips et al. 1980; Tabatabaeefar et al. 2009), but limited information is currently available on the no–till effect on energy use in rice production. In the present paper, we (1) briefly reviewed progress in rice improvement and changes in rice establishment methods in China and (2) compared yield performance and energy input between transplanted and no–till direct seeded rice production systems by using data from field experiments and production survey.

2 Progress in Rice Improvement in China

Rice is the staple food for ≈65% of the population in China (Huang et al. 2011c). Within a period of four decades from the 1960s to 1990s, rice yield has undergone two big leaps in China, primarily as the result of genetic improvement: increasing harvest index by using the semi-dwarf gene and utilization of heterosis by producing hybrids (Zhang 2007). However, rapid population growth and economic development have been posing a growing pressure for increased food production (Normile 2008). To further increase rice yield potential, China established a nationwide mega–project with the goal to develop super rice based on the ideotype concept in 1996 (Cheng et al. 1998). In 1998, Prof. Longping Yuan proposed a strategy for developing super hybrid rice by combining an ideotype approach with the use of intersubspecific heterosis (Yuan 2001). The ideotype was reflected in the following morphological traits: moderate tillering capacity (270–300 panicles m^{-2}); heavy (5 g per panicle) and dropping panicles at maturity; panicle height of 60 cm (from soil surface to the top of panicles with panicles in natural position) at maturity; and long, erect, thick, narrow, and V-shaped top three leaves. Up to 2011, 27 inbred and 56 hybrid cultivars that met super rice criteria were released by provincial or national seed boards (Fig. 1). It has been concluded that super hybrid cultivars have increased rice yield potential by more than 10% compared with ordinary hybrid and inbred cultivars under subtropical conditions (Zhang et al. 2009; Huang et al. 2011c).

Fig. 1 The first super hybrid rice cultivar (Liangyoupeijiu) of China. This cultivar is an *indica–japonica* hybrid (Peiai64S × 9311) released by Jiangsu Academy of Agricultural Sciences of China in 1999 and approved as super hybrid rice by the Ministry of Agriculture of China in 2005

3 Changes in Rice Establishment Methods in China

Transplanting is the traditional and dominant establishment method for rice production in China. However, the land preparation and seedling transplanting in this method require a large amount of labor (Chen et al. 2007). Although it is generally considered that labor is plentiful in China, limited labor availability has constrained rice production in the country in recent two decades because an increasing amount of young farmers have left for jobs in the cities leaving the older farmers behind (Derpsch and Friedrich 2009). To solve this constraint, several simplified establishment methods, such as conventional–till seedling throwing, conventional–till direct seeding, no–till seedling throwing and no–till direct seeding, have been developed for rice production in China in recent years (Huang et al. 2011a). Of these simplified establishment methods, the first three ones have become increasingly popular in the main rice production regions. For example, over 70% of rice was grown under conventional–till direct seeding in northern Hunan Province in 2009 (Fig. 2; Ao et al. 2011); no–till direct seeding is the simplest one (Fig. 3; Huang et al. 2011a). Huang et al. (2011b) reported that labor input under no–till direct seeding was 30% lower than under transplanting for super hybrid rice production in rice-oilseed rape cropping system. However, this simplified establishment method is relatively new and rarely used by rice farmers in China.

4 Yield and Energy Use of No-Till Direct Seeded Rice

4.1 Data Collection and Analysis

4.1.1 Field Experiments

Field experiments were conducted in Changsha (2004–2010) and Nanxian (2009–2010), Hunan Province, China. Both locations are situated in the sub–tropical monsoon climate zone. Changsha (28°11′ N, 113°04′ E) is at an altitude of 32 m with a mean annual temperature of ≈17.0°C, a mean annual rainfall of ≈1,355 mm and a mean annual sunshine of ≈1,677 h. Nanxian (29°21′ N, 112°23′ E) at an altitude of 30 m with a mean annual temperature of ≈16.6°C, a mean annual rainfall of ≈1,238 mm and a mean annual sunshine of ≈1,776 h. The soils of the experimental fields in Changsha and Nanxian were a tidal clay with organic matter = 15.0 g kg^{-1}, total N = 1.40 g kg^{-1}, Olsen P = 38.4 mg kg^{-1}, NH_4OAc extractable K = 113 mg kg^{-1} and a purple calcareous clay with organic matter = 24.2 g kg^{-1}, total N = 1.10 g kg^{-1}, Olsen P = 14.8 mg kg^{-1}, NH_4OAc extractable K = 80.1 mg kg^{-1}, respectively.

Two super hybrid rice cultivars, Liangyoupeijiu and Y–liangyou 1, were selected in the experiments. The former is an *indica–japonica* hybrid (Peiai64S × 9311) developed by Jiangsu Academy of Agricultural Sciences of China in 1999 and was

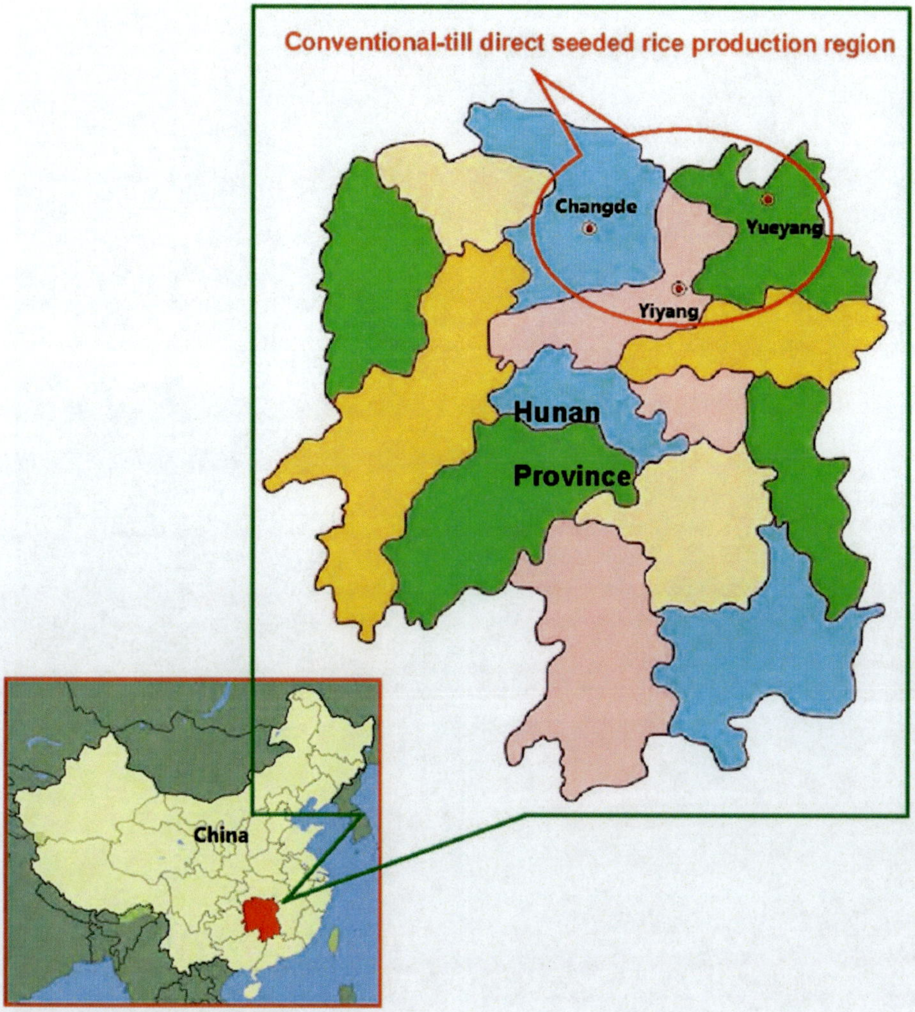

Fig. 2 Conventional–till direct seeded rice production region in Hunan Province, China. Hunan Province produces more rice than any other provinces in China. The rice–harvested area of the province is ≈3.9 Mha, accounted for ≈14% of the rice–harvested area nationally. Changde, Yueyang and Yiyang are the main rice production regions in Hunan Province. The rice–harvested area in these regions accounted for ≈37% of the total rice–harvested area in the province

used in both experimental locations, while the latter is an *indica* hybrid (Y58S × 9311) developed by Hunan Hybrid Rice Research Center of China in 2006 and was only used in Nanxian. These two cultivars have been widely grown by rice farmers in China.

Each cultivar was grown under two establishment methods: transplanting and no–till direct seeding (Fig. 4). Plots were laid out in a randomized complete block

Fig. 3 No–till direct seeding for super hybrid rice production in rice–oilseed rape rotation in Anxiang County, Hunan Province, China in 2007

design with four replications, using plot size of 30 m^2. The land preparation for transplanting was carried out by one ploughing followed by two harrowing; and for no–till direct seeding, herbicide Gramoxone (paraquat 20%) was used (diluted 5 mL L^{-1} and applied at 750 L ha^{-1}) 7 days before seeding and the plots were soaked with water 5 days after herbicide spraying. For transplanting, seedlings were raised in nursery beds, and 25–days–old seedlings were manually transplanted at a spacing of 20 × 20 cm with one seedling per hill between May 31st and June 24th. For no–till direct seeding, pre–germinated seeds were manually broadcasted onto the soil surface at a seed rate of 22.5 kg ha^{-1} between May 11th and June 1st. Urea was used

Fig. 4 Super hybrid rice Liangyoupeijiu grown under transplanting and no–till direct seeding in Changsha, Hunan Province, China in 2010

as a source of N, single superphosphate of P and potassium chloride of K with rates of 150 kg N ha^{-1}, 90 kg P_2O_5 ha^{-1} and 180 kg K_2O ha^{-1}. Nitrogen was split–applied: 90 kg N ha^{-1} as basal, 45 kg N ha^{-1} at mid–tillering, and 15 N kg ha^{-1} at panicle initiation. Phosphorus was applied as basal and K was split equally at basal and panicle initiation. Water management adopted a strategy of flooding–midseason drainage–reflooding–moist intermittent irrigation. Insects, diseases and weeds were controlled by using approved pesticides to avoid yield loss. However, the yield was detrimentally affected due to lodging caused by a typhoon in the growing season of 2005 in Changsha. Hence, the data of 2005 were excluded from the analysis.

Plants were sampled from a 0.48–m^2 area for each plot at maturity to determine yield components. Plants were separated into straw (including rachis) and grains by hand threshing. Filled spikelets were separated from unfilled spikelets by submerging in tap water. Three sub–samples of 30 g filled spikelets and all unfilled spikelets

Table 1 Information of farmers' households interviewed in rice production survey in Taoyuan, Anxiang and Loudi, Hunan Province, China in 2010

Characteristic	Range	Mean	Class distribution	
Age of household head (years old)	30–77	53	≤53 (38.5)	>53 (61.5)
Agricultural labor force (persons per household)	1–4	2	≤2 (92.3)	>2 (7.7)
Rice cultivated area (ha per household)	0.07–0.57	0.20	≤0.20 (69.2)	>0.20 (30.8)

Values in parenthesis are the percentage of the total interviewed households ($n=26$)

were taken to count the number of spikelets. Spikelets per panicle, spikelet filling percentage, and grain weight were calculated. Grain yield was determined from a 5–m^2 area in each plot and adjusted to the standard moisture content of 0.14 g H_2O g^{-1}.

4.1.2 Production Survey

Rice production survey was done in Taoyuan (29°09′ N, 111°22′ E), Anxiang (29°27′ N, 112°08′ E) and Loudi (27°31′ N, 111°23′ E), Hunan Province, China in 2010. The three locations are situated in the sub-tropical monsoon climate zone with mean annual temperature of 16.5–17.0°C, mean annual rainfall of 1,218–1,437 mm and mean annual sunshine of 1,516–1,824 h. Machinery has been widely used in land preparation and harvesting in rice production in the study areas. Super hybrid rice cultivars Liangyoupeijiu and Zhunliangyou 527 (ZhunS × Shuhui 527, *indica*) were cultivated under transplanting or no–till direct seeding in the survey areas. Since no–till direct seeding is relatively new and rarely used by rice farmers in China, only three to five pairs of adjacent fields were chosen in each survey area such that one was under no–till direct seeding and the other was under transplanting. All farmers (households, 26 in total) of the selected fields were interviewed by questionnaire. The questions concerned (a) household information (Table 1), (b) grain yield, and (c) inputs (labor, machinery, seed, fertilizers, herbicides, insecticides and fungicides).

4.1.3 Energy Assessment

Energy was defined as fossil energy measured in joule (J). Energy use for rice production includes the energy used for field operations ($EU_{field\ operations}$) by labor (EU_{labor}, e.g. sowing, fertilizing, spraying) and machinery ($EU_{machinery}$, e.g. ploughing, harrowing and harvesting), and for producing external inputs ($EU_{external\ inputs}$, e.g. seed, fertilizers and insecticides). Most rice in the study areas is irrigated by water coming from dams and its plant protection is done using hand tools, so only labor was taken into the inputs of these two operations. Energy equivalents of inputs used in the rice production were taken as the medians of the lowest and highest values found in the previous studies and given in Table 2. If D_p, I and N are respectively the total person–days, the

Table 2 Energy equivalents for inputs in rice production system

Item	Unit	Range[a]	Energy equivalent[b]	References[c]
Field operation[d]				
Labor	MJ person^{-1} day^{-1}	5–18	12	1–12
Plougher	MJ ha^{-1}	437–1,700	1,069	13–17
Harrower	MJ ha^{-1}	114–874	494	13–15, 17
Harvester	MJ ha^{-1}	364–988	676	13, 14, 17
External input				
Seed	MJ kg^{-1}	16–25	21	6, 7, 18–20
Nitrogen	MJ kg^{-1} N	43–78	61	2, 4, 6–8, 10, 12, 17, 18, 20–24
Phosphorus	MJ kg^{-1} P_2O_5	7–17	12	2, 4–8, 10, 12, 17, 18, 20–24
Potassium	MJ kg^{-1} K_2O	6–14	10	2, 4–8, 10, 17, 18, 20–24
Herbicides	MJ kg^{-1} active agent	80–460	270	5, 7, 12, 13, 24–26
Insecticides	MJ kg^{-1} active agent	58–580	319	5, 7, 10, 12, 13, 25–27
Fungicides	MJ kg^{-1} active agent	61–397	229	4, 5, 7, 10, 12, 13, 25–28

[a] Ranges indicate the lowest and highest values found in references

[b] Energy equivalents are taken as the medians of the ranges

[c] 1: Mitchell (1979); 2: Pimentel and Pimentel (1979); 3: Safa and Tabatabaeefar (2002); 4: Gezer et al. (2003); 5: Tippayawong et al. (2003); 6: Ozkan et al. (2004); 7: Bockari-Gevao et al. (2005); 8: Canakci and Akinci (2006); 9: Demircan et al. (2006); 10: Strapatsa et al. (2006); 11: Kumar (2011); 12: Tabatabaeefar et al. (2009); 13: Stout et al. (1982); 14: McFate (1983); 15: Nielsen (1989); 16: Vitlox and Pletinckx (1989); 17: Dalgaard et al. (2001); 18: Rutger and Grant (1980); 19: Singh and Mittal (1992); 20: Chamsing et al. (2006); 21: Leach (1976); 22: Lockeretz (1980); 23: Mudahar and Trignett (1987); 24: Pleanjai and Gheewala (2009); 25: Green (1987); 26: Fluck (1992); 27: Fluck and Baird (1982); 28: Hülsbergen et al. (2001)

[d] One person–day is considered to be one person working for 8 h in a day. Energy equivalent of diesel fuel used in machinery is 52 MJ L^{-1} (Austin et al. 1978)

total number of field operations by machinery and the total number of external inputs for growing rice, and EU_p, EU_i and EU_n are respectively the energy used for each person–day, each of these field operations by machinery and producing each of these external inputs, obtained or calculated according to the energy equivalents in Table 2, the total energy use (EU_{total}) for the rice production can be expressed by Eq. 1:

$$\begin{aligned} EU_{total} &= EU_{fieldoperations} + EU_{externalinputs} \\ &= \left(EU_{labor} + EU_{machinery}\right) + \sum^{N} EU_n \\ &= \left(EU_p \times D_p + \sum_{i=1}^{I} EU_i\right) + \sum_{n=1}^{N} EU_n \end{aligned} \quad (1)$$

4.1.4 Statistical Analysis

Analysis of variance (Statistix 8, Analytical software, Tallahassee, FL) was employed to determine the significance of differences between means of two establishment methods in the field experiments. A P–value < 0.05 was considered as statistically significant.

4.2 Results and Discussion

Under researcher–managed conditions, grain yield among cultivars, locations and years ranged from 8.38 to 11.25 t ha^{-1} for transplanting and 8.46 to 11.12 t ha^{-1} for no–till direct seeding, showing the means of 9.57 and 9.68 t ha^{-1} under transplanting and no–till direct seeding, respectively (Fig. 5a). Under transplanting, spikelet number ranged from 39.5×10^3 to 56.6×10^3 m^{-2}, with the mean of 48.7×10^3 m^{-2} (Fig. 5b). Under no–till direct seeding, it ranged from 42.2×10^3 to 59.8×10^3 m^{-2}, showing the mean of 49.0×10^3 m^{-2}. Average spikelet filling percentage was 78% under transplanting and 79% under no–till direct seeding (Fig. 5c). Grain weight demonstrated a very small variation depending on establishment method (Fig. 5d). Overall, there were no statistically significant differences between the two establishment methods in the yield attributes measured ($P = 0.50–0.98$), a result which is consistent with those of previous studies that reported on ordinary rice cultivars (Cho et al. 2001; Bhushan et al. 2007).

Under farmer–managed conditions, Anxiang produced the highest grain yield of 8.29 t ha^{-1} in no–till direct seeded system (Table 3). The lowest grain yield of

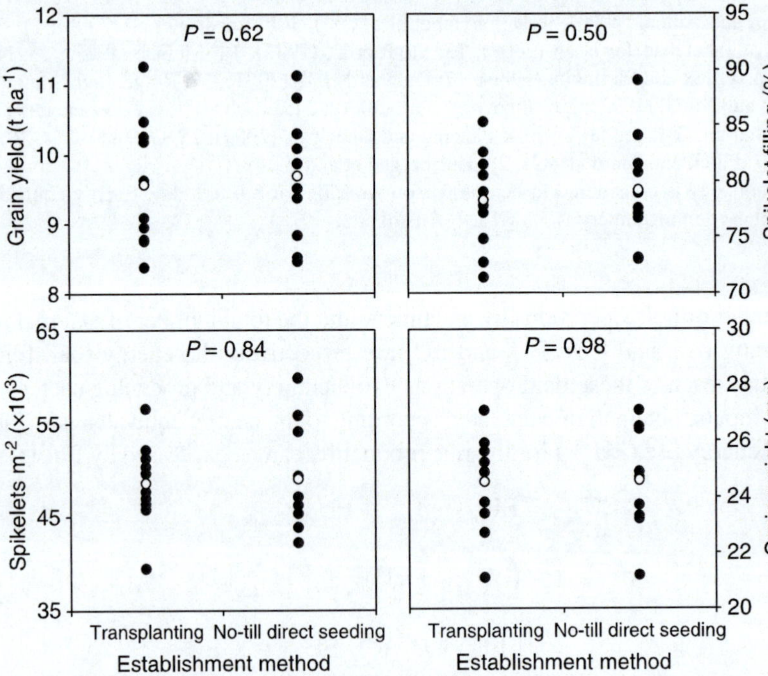

Fig. 5 Yield attributes of super hybrid rice in transplanted and no–till direct seeded systems under researcher–managed conditions. Data were obtained from field experiments in which super hybrid rice cultivar Liangyoupeijiu or Y–liangyou 1 was grown in Changsha (2004, 2006–2010) and Nanxian (2009–2010), Hunan Province, China. • represent means of four replications for each cultivar in each location and year, and ○ represent means of establishment methods. *P*–values (analysis of variance) are used to compare means of establishment methods, and statistical significance is inferred at $P < 0.05$

Table 3 Grain yield and energy use in transplanted and no–till direct seeded super hybrid rice production systems under farmer–managed conditions in Taoyuan, Anxiang and Loudi, Hunan Province, China in 2010

	Transplanting				No–till direct seeding			
Item[a]	Taoyuan	Anxiang	Loudi	Mean	Taoyuan	Anxiang	Loudi	Mean
GY (t ha^{-1})	7.50	7.88	7.13	7.50	7.13	8.29	7.50	7.64
$EU_{field\ operations}$ (GJ ha^{-1})								
Labor	0.54 (3.1)	0.55 (2.9)	0.55 (2.5)	0.55 (2.8)	0.29 (1.9)	0.36 (2.3)	0.32 (2.1)	0.32 (2.1)
Plougher	1.07 (6.2)	1.07 (5.7)	1.07 (4.8)	1.07 (5.5)	0 (0)	0 (0)	0 (0)	0 (0)
Harrower	0.49 (2.9)	0.49 (2.6)	0.49 (2.2)	0.49 (2.5)	0 (0)	0 (0)	0 (0)	0 (0)
Harvester	0.68 (4.0)	0.68 (3.6)	0.68 (3.1)	0.68 (3.5)	0.68 (4.5)	0.68 (4.4)	0.68 (4.4)	0.68 (4.5)
Total	2.78 (16.2)	2.79 (14.9)	2.79 (12.5)	2.79 (14.4)	0.97 (6.5)	1.04 (6.7)	1.00 (6.5)	1.00 (6.6)
$EU_{external\ inputs}$ (GJ ha^{-1})								
Seed	0.32 (1.9)	0.35 (1.9)	0.32 (1.4)	0.33 (1.7)	0.87 (5.8)	0.63 (4.1)	0.63 (4.1)	0.71 (4.6)
Nitrogen	11.48 (66.8)	11.07 (59.0)	15.65 (70.2)	12.73 (65.6)	10.49 (70.1)	9.86 (63.9)	10.18 (65.9)	10.18 (66.6)
Phosphorus	1.08 (6.3)	1.89 (10.1)	0.86 (3.9)	1.28 (6.6)	0.41 (2.7)	1.49 (9.7)	1.07 (6.9)	0.99 (6.5)
Potassium	0.45 (2.6)	1.13 (6.0)	0.90 (4.0)	0.83 (4.3)	0.56 (3.7)	1.01 (6.5)	1.23 (8.0)	0.93 (6.1)
Herbicides	0.04 (0.2)	0.05 (0.3)	0 (0)	0.03 (0.2)	0.18 (1.2)	0.10 (0.6)	0.16 (1.0)	0.15 (1.0)
Insecticides	1.01 (5.9)	1.43 (7.6)	1.72 (7.7)	1.39 (7.2)	1.44 (9.6)	1.26 (8.2)	1.15 (7.4)	1.28 (8.4)
Fungicides	0.03 (0.2)	0.04 (0.2)	0.05 (0.2)	0.04 (0.2)	0.04 (0.3)	0.04 (0.3)	0.03 (0.2)	0.04 (0.2)
Total	14.41 (83.8)	15.96 (85.1)	19.50 (87.5)	16.63 (85.6)	13.99 (93.5)	14.39 (93.3)	14.45 (93.5)	14.28 (93.4)
EU_{total} (GJ ha^{-1})	17.19 (100.0)	18.75 (100.0)	22.29 (100.0)	19.42 (100.0)	14.96 (100.0)	15.43 (100.0)	15.45 (100.0)	15.28 (100.0)

Values in parenthesis are the percentage of the total energy use

[a] GY, $EU_{field\ operations}$, $EU_{external\ inputs}$, and EU_{total} are grain yield, energy use for field operations, energy use for producing external inputs and total energy use, respectively

7.13 t ha^{-1} was observed in transplanted system in Loudi. Average grain yield across three locations was 7.50 and 7.64 t ha^{-1} for transplanted and no–till direct seeded systems, respectively. These results further support the result obtained from field experiments that no–till direct seeded system does not reduce grain yield of super hybrid rice compared with transplanted system. On the other hand, consistently lower total energy use was observed in no–till direct seeded system than in transplanted system (Table 3). Averaged across three locations, no–till direct seeded system had a total energy use of 15.28 GJ ha^{-1}, which was 21% lower than transplanted system. This magnitude of energy use reduction is large compared to that reported in corn (7%), wheat (13%) and soybean (18%) production (Phillips et al. 1980; Tabatabaeefar et al. 2009).

The reduced total energy use in no–till direct seeded system was attributed to lower energy use in both field operations and producing external inputs (Table 3). For field operations, mean energy use by labor and machinery (plougher and harrower) across three locations were 0.23 and 1.56 GJ ha^{-1} lower in no–till direct seeded system than in transplanted system, respectively, which contributed ≈6% and 38% of the reduced total energy use. For producing external inputs, no–till direct seeded system had 0.60 GJ ha^{-1} higher energy use in seed (0.38 GJ ha^{-1}), K fertilizer (0.10 GJ ha^{-1}) and herbicides (0.12 GJ ha^{-1}) but 2.95 GJ ha^{-1} lower energy use in N fertilizer (2.55 GJ ha^{-1}), P fertilizer (0.29 GJ ha^{-1}) and insecticides (0.11 GJ ha^{-1}) than transplanted system averaged across three locations, which resulted in a reduced average energy use of 2.35 GJ ha^{-1} in no–till direct seeded system than in transplanted system.

It is not difficult to understand that no–till direct seeded system had lower energy use in field operations, since both crop establishment and land preparation are greatly simplified in this system (Huang et al. 2011a). The lower energy use for producing external inputs in no–till direct seeded system was mainly driven by reduced N fertilizer application. Our field experiments in Changsha showed that there was no significant difference in internal N efficiency (kg grain per kg N in aboveground plant dry matter) between no–till direct seeded and transplanted super hybrid rice (data not shown). In other words, no–till direct seeded system does not require more N than transplanted system to produce equal grain yield of super hybrid rice. On the other hand, it has been generally accepted that no–till has improving effect on soil fertility (Chen et al. 2007; Triplett and Dick 2008). In the field experiments in Changsha, NaOH hydrolysable N content in 0–5 cm soil layer was increased by ≈15% in no–till direct seeded system than in transplanted system (data not shown). In the production survey, we found that the farmers had recognized the positive effect of no–till on soil fertility. This might be why less N fertilizer was applied in no–till direct seeded system than in transplanted system. Moreover, our study showed that energy use in N fertilizer remained the highest contributor to total energy use in both establishment methods and all three locations (Table 3). Although similar result was observed in rice production in Japan and Thailand (Pimentel and Pimentel 1996; Chamsing et al. 2006), the contribution of energy use in N fertilizer to total energy use in our survey areas (59–70%) is much higher than in Japan (34%) and Thailand (40%). Averaged across two establishment methods and three locations, the energy use in N fertilizer was 11.46 GJ ha^{-1} and was equivalent to a N fertilizer application rate of 188 kg ha^{-1} (Table 3), which was close to the average N fertilizer application rate for rice production in China (180 kg ha^{-1}) but 83% higher than the world average (103 kg ha^{-1}) (Peng et al. 2002). Overapplication of N fertilizer does not only directly waste energy, but may also increase crop damage from pests and diseases (Peng et al. 2002; Peng et al. 2009). Consistent with this, in the present study, both maximum energy use in N fertilizer and insecticides were observed in Loudi in transplanted system (Table 3). These results suggest that a reduction in N fertilizer application rates could be an effective way to save energy in rice production in China. It has been recognized that improper N management practices, including improper rates and timing of N application, are responsible for the overuse of N fertilizer in rice production (Peng et al. 2006). In recent years, a number of researchers have focused on this issue, and have demonstrated that improved nutrient management strategies, such as real–time N management and fixed–time adjustable dose N management, can significantly reduce N fertilizer application rates without decreasing rice yields (Hussain et al. 2000; Singh et al. 2002; Peng et al. 2006). However, these studies were usually conducted under transplanted conditions. There are very few reports on optimizing N management for no–till direct seeded super hybrid rice. At present, the farmers basically follow their experience to apply fertilizers in no–till direct seeded super hybrid rice production in China. Thus, in future studies, great attention needs to be paid to establish knowledge–based N management practices for no–till direct seeded super hybrid rice to achieve further energy saving.

5 Conclusion

Significant progress has been made in super hybrid rice breeding in China in recent years. Meanwhile, simplified rice establishment methods have become increasingly attractive in China because of their labor–saving benefits. No–till direct seeding is a relatively new simplified rice establishment method, it does not reduce grain yield but reduces total energy use by 21% compared to the traditional rice establishment method (transplanting) in super hybrid rice production.

Acknowledgements This work was supported by the International Rice Research Institute, the German Federal Ministry of Economic Cooperation and Development, the Ministry of Agriculture of China, and the China National Rice Research Institute.

References

Allmaras RR, Dowdy RH (1985) Conservation tillage systems and their adoption in the United States. Soil Tillage Res 5:197–222

Ao H, Xiao A, Huang M, Zou Y (2011) Development strategies for direct–seeded rice production in Hunan Province, China. China Rice 17:28–31

Austin RB, Kingston G, Longden PC, Donovan PA (1978) Gross energy yields and the support energy requirements for the production of sugar from beet and cane: a study of four production areas. J Agric Sci 91:667–675

Bhushan L, Ladha JK, Gupta RK, Singh S, Tirol-Padre A, Saharawat YS, Gathala M, Pathak H (2007) Saving of water and labor in a rice–wheat system with no–tillage and direct seeding technologies. Agron J 99:1288–1296

Bockari-Gevao SM, Wan Ishak WI, Azmi Y, Chan CW (2005) Analysis of energy consumption in lowland rice–based cropping system of Malaysia. Songklanakarin J Sci Technol 27:819–826

Brown LR (2006) Plan B 2.0: rescuing a planet under stress and a civilization in trouble. W.W. Norton & Company, New York

Canakci M, Akinci I (2006) Energy use pattern analyses of greenhouse vegetable production. Energy 31:1243–1256

Chamsing A, Salokhe VM, Singh G (2006) Energy consumption analysis for selected crops in different regions of Thailand. Agric Eng Int: The CIGRE journal, Manuscript EE 06013

Chen S, Xia G, Zhao W, Wu F, Zhang G (2007) Characterization of leaf photosynthetic properties for no–tillage rice. Rice Sci 14:283–288

Cheng S, Liao X, Min S (1998) China's super rice research: background, goals and issues. China Rice 1:3–5

Cho YS, Lee BZ, Choe ZR, Ockerby SE (2001) An evaluation of a no–tillage, unfertilised, direct–sown, wheat–rice cropping system in Korea. Aust J Exp Agric 41:53–60

Dalgaard T (2000) Farm types—how can they be used to structure, model and generalize farm data? In: Weidema BP, Meeusen MHG (eds) Agricultural data for life cycle assessments. Agricultural Economics Research Institute, Hague, pp 98–114

Dalgaard T, Halberg N, Poter JR (2001) A model for fossil energy use in Danish agriculture used to compare organic and conventional farming. Agric Ecosyst Environ 87:51–65

Demircan V, Ekinci K, Keener HM, Akbolat D, Ekinci C (2006) Energy and economic analysis of sweet cherry production in Turkey: a case study from Isparta province. Energy Convers Manage 47:1761–1769

Derpsch R, Friedrich T (2009) Development and current status of no–till adoption in the world. In: Proceedings on CD, 18th triennial conference of the International Soil Tillage Research Organization (ISTRO), Izmir, pp 15–19

Flavin C, Dunn S (1998) Responding to the threat of climate change. In: Brown LR, Flavin C, French H (eds) State of the world. W.W. Norton & Company, New York, pp 113–130

Fluck RC (1992) Energy analysis in agricultural systems. In: Fluck RC (ed) Energy in farm production. Elsevier Science Publishers B.V, Amsterdam, pp 45–51

Fluck RC, Baird CD (1982) Agricultural energetics. AVI Publications, Westport

Gezer I, Acaroglu M, Haciseferogullari H (2003) Use of energy and labour in apricot agriculture in Turkey. Biomass Bioenerg 24:215–219

Green M (1987) Energy in pesticide manufacture, distribution and use. In: Helsel ZR (ed) Energy in plant nutrition and pest control. Elsevier Science Publishers B.V, Amsterdam, pp 165–177

Huang M, Md I, Xia B, Zou Y (2011a) Significance, progress and prospects for research in simplified cultivation technologies for rice in China. J Agric Sci 149:487–496

Huang M, Zou Y, Feng Y, Cheng Z, Mo Y, Md I, Xia B, Jiang P (2011b) No–tillage and direct seeding for super hybrid rice production in rice–oilseed rape cropping system. Eur J Agron 34:278–286

Huang M, Zou Y, Jiang P, Xia B, Md I, Ao H (2011c) Relationship between grain yield and yield components in super hybrid rice. Agric Sci China 10:1537–1544

Hülsbergen KJ, Feil B, Biermann S, Rathke GW, Kalk WD, Diepenbrock W (2001) A method of energy balancing in crop production and its applications in a long–term fertilizer trail. Agric Ecosyst Environ 86:303–321

Hussain F, Bronson KF, Yadvinder S, Bijay S, Peng S (2000) Use of chlorophyll meter sufficiency indices for nitrogen management of irrigated rice in Asia. Agron J 92:875–879

Kumar M (2011) Yield production and energy budget of traditional agricultural crops in Garhwal Himalaya. Agric Sci China 10:78–85
Leach G (1976) Energy and food production. IPC Science and Technology Press Ltd, Guilford
Li C, Salas W, DeAngelo B, Rose S (2006) Assessing alternatives for mitigating net greenhouse gas emissions and increasing yields from rice production in China over the next 20 years. J Environ Qual 35:33–44
Lockeretz W (1980) Energy inputs for nitrogen, phosphorus and potash fertilizers. In: Pimentel D (ed) Handbook of energy utilization in agriculture. CRC Press, Boca Raton, pp 23–24
McFate KL (1983) Farm energy analysis: a program aimed at fostering a team approach for implementing wise energy management and efficient energy use practices on American farms. National Food and Energy Council, Columbia
Mitchell R (1979) An analysis of India agroecosystem. Interprint, New Delhi
Mudahar MS, Trignett TP (1987) Energy requirements, technology and resources in fertilizer sector. In: Helsel ZR (ed) Energy in plant nutrition and pest control: energy in world agriculture. Elsevier Science Publishers B.V, Amsterdam, pp 25–61
Nielsen V (1989) Specific fuel consumption in European countries: Denmark. In: Pick E, Noren O, Nielsen V (eds) Energy consumption and input–output relations of field operations. Food and Agriculture Organization of the United Nations, Rome, pp 36–52
Normile D (2008) Reinventing rice to feed the world. Science 321:330–333
Ozkan B, Akcaoz H, Cemal F (2004) Energy input–output analysis in Turkish agriculture. Renew Energy 29:39–51
Peng S, Huang J, Zhong X, Yang J, Wang G, Zou Y, Zhang F, Zhu Q, Buresh R, Witt C (2002) Challenge and opportunity in improving fertilizer–nitrogen use efficiency of irrigated rice in China. Agric Sci China 1:776–785
Peng S, Buresh RJ, Huang J, Yang J, Zou Y, Zhong X, Wang G, Zhang F (2006) Strategies for overcoming low agronomic nitrogen use efficiency in irrigated rice systems in China. Field Crops Res 96:37–47
Peng S, Tang Q, Zou Y (2009) Current status and challenges of rice production in China. Plant Prod Sci 12:3–8
Phillips RE, Thomas GW, Blevins RL, Frye WW, Phillips SH (1980) No–tillage agriculture. Science 208:1108–1113
Pimentel D, Pimentel M (1979) Food, energy and society. Edward Arnold Publishers Ltd, London
Pimentel D, Pimentel M (1996) Food, energy and society. University Press of Colorado, Niwot
Pleanjai S, Gheewala SH (2009) Full chain energy analysis of biodiesel production from palm oil in Thailand. Appl Energy 86(Suppl):209–214
Rutger JN, Grant WR (1980) Energy use in rice production. In: Pimentel D (ed) Handbook of energy utilization in agriculture. CRC Press, Boca Raton, pp 93–98
Safa M, Tabatabaeefar A (2002) Energy consumption in wheat production in irrigated and dry land farming. In: Proceedings of international agricultural engineering conference. Wuxi, pp 28–30
Singh S, Mittal JP (1992) Energy in production agriculture. Mittal Publications, New Delhi
Singh B, Singh Y, Ladha JK, Bronson KF, Balasubramanian V, Singh J, Khind CS (2002) Chlorophyll meter- and leaf color chart based nitrogen management for rice and wheat in northwestern India. Agron J 94:821–829
Stout BA, Myers C, Schwab G, Helsel Z (1982) Energy management in US production agriculture. In: Robinson DW, Mollan RC (eds) Energy management and agriculture. Royal Dublin Society, Dublin, pp 67–84
Strapatsa AV, Nanos GD, Tsatsarelis CA (2006) Energy flow for integrated apple production in Greece. Agric Ecosyst Environ 116:176–180
Tabatabaeefar A, Emamzadeh H, Varnamkhasti MG, Rahimizadeh R, Karimi M (2009) Comparison of energy of tillage systems in wheat production. Energy 34:41–45

Tippayawong N, Pittayapak A, Jompakdee W (2003) Analysis of energy requirement for vegetable oil production in Northern Thailand's farms. CMU J 2:37–47

Triplett GB Jr, Dick WA (2008) No–tillage crop production: a revolution in agriculture! Agron J 100(Suppl):153–165

Vitlox M, Pletinckx A (1989) Specific fuel consumption in European countries: Belgium. In: Pick E, Noren O, Nielsen V (eds) Energy consumption and input–output relations of field operations. Food and Agriculture Organization of the United Nations, Rome, pp 12–17

Yuan L (2001) Breeding of super hybrid rice. In: Peng S, Hardy B (eds) Rice research for food security and poverty alleviation. International Rice Research Institute, Los Baños, pp 143–149

Zhang Q (2007) Strategies for developing green super rice. Proc Natl Acad Sci USA 104:16402–16409

Zhang Y, Tang Q, Zou Y, Li D, Qin J, Yang S, Chen L, Xia B, Peng S (2009) Yield potential and radiation use efficiency of "super" hybrid rice grown under subtropical conditions. Field Crops Res 114:91–98

Agricultural Water Poverty Index for a Sustainable World

Masoumeh Forouzani, Ezatollah Karami, Mansour Zibaei, and Gh. Hossein Zamani

Abstract Water is a major constraint to development of all countries with low water access, especially those located in arid and semi-arid areas with heavy dependence on agriculture. Water scarcity due to long-term droughts, increased demand, and mismanagement of the available water resources has threaten the sustainability of agricultural development. The need of more food production for growing population drives the agricultural sector to manage the water resources in a sustainable way. Water sustainability encompasses different dimensions which need to be addressed clearly by appropriate indices. The Agricultural Water Poverty Index as an assessment tool measures the level of agricultural water poverty as the most important construct that influences agricultural water management. It encompasses a variety of water-related aspects including availability and access to water resources, capacities to manage available water usage and finally environmental factors affecting availability of water resources. Each aspect needed to be translated into practical and subtle indicators to demonstrate what it is intended to measure, and also to jointly measure the Agricultural Water Poverty Index. Accurately and reasonable selection of indicators needs to apply a suitable framework which increases scientific credibility of water assessment.

M. Forouzani (✉)
Department of Agricultural Extension and Education, Khuzestan Ramin Agricultural and Natural Resources University, Mollasani, 63417-73637, Ahvaz, Iran
e-mail: m.forouzani@yahoo.com

E. Karami • G.H. Zamani
Department of Agricultural Extension, College of Agriculture, Shiraz University, Shiraz 71441-65186, Iran
e-mail: ezatkarami@gmail.com; Ghh_zamani@yahoo.com

M. Zibaei
Department of Agricultural Economics, College of Agriculture, Shiraz University, Shiraz 71441-65186, Iran
e-mail: zibaei@shirazu.ac.ir

E. Lichtfouse (ed.), *Farming for Food and Water Security*, Sustainable Agriculture Reviews 10, DOI 10.1007/978-94-007-4500-1_6,

This report develops indicators of Agricultural Water Poverty for assessing the agricultural water situation at farm level. To reach this goal, firstly different current conceptual frameworks suggested to frame the indicator selection process, including causal chain, hierarchical, and integrated framework, are shortly described and criticized. This review reveals that as the causal chain frameworks were improved over time, due to introducing the additional concepts and lack of clear agenda to build such models, the development of their structures has become complicated. Although the hierarchical frame is more simplified, it includes general concepts which are not robust enough to identify practical and useful indicators. Thus, our report focuses on using the Integrated Causal Network and Ecological Hierarchy Network framework to develop the indicators of Agricultural Water Poverty, because it encompasses all the special issues which may be the concern of indicator construction. To modify this framework a conceptual path to develop the indicators of Agricultural Water Poverty is offered, in which a sequence of steps are proposed. This conceptual path leads to the construction of abstract and concrete indicators of Agricultural Water Poverty and applies a quantitative method (Binary Integer Programming) for selecting the final indicators. An Analytic Hierarchy Process is conducted to assign indicator weights and the model criteria requirements are considered. Finally, a total of 25 indicators are determined by translating the requirements into quantitative constraints and resolving the model.

Keywords Indicator selection • Agricultural Water Poverty Index (AWPI) • Causal Network Frameworks • Analytic Hierarchy Process (AHP) • Binary Integer Programming

Abbreviations

AWP	Agricultural Water Poverty
AWPI	Agricultural Water Poverty Index
AHP	Analytic Hierarchy Process

1 Introduction

Water is a constraint to economic development of all countries located in arid and semi-arid areas with low water access, especially those depend heavily on agriculture (The emerging drought combined with the complexity of water scarcity issue has become an increasingly important issue in many areas of Iran as an arid country, See Fig. 1). Therefore, a reliable supply of water is essential to foster social, economic and environmental developments. There is evidence that water scarcity already affects more than 40% of the people on our planet and by 2025, 1.8 billion people will be living in countries or regions with absolute water scarcity. Further, two-thirds of the world's population could be living under water stressed conditions

Fig. 1 The appearing cracks in ground surface and ground subsidence (**a**); Drying springs (**b**); and abandoned villages (**c**) in Darab plain, Fars Province, Iran (Source: Torabi Haghighi and Keshtkaran 2008, pp. 7–8)

(Food and Agriculture Organization (FAO) 2007). Continued population growth calls for increased allocations of water for multiple uses, such as domestic, agriculture and industrial consumption. Thus worldwide concerns are growing related to water scarcity and its impact on food security (Yang et al. 2003).

In water scarce countries irrigation is vital to enhance crop production (Jhorar et al. 2009), but growing water scarcity puts pressure on irrigation systems, as the main consumptive user, to release water for other uses (Malano et al. 2004). Imbalances between supply and demand, inter-sectorial competition, inter-regional conflicts, and degradation of surface and groundwater quality often occur in these regions (Pereira et al. 2002). Therefore, it is expected that the agricultural sector uses less water without decreases in food production for the growing population. This means that water is an important factor in agriculture, it can affect the food production, and it should be managed in a way that it can be sustained.

Generally speaking, sustainable management of water is a priority for agriculture in the world, especially in those regions which face water scarcity. Water sustainability encompasses different dimensions which need to be addressed clearly by appropriate indices. Hence, development of appropriate tools to help the assessment of water scarcity is a main requirement. In this regard, Forouzani and Karami (2010) developed an Agricultural Water Poverty Index (AWPI) to measure water situation at farm level. The most fundamental function of this index is to measure the level of agricultural water poverty as the most important construct that influences agricultural water management. The AWPI is made of a various range of appropriate components; each which can be measured by a group of indicators. Agricultural water poverty indicators can help to assess the situation of agricultural water in a region, to monitor trends in water conditions over time or to diagnose the causes of water poverty in a region over a specific time period (Forouzani and Karami 2010).

Indicators are constructed to simplify managing and understanding complex systems like the environment (Turnhout et al. 2007). Indicator characteristics are helpful in reducing complexity, and making meaningful multifaceted concepts can assist to make decisions and planning actions (Doody et al. 2009; Ronchi et al. 2002). Development of such indicators is useful and provides well-suited tools to apply research results to a broader field of utilization (Muller and Lenz 2006). Development of measurable indicators allows assessment of entities that originally may not seemed measurable (Turnhout et al. 2007).

Reasonable selection of an indicator is a prerequisite for effectively using it (Lin et al. 2009). Trade-offs between science and utility often influence the process of indicator selection (Turnhout et al. 2007), and dependency on poor indicators may prevent accurate assessment of complex systems (i.e. spatial and temporal interactions). As a result, this hampers the use of such indicators as a resource management tool (Dale and Beyeler 2001).

Absence of a properly documented indicator selection process is not a minor issue (Sullivan 2002) because without a defined protocol with scientific rigor to identify indicators, the selection process will be subjected to arbitrary decisions by different researchers (Lin et al. 2009). It is, therefore, very important to have a well-defined and transparent conceptual framework leading from problem definition to indicator set to interpretation of the indicator values. Standard procedures for selecting indicators allow repeatability, avoid bias, and impose discipline upon the selection process (Dale and Beyeler 2001).

Given the aforementioned explanations, this paper aims to develop the AWPI indicators based on a proper framework. In this case, the conceptual framework introduced by Lin et al. (2009), the Integrated Causal Network and Ecological Hierarchy Network framework, was applied and consequently, the most appropriate indicators for measuring the AWPI according to a Binary Integer Programming Method were determined. Therefore, in the following sections, a brief description of AWPI is provided, for selection the appropriate framework for AWPI the current conceptual frameworks related to indicator construction are described. The final section presents the process for determining AWPI indicators based upon the selected conceptual framework.

2 The Agricultural Water Poverty Index

Before any progress can be made on the construction of an AWPI, certain confusions about words and concepts regarded to its concept and related components must be cleared up. The theoretical framework of the AWPI has been explained in detail in another article of the first two authors of this paper (Forouzani and Karami 2010).

They argued that growing water scarcity due to natural phenomena such as long-term droughts, increased demand, and mismanagement of the available water resources is the major threat to sustainable agricultural development. Without any doubt, this is likely to cause serious predicaments in the agricultural sector. While water is the most important construct influencing agricultural development, the precise examination of agricultural sustainability indicators indicates that it provides little to no contribution to the indicators developed for agricultural sustainability. One strand of this problem suggests that the water management in more effective ways should be considered as a proper approach to use the limited available water, while another strand implies the urgency of the need to assess the water resources situation as an important indicator to sustainable agricultural development. Forouzani and Karami (2010) concluded that there is a serious need to develop an appropriate assessment tool to identify the water situation in the agricultural sector. Consequently, they introduced the Agricultural Water Poverty Index as a new index to practically determine the extent to which a farmer or an agricultural community is water poor (Forouzani and Karami 2010). (The calculation method of the AWPI including methodology and formulas has been provided in detail in another paper of these two authors. Using the indicators developed in this paper, they have measured the AWPI for a specific region).

The AWPI provides the ability to aggregate the main aspects of agricultural water into an index. It combines various dimensions associated with water management issues; including agricultural, social, economical and environmental information. It assists farmers and/or communities to assess agricultural water poverty by measuring five components which include: resources, access, use, capacity and environment (Forouzani and Karami 2010). Brief definitions of these components are as follow (Forouzani and Karami 2010):

Resources: the amount of agricultural water (surface and groundwater) that is currently available in a given region. Resource availability is related to the quality of resources; therefore, the quality of water resources will be included as a sub-component of the "Environment" component.

Access: the extent to which farmers have access to agricultural water resources in the region. This component can be divided into two parts: (a) farmers' access to water; and (b) potential and quality of land to receive available water.

Use: estimated physical water use efficiency of available agricultural water. This component is influenced by the ability to use agricultural water effectively.

Capacity: this component points to the current potentials in managing agricultural water at farm level. The AWPI considers this component as having capacity for sustaining access and optimal use of agricultural water and implies an ability to manage water for the sake of effective water usage. Capacity can be divided into three categories: a) human capital, mostly in the form of farmer's water management knowledge, education and other abilities; b) real capital, mainly technological and financial (savings and investments); and c) social capital which interacts with real capital to provide a capacity to improve water use efficiency.

Environment: environmental factors influencing the quality and quantity of agricultural water.

The AWPI as a set of indicators allows comparisons to be made between individuals and/or communities, and enables decision makers to prioritize actions in the water sector. In order to reach this goal, appropriate indicators are needed to demonstrate what it is intended to measure.

In general, a variety of hydrological, agricultural, economic and social indicators should be taken into consideration to measure the main abstract components of AWPI. In other words, each component should be composed of specific practical indicators to combine different water related elements of an agricultural system, and also to jointly measure the AWPI.

3 Conceptual Frameworks for Selecting Indicators

There are two perspectives for indicator selection: (1) based on the developers' perspective and (2) a specific framework. Indicators which are selected based on the developers' prior knowledge and experience implicitly reflect the values of those who develop and select them (Manoliadis 2002; Niemeijer and de Groot 2008). Niemeijer and de Groot (2008) noted that "in many current practices, indicators are often selected based on the degree to which experts meet a number of criteria individually, rather than on the basis of how they jointly provide an answer to environmental questions". In short, developer's based perspective indicator selection is based on developer interests.

Working on the basis of a concrete framework directs indicator selection through an analytical logic rather than developer's preferred characteristics. However, a more rigorous and transparent indicator selection process will increase both the value and scientific credibility of environmental assessments. The analytical logic insists on the premise that each indicator in an index set should have a particular function in the analytical problem solving logic of the environmental issues that are to be addressed with the use of indicators (Dale and Beyeler 2001; Niemeijer and de Groot 2008). Therefore, different conceptual frameworks were suggested that should be used to frame the indicator selection process. Three kinds of conceptual frameworks have been identified: (1) causal chain frame, (2) hierarchical frame, and (3) integrated framework. In the following sections a short description of these frameworks is provided.

3.1 Causal Chain Frameworks

The first set of frameworks is the causal chain frameworks. They are established based on the causal links (chains) which share this main logic: human and economic activities exert pressure on the environment, and this makes changes in the state or environmental conditions. As a result, society may be provoked to respond and change the pressures and state of the environment (Niemeijer and de Groot 2008).

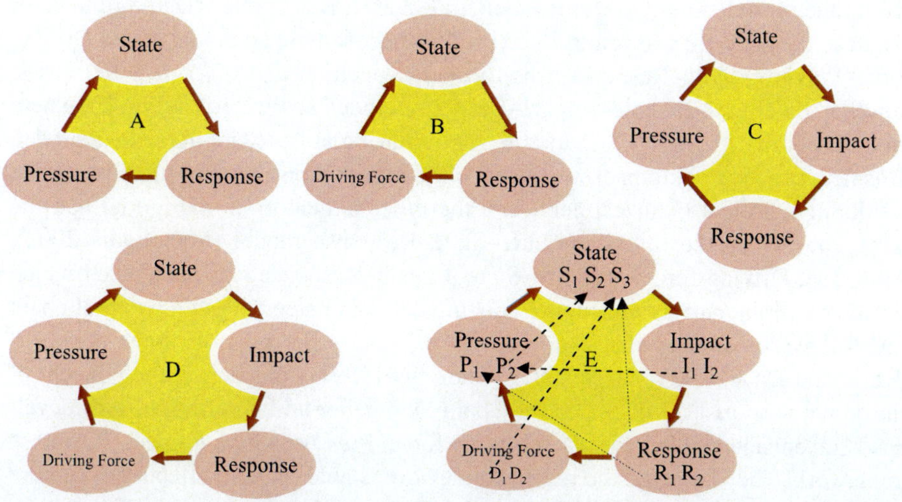

Fig. 2 The causal chain frameworks including (**a**) Pressure–State–Response, (**b**) Driving force–State–Response, (**c**) Pressure–State–Impact–Response, (**d**) Driving force–Pressure–State–Impact–Response, and (**e**) Enhanced Driving force–Pressure–State–Impact–Response frameworks (Adapted from: Niemeijer and de Groot 2008, p. 16; Ronchi et al. 2002, p. 199). The causal chain frameworks are established based on the causal links which show the pressure of human and economic activities on the environment, followed by changes in the state or environmental conditions, and provoking the society responses to mitigate the driving forces and pressures. As development occurs over time they include not only the more components, but also reflect the inter-relations among those components

The degree in which researchers subdivide steps in the causal chain is the main difference between frameworks. The most common conceptual frameworks used in indicator construction are the (1) Driving force–Pressure–State–Impact–Response, (2) Pressure–State–Impact–Response, (3) Pressure–State–Response, or (4) Driving force–State–Response; which organize and structure indicators in the context of a so-called causal chain (see Fig. 2) (Niemeijer and de Groot 2008; Ronchi et al. 2002).

Prior to the Rio de Janeiro summit, the Pressure–State–Response model, as a new conceptual framework for environmental analysis, was established by the Organization for Economic Cooperation and Development (Ronchi et al. 2002) (Fig. 2a). The Pressure–State–Response model includes indicators of pressure which point to burden of activities on the environment; indicators of state which show the conditions of the environment and development; and indicators of response which direct to success of physical, behavioral, or institutional programs to causes, pressures or states (Golusin and Ivanovic 2009; Manoliadis 2002). Because of the incompatibility of this primary model with social and economic systems, the concept of "driving force" was introduced and the "pressure" component was replaced with this new concept. As a result, the Driving force–State–Response model was formed (Fig. 2b), in which "driving force" is the area of human activities and recognizes both beneficial and harmful impacts of an ecological system on the environment

(Niemeijer and de Groot 2008; Ronchi et al. 2002). But in many leading indicator projects, the pressure was reintroduced and somehow a mixed model called Driving force–Pressure–State–Response was formed (Ronchi et al. 2002). In other cases, however, the Pressure–State–Impact–Response model was preferred (Fig. 2c) where the concept of "impact" was added to the initial Pressure–State–Response model. This new concept, i.e. impact, was used to monitor the long-term or more pervasive results of a project or investment; and the other indicators in the model had the same concept as the initial Pressure–State–Response model (Manoliadis 2002). After that, Driving force–Pressure–State–Impact–Response was introduced which combines all the current concepts to distinguish more steps through a causal chain and follows essentially the same general pattern as the previous ones (Fig. 2d) (Borja and Dauer 2008; Smeets and Weterings 1999). Considering the concept of causal network instead of the causal chain, Niemeijer and de Groot (2008) developed an enhanced form of the Driving force–Pressure–State–Impact–Response framework. They believed the recent framework is able to more effectively capture the whole range of causes and effects and their inter-relations. Since this framework encompasses the interconnections between different components and processes, it is called the Causal Network framework, which graphically represents a series of causal and feedback loops (Lin et al. 2009).

In the last two decades causal chain frameworks have become an increasingly important issue in the field of indicator construction. As frameworks were improved, due to introducing the additional concepts, the development of framework structures has become complicated. Niemeijer and de Groot (2008), for example, proposed several detailed steps to form a Causal Network framework so that one must pay much more time to build the resulting complex framework. However, its complexity makes it less practical and useful.

Moreover, in most cases a clear agenda is not presented to build such models in which complex ecological systems are addressed. Rather, the equal values are not considered by users for each concept. Borja and Dauer (2008), for example, using Driving force–Pressure–State–Impact–Response in their study on "assessing the environmental quality status in estuarine and coastal systems" attempted to develop the Impact and Response concepts in more detail, so they placed more emphasis on these components rather than others, i.e. Driving Force, Pressure and State.

3.2 *Ecological Hierarchy Network*

The complexity of ecological systems forms a bottleneck for appropriate indicator selection. In order to consider more appropriate indicators, it is suggested that a hierarchical model can better address such difficulties by choosing the indicators which represent the structure, function, and composition of an ecological system. These three elements define a system. If the linkages between underlying processes, composition and structural elements are broken, then sustainability and integrity are endangered and restoration may be difficult and complex (Dale and Beyeler 2001).

Although the Ecological Hierarchy Network framework provides valuable information, it is more simplified in contrast to causal chain frameworks. While it offers an easy-to-understand picture of issues vital for indicator development, the picture may be distorted due to unclear procedures for building the framework. The Ecological Hierarchy Network framework includes general concepts and is not robust enough to identify practical and useful indicators.

3.3 Integrated Causal Network and Ecological Hierarchy Network Frameworks

Lin et al. (2009) combined the two recent frameworks to build up a new one. They identified three kinds of ecological indicators under the name of Integrated Network framework. The synthetic or general indicators, which represent the general objective of a plan or program, are at the top of the framework. The abstract indicators, in the middle, are the logical extensions of the general indicator and may be multi-layered to interpret the hierarchy relationships. Finally, at the bottom, concrete indicators are clear and specific enough to be put into practice directly.

The emphasis of this merged framework is that both Ecological Hierarchy Network and Causal Network frameworks can be integrated in an ecological indicator selection process. The former can be first used to collect the relevant ecological elements around the general indicators and arrange them along the scale gradients of composition, structure and function. Then, the latter can be used as a guideline to identify the rational indicators from the general objective within the Ecological Hierarchy Network and connect those through the Driving force–Pressure–State–Impact–Response chains. As a result, a series of ecological indicators with a network structure is formed (Lin et al. 2009).

Using this combined framework helps to overcome limitations in applying Causal Chain and Ecological Hierarchy Network frameworks individually. Because each one of those frameworks emphasizes on special issues which may all together be the concern of indicator construction. In order to develop the AWPI indicators, therefore, this study is intended to use Integrated Network framework which is modified according to the nature of issue at hand. Following provides a more detailed description of this framework by developing the indicators of the AWPI.

4 The Conceptual Path to Develop Indicators

AWPI is a new index intended to assess the water situation in the agricultural sector and is developed based on a rigorous structure. According to Lin et al. (2009), four steps should be followed to construct indicators based on the

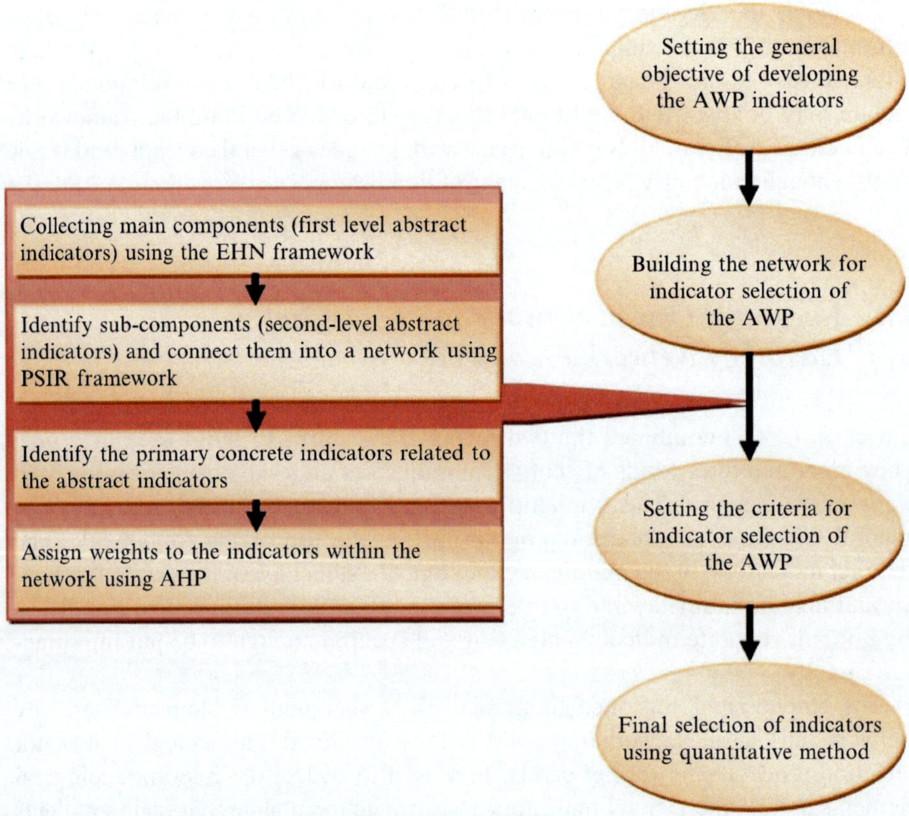

Fig. 3 The extended framework used for selection of the Agricultural Water Poverty indicators (Where: *AWP* Agricultural Water Poverty, *EHN* Ecological Hierarchy Network, *PSIR* Pressure–State–Impact–Response, *AHP* Analytic Hierarchy Process). Using the integrated Causal Network and Ecological Hierarchy Network, an extended framework was provided to develop indicators of the Agricultural Water Poverty. It conceptually clarifies the sequence of steps which lead to build the Agricultural Water Poverty Index

Integrated Causal Network and Ecological Hierarchy Network framework which include:

(1) Identifying the problem and establishing the general objective; (2) Building a network for indicator selection; (3) Setting the criteria and requirements of indicator selection; and (4) Final selection of indicators.

The AWPI was developed by compounding diverse components with specific regard to the integrated framework. Figure 3 represents an extended framework as a conceptual path to develop the AWPI indicators in which a sequence of steps are proposed. In other words, a process for indicator selection was offered which eventually will apply a quantitative method to select the final AWP indicators. This process will be illustrated in the following diagram.

4.1 Setting the General Objective for Developing Indicators of Agricultural Water Poverty

The general objective determines the study boundary and is core stage for ecological indicator selection (Lin et al. 2009). The establishment of objectives must be as detailed as possible to enable indicator selection (Manoliadis 2002).

Since assessing water situation in an agricultural system is the base of sustainability for agricultural water management, the general objective of the AWPI is to assess the agricultural water poverty situation at farm level (i.e. for a farmer). Consequently, the main research question of this study can be formulated as "How can one measure a farmer's agricultural water poverty?"

4.2 Building the Network for Indicator Selection of Agricultural Water Poverty

4.2.1 Collecting and Arranging the Main Components Related to the General Objective

One of the key issues which must be addressed in developing any index is the choice of components. Whereas, Forouzani and Karami (2010) considered five components: (1) Resources, (2) Access, (3) Use, (4) Capacity and (5) Environment, as main elements of the AWPI, Lin et al. (2009) argued that if one wishes to identify the relevant ecological elements around the general objective, an Ecological Hierarchy Network framework can help to achieve this goal. Although, the main components of the AWPI (i.e. Resources, Access, Use, Capacity and Environment) have been predetermined, it is also possible to correspond to each of the main elements of an Ecological Hierarchy Network framework. Thus, by considering the point that each element of an Ecological Hierarchy Network framework denotes to an agricultural water system, it can be concluded that "Resources" and "Environment" are considered as compositional components; "Access" and "Capacity" as structural components; and "Use" as the functional component (Fig. 4). Since some environmental factors influencing the quality of agricultural water are outcomes of agricultural functions in rural areas, the element of "Environment" can be considered as an additional impressive element pertaining to the function of an agricultural water system. A combination of these five components define an agricultural water system and provide a means to select an appropriate set of indicators representative of the key characteristics of the AWPI. In other words, these components are logical extensions of the general indicator (AWPI), and as insisted by Lin et al. (2009), they compose the first level abstract indicators of the AWPI indicators.

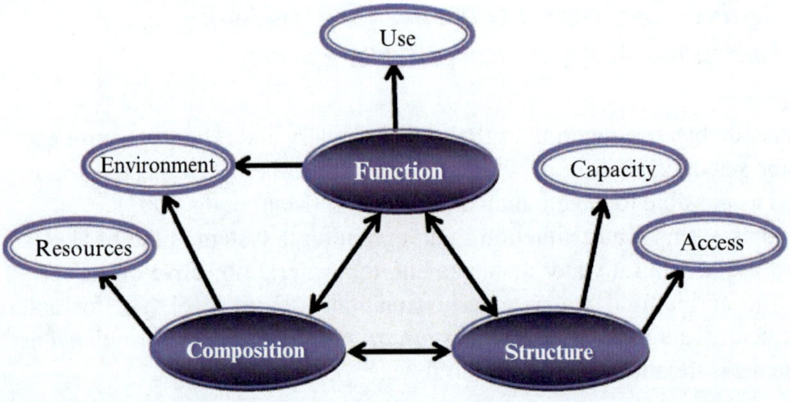

Fig. 4 The key components of the Agricultural Water Poverty Index (AWPI) and their relation to the Ecological Hierarchy Network framework in an agricultural water system. The three elements of an Ecological Hierarchy Network (Composition, Function and Structure) provide a basis to select the first level of abstract indicators of the Agricultural Water Poverty Index. They are general enough to capture all facets of an agricultural system with specific regard to water

4.2.2 Identifying the Sub-Components and Connecting them into a Network

As proposed in the Integrated Causal Network and Ecological Hierarchy Network framework, after identifying the first level of abstract indicators, this framework guides to draw the second level of abstract indicators by building a Causal Network using a specific causal chain model. According to Forouzani and Karami (2010), each main proposed component of the AWPI includes sub-components as the most relevant and rational abstract indicators (Table 1). It also can be assumed that cause and effect relationships between components and sub-components exist. Instead of using the Driving force–Pressure–State–Impact–Response model applied by Lin et al. (2009) for constructing a causal network model, this study used a Pressure–State–Impact–Response model to better structure and more easily identify relationships between the AWPI abstract indicators. Thus, a causal network for AWPI was built.

Considering the aforementioned descriptions of the AWPI components and the Pressure–State–Impact–Response framework, it can be said that current increased demand on water supply due to rapidly expanding population and the need to secure water to cover food demands for the people, coupled with environmental factors (for example, drought or contamination) are noticeable factors that exert pressures on available water resources. These pressures lead to changes and decline in quality and quantity of the current state of agricultural water resources. Decreasing ground and surface water resources, as well as declining in water quality, contribute to reduced farmer access to water. In other words, the water amount is constant in the globe and the growing population of users makes the proportion per user less available. These factors may limit water access to farmers in the future. Indeed, some factors like soil texture, land slope, distance between water source and farm,

Table 1 The first and second level suggested abstract indicators for the Agricultural Water Poverty Index

Abstract indicators	
First level	Second level
Resources (R)	R_1: Surface water resources
	R_2: Groundwater resources
Access (A)	A_1: Farmers' access to water
	A_2: Potential and quality of land
Use (U)	U_1: Efficient water use
Capacity (C)	C_1: Human capital
	C_2: Real capital (technological and financial)
	C_3: Social capital
Environment (E)	E_1: Water quality

Note: In order to measure Agricultural Water Poverty Index, *first level* indicators which reflect the most abstract components of the index must be translated into the more illustrative indicators at the *second level*

and other factors relating to the potential of farmlands for perfectly using available water, have influences on the accessibility of land to receive available water. Consequently, declining available water resources coupled with the low accessibility of lands to receive available water can contribute to the long term impact on farmer's access to water for agricultural use (Fig. 5).

These conditions allow factoring of the results of those pressures, and may provoke responses in order to better use of available agricultural water in more efficient ways by improving farmer human, real, and social capitals related to agricultural water management. The response component of the Pressure–State–Impact–Response, therefore, constitutes management strategies to reduce and mitigate water poverty in agricultural sector while promoting sustainable water management. A couple of fundamental factors are considered to be important for better management of water through increased farmer water use efficiency and eventually raise their access to available water resources. They are related to improved human capitals such as education and water management knowledge; real capitals such as modern irrigation methods; and social capitals such as membership in water user associations, attending water management classes, and farmer's social and economic status.

Not surprisingly, low water availability in agriculture against high water demand, leads to unequal proportion of farmers to available water. In this regard, some farmers who have more capacities than others can take more advantageous from limited water resources. However, as the resources become scarce, an increasing number of rural people see their income sources decline or disappear (Food and Agriculture Organization (FAO) 2007). Income poverty according to the famous vicious circle of "poverty and environmental degradation" may lead to less efficient or even inappropriate use of available water resources, and make the situation worse for the community. Therefore, poor capacities can be observed as factors directly influence efficient use of available water resources in agriculture.

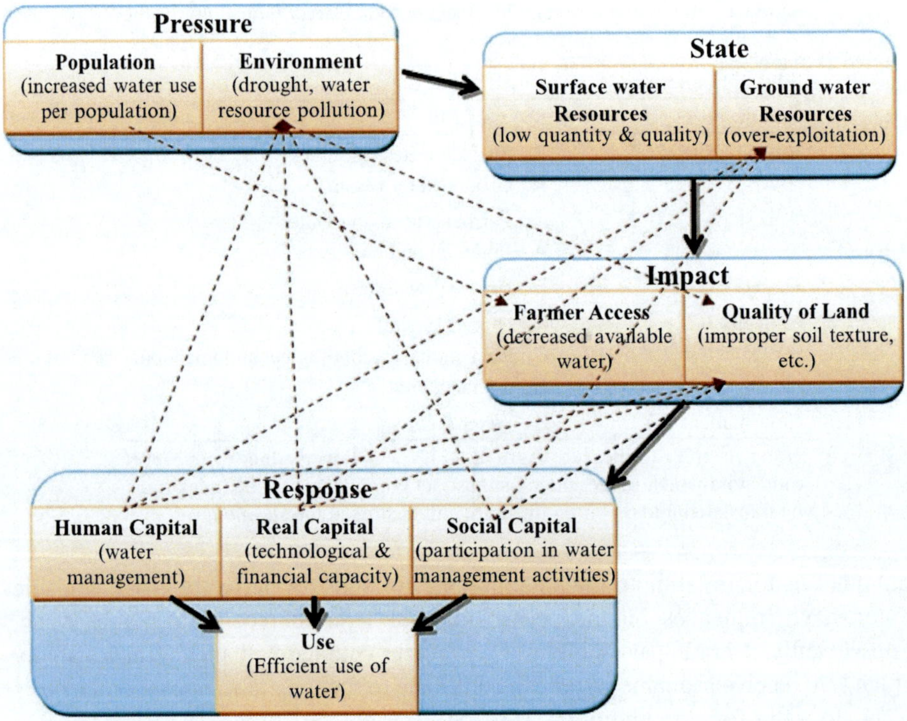

Fig. 5 The schematic representation of the conceptual Causal Network framework for Agricultural Water Poverty Index. Specific causal links are drawn with *black arrows* showing the direct causal relations between abstract indicators and the *dotted arrows* indicate the interrelations among them

Whereas the direct causal relations between the AWPI abstract indicators are identified by black arrows in Fig. 5, dotted arrows show the interrelation among them. For example, although the growing population of users directly causes over-exploitation of water resources, the impact of mismanagement of available water resources may emanate from ignorance, a lack of technical knowledge and ability among farmers (human capital), poor operational and management structures, absence of people's involvement and participation (social capital), infrastructure deficiencies due to inappropriate planning (Perret and Touchain 2002) or income poverty (real capital).

As illustrated by Niemeijer and de Groot (2008), a causal network may include both direct and indirect cause and effect relationships between abstract indicators (Fig. 5). A map of involved indicators in directional causal networks helps to easily identify key nodes among the indicators and determination of the most robust indicators (Lin et al. 2009; Niemeijer and de Groot 2008). However, a precise examination of the relationship between indicators reveals that the "capacity" component has many outgoing arrows and the "Access" component has many incoming arrows (see Fig. 5). This situation reflects the more relative importance of these two components than the others. Niemeijer and de Groot (2008) identify these indicators as root-nodes

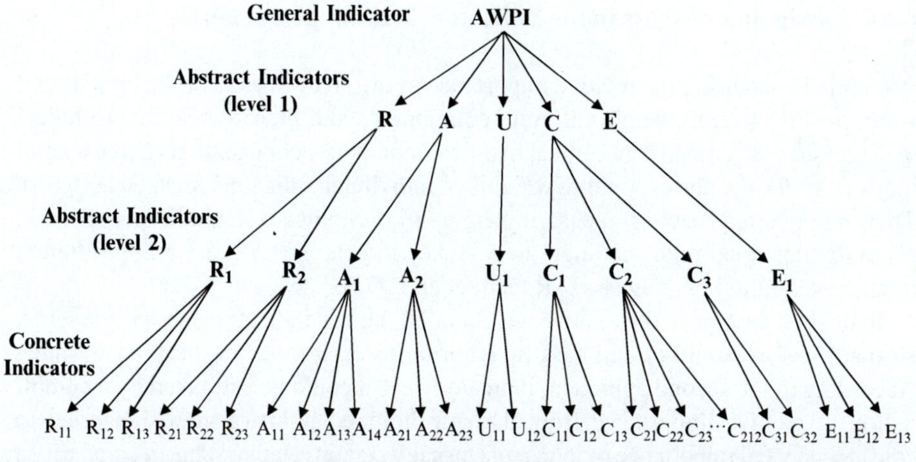

Fig. 6 A schematic model of integrated Causal Network and Ecological Hierarchy Network framework in the case of Agricultural Water Poverty Index (*AWPI*) (Where: *R* Resources, *A* Access, *U* Use, *C* Capacity, *E* Environment)

and end-of-chain nodes, respectively. They believe root nodes are important because their related indicators typically provide information about the source of multiple environmental issues or problems. End-of-chain nodes are important, as well, because they are located at the end of a series of cause and effect chains and allow their associated indicators to gauge the impact of management processes or issues.

Lawrence et al. (2002), in their international comparison of countries by using the Water Poverty Index (WPI), found that "access" and "capacity" components are relatively highly correlated with the whole index. Therefore it may be concluded that these two components play a major role in forming the Agricultural Water Poverty Index, as well.

In short, the outcome of the two previous steps is to specify abstract indicators that need to be stated in the form of concrete indicators to measure accurately the agricultural water poverty.

4.2.3 Identifying the Primary Concrete Indicators

In the integrated network framework, after abstract indicators, at the lowest level, there are concrete indicators which must be clear and practical enough to measure the whole index. Indicators suggested as primary concrete indicators warrant practical considerations through field observations and reviewing the accomplishments in the field of water management by authors. They are indicated in Fig. 6 and a brief description of each is represented in Table 2. The authors considered them as the primary list of concrete indicators that must serve as a basis for measuring the AWPI. Figure 6 also demonstrates a schematic representation of the abstract indicators that are bi-layered with respect to their associated concrete indicators.

4.2.4 Assigning Weights to the Indicators Within the Network

Perceptions regarding the relative importance of different aspects of the agricultural water poverty lead to weight different components and their indicators. To reflect the significance or importance of each indicator and their contributions to the overall index (AWPI), the corresponding weights of individual indicators must be assigned. There are two approaches to assign weights to the indicators within the network: (1) assigning equal relative weight to each key indicator, or (2) assigning different relative weight to key indicators (Ronchi et al. 2002).

In the first approach, the relative weight of all the key indicators may be the same, so that they are simply combined by addition to each other in the final formula. According to the second approach, indicator weight could be estimated by a combination of relative importance from the stakeholders' point of views. In order to consider unequal importance of indicators, usually certain relative value in comparison to the others is given to each indicator. These relative indicator values are the representation of their weights in the index as a whole.

The weighing mechanism can be based on Delphi techniques, multi-criteria analysis or public opinion polls (Hope et al. 1992, cited in: Malkina-Pykh and Pykh 2008, p. 857). The Analytic Hierarchy Process (AHP) as one of the most effective methods in assigning weight can also be applied, in which the relative importance of indicators is assessed and compared (Lin et al. 2009).

Generally, it would be more efficient to conduct an AHP analysis to assign the weights to the proposed abstract and concrete indicators of the AWPI. The AHP explicitly deals with a hierarchy structure and is essentially a methodology of decision making used for combining or synthesizing quantitative as well as qualitative criteria (Sadeghi Niaraki and Kim 2009).

In this study, a group of practitioners and researchers including the first two authors evaluated the AWPI which is divided into several components in a hierarchical structure, to assign weight in an AHP process. In this regard, by using Expert Choice software (a multi-objective decision support tool based on the Analytic Hierarchy Process), a hierarchical decision structure was established, then, both abstract and concrete indicators were weighted individually at every level relative to each other through a comparative judgment process and consistency index. Table 2 displays the indicators coupled with their weights assigning results. It also helps to understand alternative concrete indicators since some of the concrete indicators are related to the AWPI positively and some negatively. A positive relationship (represented by positive sign) means that as the concrete indicator increases, the agricultural water poverty will increase, too. These indicators have a reverse relation with one's agricultural water wealth situation. Conversely, a negative relationship (represented by negative sign) means that as the concrete indicator increases, the agricultural water poverty will decrease, and it relates to agricultural water wealth positively. Increasing in fertilizer consumption, for example, declines the water quality, and consequently exacerbates the agricultural water poverty.

Table 2 The abstract and primary suggested concrete indicators for the Agricultural Water Poverty Index

Abstract indicators			
First level	Second level	Concrete indicators	Weight
Resources (0.238)	R_1: Surface water resources (0.1085)	R_{11}: Purchased canal water (L/s/ha) (–)	0.0426
		R_{12}: River water (L/s/ha) (–)	0.0284
		R_{13}: Rain (mm) (–)	0.0374
	R_2: Groundwater resources (0.1295)	R_{21}: Well water (borehole, deep and semi-deep) (L/s/ha) (–)	0.0618
		R_{22}: Qanat water (L/s/ha) (–)	0.0414
		R_{23}: Spring water (L/s/ha) (–)	0.0263
Access (0.223)	A_1: Farmers' access to water (0.1255)	A_{11}: Water right (–)	0.0359
		A_{12}: Uncultivated lands due to water scarcity (ha) (+)	0.0433
		A_{13}: Reported conflicts over water use (+)	0.0189
		A_{14}: Upstream lands in water allocation and distribution (ha) (–)	0.0274
	A_2: Potential and quality of land (0.0975)	A_{21}: Distances between water source and farm (km) (+)	0.0406
		A_{22}: Soil texture (water maintaining quality) (–)	0.0274
		A_{23}: Land slope (degree) (+)	0.0295
Use (0.276)	U_1: Efficient water use (0.276)	U_{11}: Physical water use efficiency (–)	0.164
		U_{12}:Ploughed down lands after cultivation due to water scarcity (ha) (+)	0.1121
Capacity (0.165)	C_1: Human capital (0.0498)	C_{11}: Educational level (year) (–)	0.0185
		C_{12}: Water management knowledge (–)	0.0164
		C_{13}: Being model farmer in agriculture (times/each 3 years) (–)	0.0149
	C_2: Real capital (technological and financial) (0.0829)	C_{21}: Lands under modern irrigation (ha) (–)	0.0099
		C_{22}: Storage water pool (m^3) (–)	0.0066
		C_{23}: Subsurface drainage system (km) (–)	0.0071
		C_{24}: Land leveling (ha) (–)	0.0085
		C_{25}: Canal lining (km) (–)	0.0083
		C_{26}: Using pipe system for water delivery (km) (–)	0.0083
		C_{27}: Cultivation of water-saving crops or varieties (ha) (–)	0.006
		C_{28}: Monetary value of land (Rial) (+)	0.0034
		C_{29}: Loss in land's annual revenue due to water scarcity (Rial/ha) (+)	0.0056
		C_{210}: Invested income for improving irrigation system (Rial/ha) (–)	0.0059
		C_{211}: Devoted credits and loans for improving irrigation system (Rial/ha) (–)	0.0065
		C_{212}: Drought insurance (ha) (–)	0.0069

(continued)

Table 2 (continued)

Abstract indicators			
First level	Second level	Concrete indicators	Weight
	C_3: Social capital (0.0323)	C_{31}: Attending water management classes (–)	0.0202
		C_{32}: Participation in formal and/or informal activities related to agricultural water (–)	0.0121
Environment (0.098)	E_1: Water quality (0.098)	E_{11}: Water quality (electrical conductivity) (μs/cm) (+)	0.0394
		E_{12}: Fertilizer consumption (kg/ha) (+)	0.0322
		E_{13}: Pesticide and herbicide use (g or mL/ha) (+)	0.0263

Note: Agricultural Water Poverty Index eventually is measured by concrete indicators reflecting the most practical indicators in the field. The overall impact of each concrete indicator on Agricultural Water Poverty Index depending on *positive* or *negative* effect is demonstrated by a sign in *parenthesis*. In the two first columns, the numbers in the parenthesis indicate the weight each abstract indicator has obtained through an Analytic Hierarchy Process. The more weight is assigned, the more importance it reveals

4.3 *Setting the Criteria for Indicator Selection of Agricultural Water Poverty*

Limitations in money and time resources lead to a concentration on selecting "effective" indicators for measuring agricultural water poverty. Effective indicators should provide users with the information they need in an understandable form according to their tasks, responsibilities, and values (Bouleau et al. 2009). Therefore, selection of the most suitable indicators provides great value. The "purpose" usually influences the choice of indicators, but what is more important is trade-offs between desirable features, costs, and feasibility options which often determine the choice of indicators. This point indicates that the precise and various criteria should be noticed in the indicator selection process (Dale and Beyeler 2001).

There are different kinds of criteria as well as different sorts of approaches to offer these criteria. Some researchers prefer to introduce the criterion itself (i.e. Dale and Beyeler 2001; Elle et al. 2010; Lin et al. 2009; Manoliadis 2002), while others, like Kurtz et al. (2001) take a different approach by focusing on a hierarchical evaluation guideline for indicators instead of selection criteria.

Table 3 presents criteria identified as appropriate in most studies related to environmental indicator selection. Although some used in different studies have similar meaning, various researchers use different terminologies. "Understandable" defined by Elle et al. (2010) as "to be clear and must be made sense to all relevant actors", for example, has the similar meaning in comparison with the concept of "Accessibility" that was applied by Ravetz et al. (2001, cited in: Doody et al. 2009, p. 1132) showing "understandable by different users". In Table 3 such similar cases are modified and summarized. As shown in Table 3, some criteria are emphasized

Table 3 Common criteria in indicator selection

Criterion	Description	Reference
Measurable	Be measurable in qualitative or quantitative terms	Schomaker (1997) cited in Niemeijer and de Groot 2008, p. 16; Yli-Viikari et al. 2007; Ioris et al. 2008; Lin et al. 2009
Achievable	Be achievable in terms of the available resources (data)	Schomaker (1997) cited in Niemeijer and de Groot 2008, p. 16; Yli-Viikari et al. 2007
Practical	Be straightforward and relatively inexpensive to measure	World Bank 1999 cited in: Manoliadis 2002, p. 170; Bell and Morse 2003; Niemeijer and de Groot 2008; Yli-Viikari et al. 2007
Necessary skills	Not require excessive data collection skills	Niemeijer and de Groot 2008
Relevant	Closely linked to the environmental problems and audiences being addressed	Dale and Beyeler 2001; Doody et al. 2009; Elle et al. 2010; Niemeijer and de Groot 2008; Yli-Viikari et al. 2007
Quality	Produce trustable and valid data	World Bank 1999, cited in: Manoliadis 2002, p. 170; Ravetz et al. 2001, cited in Doody et al. 2009, p. 1132
Disaggregation	Produce data that represent peaks, troughs, averages, distributions, etc.	Ravetz et al. 2001, cited in Doody et al. 2009, p. 1132
Comprehensibility (Understandable)	Comprehensible to all relevant actors, not only to the experts of the monitored areas	Golusin and Ivanovic 2009; Elle et al. 2010; Yli-Viikari et al. 2007
Representation	Reflect policy makers' and managers' concerns and allow their involvement in the assessment of the issue at hand	Ioris et al. 2008
Usable	Be practical enough and employable in the decision-making situations	Bell and Morse 2003
Controllable	Represent changes that can be controlled by management actions	Lin et al. 2009
Responsive	Have a known response to stresses in a clear and predictable manner	Dale and Beyeler 2001; Lin et al. 2009; Yli-Viikari et al. 2007
Stable	Have low variability in response	Lin et al. 2009
Sensitive	Must readily change as circumstances change	Dale and Beyeler 2001; Bell and Morse 2003; Lin et al. 2009; Yli-Viikari et al. 2007

(continued)

Table 3 (continued)

Criterion	Description	Reference
Anticipatory	Change in the indicator should be measurable before substantial change in ecological system integrity occurs.	Dale and Beyeler 2001
Appropriate scale	Meet appropriate spatial and temporal scale	Bell and Morse 2003; Yli-Viikari et al. 2007
Portability	Be repeatable and reproducible in different levels	Riley 2000, cited in Niemeijer and de Groot 2008, p. 17

Note: Due to money and time restrictions, a manageable number of concrete indicators must be selected to measure the Agricultural Water Poverty Index in practice. Setting the criteria directs to the final selection of effective indicators

by researchers more than others. This indicates that those criteria are more accepted by various researchers to evaluate the suitability of indicators.

The outcome of this section is to identify a couple of appropriate criteria to assess and select between alternative concrete indicators. Therefore, it is necessary, first to determine the most appropriate criteria for individual selection of indicators. Niemeijer and de Groot (2008) noted that the most common criteria are measurability, low resource demanding, analytical soundness, policy relevance, and sensitivity to changes within policy time frames. As it was emphasized on the previous sections, this paper aims to select appropriate indicators for the AWPI which has been proposed as an index to assess the agricultural water poverty specifically at farm level (i.e. farmer level). According to the nature of the issue at hand, five criteria were considered to be relevant enough to assess the concrete indicators. They include: (1) Measurable, (2) Practical, (3) Relevant, (4) Comprehensibility, and finally (5) Controllable.

4.4 *Final Selection of the Agricultural Water Poverty Indicators*

Assigning weights to indicators makes it possible to select final indicators using special methods. A couple of quantitative and qualitative methods like Compromise programming (Manoliadis 2002), Q-method (Barry and Proops 1999; Doody et al. 2009), Binary Integer Programming method (Lin et al. 2009), have been used by different researchers to construct and select indicators. A quantitative method similar to that applied by Lin and his colleagues (Lin et al. 2009) was adopted for this study.

In order to select appropriate concrete indicators, it is crucial to identify those concrete indicators which satisfy the selection criteria in stage 3. Therefore, each alternative concrete indicator was assessed to determine if it met the selected criteria: i.e. Measurable, Practical, Relevant, Comprehensibility, and Controllable. Table 4, in which the criteria are represented by a letter, shows the concrete indicators which are satisfied by each criterion individually. In other words, when an

Table 4 The final selected concrete indicators after assessing based on the chosen criteria

Abstract indicators (2nd level)	Concrete indicators	Criteria
R_1: Surface water resources	R_{11}: Purchased canal water (x_{31})*	MPRCT
	R_{12}: River water (x_{32})*	MPRCT
	R_{13}: Rain (x_{33})*	MPRC
R_2: Groundwater resources	R_{21}: Well water (borehole, deep and semi-deep) (x_{34})*	MPRCT
	R_{22}: Qanat water (x_{35})*	MPRCT
	R_{23}: Spring water (x_{36})*	MPRCT
A_1: Farmers' access to water	A_{11}: Water right (x_{37})*	MPRC
	A_{12}: Uncultivated lands due to water scarcity (x_{38})*	MPRCT
	A_{13}: Reported conflicts over water use (x_{39})	MRCT
	A_{14}: Upstream lands in water allocation and distribution (x_{310})*	MPRC
A_2: Potential and quality of land	A_{21}: Distances between water source and farm (x_{311})*	MPRC
	A_{22}: Soil texture (water maintaining quality) (x_{312})	MRCT
	A_{23}: Land slope (x_{313})	MRCT
U_1: Efficient water use	U_{11}: Physical water use efficiency (x_{314})*	MPRCT
	U_{12}: Ploughed lands after cultivation due to water scarcity (x_{315})*	MPRCT
C_1: Human capital	C_{11}: Educational level (x_{316})*	MPRCT
	C_{12}: Water management knowledge (x_{317})*	MPRCT
	C_{13}: Being model farmer in agriculture (x_{318})	MPRC
C_2: Real capital (technological and financial)	C_{21}: Lands under modern irrigation (x_{319})*	MPRCT
	C_{22}: Storage water pool (x_{320})	MRT
	C_{23}: Subsurface drainage system (x_{321})	MRT
	C_{24}: Land leveling (x_{322})*	MPRCT
	C_{25}: Canal lining (x_{323})*	MPRCT
	C_{26}: Using pipe system for water delivery (x_{324})*	MPRCT
	C_{27}: Cultivation of the water-saving crops or varieties (x_{325})	MRT
	C_{28}: Monetary value of land (x_{326})	MRC
	C_{29}: Loss in land's annual revenue due to water scarcity (x_{327})	MRC
	C_{210}: Invested income for improving irrigation system (x_{328})*	MPRCT
	C_{211}: Devoted credits and loans for improving irrigation system (x_{329}) *	MPRCT
	C_{212}: Drought insurance (x_{330})*	MPRCT
C_3: Social capital	C_{31}: Attending water management classes (x_{331})*	MPRCT
	C_{32}: Participation in formal and/or informal activities related to agricultural water (x_{332})	MRCT
E_1: Water quality	E_{11}: Water quality (EC) (x_{333})*	MPRC
	E_{12}: Fertilizer consumption (x_{334})*	MPRCT
	E_{13}: Pesticide and herbicide use (x_{335})*	MPRCT

M Measurable, *P* Practical, *R* Relevant, *C* Comprehensibility, *T* Controllable

Note: The first number in the *parenthesis* refers to the level of the indicator which in this case 3 means the third level indicators (*concrete indicators*), while the other following number(s) refers to the number of indicator of total 35. The indicators marked by asterisk are those satisfy both the criteria and constraints of the mathematical model used to final selection of indicators alternative indicator meets an individual criterion, a letter representing the criteria is assigned to the indicator (see Table 4).

However, different criteria do not have equal importance in the selection process. This means some criteria are more effective for selecting indicators than others. By using Expert Choice software, a group of practitioners and researchers (including the first two authors) assigned the weights to the selected criteria through an Analytic Hierarchy Process. So, Measureable, Practical, Relevant, Comprehensible and Controllable received weights of 0.239, 0.172, 0.336, 0.099, and 0.154 from 1, respectively. Relevant and Measurable, for example, in this study outweigh the others at indicator selection process from a scientific and practical point of views. But, as it is emphasized by Lin et al. (2009), besides the selection criteria, there are always some practical requirements for an ecological indicator's utility. There are two requirements that should be considered for selecting the AWPI indicators in this study. First, this study attempts to determine an indicator set with minimal components to assess Agricultural Water Poverty. This means that the final indicators should satisfy the five selected criteria, Measureable, Practical, Relevant, Comprehensible and Controllable, at the same time. Second, indicators should meet Measureable, Practical and Relevant criteria, which received the highest importance among the others, at the same time. It means that the final selected indicators should be closely linked to the agricultural water management problems and be measurable in qualitative or quantitative methods which are relatively easy and inexpensive. In other words, the indicators not satisfying the criteria of being Measureable, Practical, and Relevant, should be discarded.

Based on the Binary Integer Programming method, steps should be followed for selecting final concrete indicators. First it is needed to simulate abstract and concrete indicators into a framework by using matrixes. The relationships between any two neighboring levels of indicators in the network framework can be represented as a matrix (Lin et al. 2009). Using Fig. 6 and Table 4, a relation matrix R between abstract layer on second level and concrete layer is constructed. This matrix R has 9 rows and 35 columns (a 9×35 matrix), representing the number of abstract indicators in the second layer multiplied by the number of concrete indicators. The value of 1 is assigned in the matrix, if the both the abstract and concrete indicators have a connection. Otherwise, 0 is assigned.

Then, it is necessary to construct another matrix that denotes the relationships between individual selection criteria and concrete indicators. This matrix, called matrix D, is constructed according to Table 4. In this matrix, the rows are related to the selection criteria M, P, R, C, T, and the columns to the alternative concrete indicators. The value of 1 indicates the criteria being satisfied by the relevant indicator, and otherwise, a 0 is denoted.

After that, the two vectors $I_1 = [1,1,1,1,1]$ and $I_2 = [1,1,1,0,0]$ are defined, while they are satisfying MPRCT (Measureable, Practical, Relevant, Comprehensible and Controllable) and MPR (Measureable, Practical and Relevant), respectively.

$$
R = \begin{pmatrix}
1&1&1&0&0&0&0&0&0&0&0&0&0&0&0&0&0&0&0&0&0&0&0&0&0&0&0&0&0&0&0&0&0&0&0\\
0&0&0&1&1&1&0&0&0&0&0&0&0&0&0&0&0&0&0&0&0&0&0&0&0&0&0&0&0&0&0&0&0&0&0\\
0&0&0&0&0&0&1&1&1&1&0&0&0&0&0&0&0&0&0&0&0&0&0&0&0&0&0&0&0&0&0&0&0&0&0\\
0&0&0&0&0&0&0&0&0&0&1&1&1&0&0&0&0&0&0&0&0&0&0&0&0&0&0&0&0&0&0&0&0&0&0\\
0&0&0&0&0&0&0&0&0&0&0&0&0&1&1&0&0&0&0&0&0&0&0&0&0&0&0&0&0&0&0&0&0&0&0\\
0&0&0&0&0&0&0&0&0&0&0&0&0&0&0&1&1&1&0&0&0&0&0&0&0&0&0&0&0&0&0&0&0&0&0\\
0&0&0&0&0&0&0&0&0&0&0&0&0&0&0&0&0&0&1&1&1&1&1&1&1&1&1&1&1&1&0&0&0&0&0\\
0&0&0&0&0&0&0&0&0&0&0&0&0&0&0&0&0&0&0&0&0&0&0&0&0&0&0&0&0&0&1&1&1&0&0\\
0&0&0&0&0&0&0&0&0&0&0&0&0&0&0&0&0&0&0&0&0&0&0&0&0&0&0&0&0&0&0&0&0&1&1
\end{pmatrix} = \begin{pmatrix} R_1\\ R_2\\ R_3\\ R_4\\ R_5\\ R_6\\ R_7\\ R_8\\ R_9 \end{pmatrix}
$$

(9×35)

$$
D = \begin{pmatrix}
1&1&1&1&1&1&1&1&1&1&1&1&1&1&1&1&1&1&1&1&1&1&1&1&1&1&1&1&1&1&1&1&1&1&1\\
1&1&1&1&1&1&1&1&0&1&1&0&0&1&1&1&1&1&1&0&0&1&1&1&0&0&0&1&1&1&1&0&1&1&1\\
1&1&1&1&1&1&1&1&1&1&1&1&1&1&1&1&1&1&1&1&1&1&1&1&1&1&1&1&1&1&1&1&1&1&1\\
1&1&1&1&1&1&1&1&1&1&1&1&1&1&1&1&1&1&1&0&0&1&1&1&0&1&1&1&1&1&1&1&1&1&1\\
1&1&0&1&1&1&0&1&1&0&0&1&1&1&1&1&1&0&1&1&1&1&1&1&1&0&0&1&1&1&1&1&0&1&1
\end{pmatrix} =
$$

[D_1, D_2, D_3, …. , D_{35}] (5×35)

In the next step, it is imperative to formulate a mathematical model for final indicator selection. Actually, an objective function to minimize the number of selected indicators should be formulated; however, all the requirements already were specified for this study are defined as the constraints of this function.

$$\text{Objective : } \min Z \sum_{j=1}^{35} x_{3j} \tag{1}$$

Subject to:

$$x_{3j} = 1 \text{ if } I_1 \bullet D = 5 \tag{2}$$

$$x_{3j} = 0 \text{ if } I_2 \bullet D \neq 3 \tag{3}$$

$$R_k \bullet X_3' > 0 \tag{4}$$

$$X_3 \bullet W_3' \geq 0.85 \tag{5}$$

$$x_{3j} = 0 \text{ or } 1 \tag{6}$$

$$j = 1, 2, \ldots, 35$$

$$k = 1, 2, \ldots, 9$$

Where Z is the number of selected indicators; $X_3 = [x_{31}, x_{32}, \ldots, x_{3j}, \ldots, x_{335}]$ is a vector which indicates the variables to be decided (number 3 on the subscript shows the third layer of the AWPI indicators, or concrete indicators), in which $x_{3j} = 1$ means the related alternative indicator is selected and 0 means not be selected; and $W_3 = [w_{31}, w_{32}, \ldots, w_{3j}, \ldots, w_{335}]$, as a vector, is the assigned weight values to the alternative concrete indicators. As it is indicated in Table 2, W_3 = [0.0426, 0.0284, 0.0374, 0.0618, 0.0414, 0.0263, 0.0359, 0.0433, 0.0189, 0.0274, 0.0406, 0.0274, 0.0295, 0.164, 0.1121, 0.0185, 0.0164, 0.0149, 0.0099, 0.0066, 0.0071, 0.0085, 0.0083, 0.0083, 0.006, 0.0034, 0.0056, 0.0059, 0.0065, 0.0069, 0.0202, 0.0121, 0.0394, 0.0322, 0.0263].

In the function (1), constraints (2) and (3) represent the individual selection criteria. Formula (2) means the indicator satisfying MPRCT (Measureable, Practical, Relevant, Comprehensible and Controllable) is selected and (3) means the indicator not satisfying MPR (Measureable, Practical and Relevant,) at the same time is discarded. These two recent constraints indicate that the selected indicator must satisfy them at the same time. Formula (4) means at least one indicator connecting to an indicator of the second level must be selected. Formula (5) means the sum weight of selected indictors should not be less than 0.85.

For resolving the model, a simple way was followed:

Step (1): Considering the constraints in formulas (2) and (3), some variables, firstly, can be determined to select from the primary list of concrete variables as follows:

$$S_1 = I_1 \bullet D = \begin{bmatrix} 5, 5, 4, 5, 5, 5, 4, 5, 4, 4, 4, 4, 4, 5, 5, 5, 5, 4, 5, 3, 3, 5, 5, 5, \\ 3, 3, 3, 5, 5, 5, 5, 4, 4, 5, 5 \end{bmatrix}$$

So, $x_{3j} = 1$, when j = 1, 2, 4, 5, 6, 8, 14, 15, 16, 17, 19, 22, 23, 24, 28, 29, 30, 31, 34, 35. It means among the total 35 indicators, the indicators whose associated subscript numbers are matched with these numbers will be selected.

$$S_2 = I_2 \bullet D = \begin{bmatrix} 3, 3, 3, 3, 3, 3, 3, 3, 2, 3, 3, 2, 2, 3, 3, 3, 3, 3, 3, 2, 2, 3, 3, 3, 2, \\ 2, 2, 3, 3, 3, 3, 2, 3, 3, 3 \end{bmatrix}$$

Considering S_1 and S_2 at the same time, it will be revealed that $x_{3j} = 0$, when j = 9, 12, 13, 20, 21, 25, 26, 27, 32. It means among the total 35 indicators, nine indicators whose associated subscript numbers are matched with these numbers will be discarded.

Findings of this stage show that $X_3 = [1, 1, x_{33}, 1, 1, 1, x_{37}, 1, 0, x_{310}, x_{311}, 0, 0, 1, 1, 1, 1, x_{318}, 1, 0, 0, 1, 1, 1, 0, 0, 0, 1, 1, 1, 1, 0, x_{333}, 1, 1]$. Up to this stage nine variables were discarded from the primary concrete indicators listed in Table 4. They include x_{39}, x_{312}, x_{313}, x_{320}, x_{321}, x_{325}, x_{326}, x_{327} and x_{332} (i.e. A_{13}, A_{22}, A_{23}, C_{22}, C_{23}, C_{27}, C_{28}, C_{29}, C_{32}) and six variables yet need to be decided which include x_{33}, x_{37}, x_{310}, x_{311}, x_{318}, and x_{333}. Hence, the model will be followed to solve the problem in the next step.

Step (2): Substituting the recent vector X_3 into the formulas of constraint (4) yields:

$$R_k \bullet X_3' = \begin{pmatrix} R_1 \\ R_2 \\ R_3 \\ R_4 \\ R_5 \\ R_6 \\ R_7 \\ R_8 \\ R_9 \end{pmatrix} \bullet [11x_{33}111x_{37}10x_{310}x_{311}001111x_{318}10011100011110x_{333}11]' = \begin{pmatrix} 2 + x_{33} \\ 3 \\ x_{37} + 1 + x_{310} \\ x_{311} \\ 2 \\ 2 + x_{318} \\ 7 \\ 1 \\ x_{333} + 2 \end{pmatrix} > 0$$

The expression $2+x_{33}$ of the recent matrix, for example, indicates that from row R_1,which included three concrete indicators (x_{31}, x_{32} and x_{33}), two indicators have been selected and one (x_{33}) needs to be decided, and as the same for the other rows.

Considering constraints (4) and (6) at the same time, it can be concluded that $x_{311}>0$ when $j=11$, because x_{3j} can only be valued as 0 or 1. Keeping this point in mind, therefore, the other four expressions, $2+x_{33}>0$, $X_{37}+1+x_{310}>0$, $2+x_{318}>0$, and $x_{333}+2>0$ can also be satisfied. Moreover, if it is supposed that at least two indicator connecting to an indicator of the second level should be selected ($R_k \bullet X_3' >1$ instead of $R_k \bullet X_3' >0$), all of those expressions except the forth ($x_{311}>1$) can be satisfied.

So, it can be concluded that, up to now, X_3=[1, 1, x_{33}, 1, 1, 1, x_{37}, 1, 0, x_{310}, 1, 0, 0, 1, 1, 1, 1, x_{318}, 1, 0, 0, 1, 1, 1, 0, 0, 0, 1, 1, 1, 1, 0, x_{333}, 1, 1], in which five variables yet need to be decided. In order to solve this problem, the model will be followed through the step 3.

Step (3): In this stage, substituting X_3 with the five undetermined variables into the formula of constraint (5) makes it possible to resolve the problem. Thus, we have:

$X_3 \bullet W'_{3j}$ = [1, 1, x_{33}, 1, 1, 1, x_{37}, 1, 0, x_{310}, 1, 0, 0, 1, 1, 1, 1, x_{318}, 1, 0, 0, 1, 1, 1, 0, 0, 0, 1, 1, 1, 1, 0, x_{333}, 1, 1] • [0.0426, 0.0284, 0.0374, 0.0618, 0.0414, 0.0263, 0.0359, 0.0433, 0.0189, 0.0274, 0.0406, 0.0274, 0.0295, 0.164, 0.1121, 0.0185, 0.0164, 0.0149, 0.0099, 0.0066, 0.0071, 0.0085, 0.0083, 0.0083, 0.006, 0.0034, 0.0056, 0.0059, 0.0065, 0.0069, 0.0202, 0.0121, 0.0394, 0.0322, 0.0263]′

$$\begin{aligned}\Rightarrow [&0.0426+0.0284+0.0374x_{33}+0.0618+0.0414+0.0263+0.0359x_{37}+0.0433\\ &+0+0.0274x_{310}+0.0406+0+0+0.164+0.1121+0.0185+0.0164\\ &+0.0149x_{318}+0.0099+0+0+0.0085+0.0083+0.0083+0+0+0+0.0059\\ &+0.0065+0.0069+0.0202+0+0.0394x_{333}+0.0322+0.0263] \geq 0.85\end{aligned}$$

$$\Rightarrow [0.0374x_{33}+0.0359x_{37}+0.0274x_{310}+0.0149x_{318}+0.0394x_{333}+0.7284] \geq 0.85$$

The recent equation indicates that the sum of the weights of determined variables (indicators) only provide 0.7248 of the total amount of 0.85. It needs, therefore, to select more indicators which can help to increase the sum weight of selected indictors to about 0.1216, to satisfy the constraint of ≥0.85.

$$\Rightarrow [0.0374x_{33}+0.0359x_{37}+0.0274x_{310}+0.0149x_{318}+0.0394x_{333}] \geq 0.1216$$

In other words, the above expression is satisfied when the sum of weights on the left hand would be equal or larger than 0.1216. It means only those variables which can satisfy this amount of weight must be selected. Thus, the process will be continued through the next step.

Step (4): In this stage, the process should be followed by using the objective function which aims to minimize the number of selected indicators.

$$\text{Objective}: \min Z \sum_{j=1}^{35} x_{3j}, \text{therefore}$$

min $Z = x_{33} + x_{37} + x_{310} + x_{318} + x_{333} + 30$, in which the constant number 30 indicates the number of determined indicators. It can also be any other arbitrary number.

Subject to:

1. $0.0374x_{33} + 0.0359x_{37} + 0.0274x_{310} + 0.0149x_{318} + 0.0394x_{333} \geq 0.1216$
2. $2 + x_{33} > 0$
3. $X_{37} + 1 + x_{310} > 0$
4. $2 + x_{318} > 0$
5. $1 + x_{333} > 0$

Since x_{3j} can only be valued as 0 or 1, the constraints 2, 3, 4, and 5 would be satisfied. The problem can be solved simply by substitution of 1 or 0 instead of each variable of x_{33}, x_{37}, x_{310}, x_{318}, and x_{333} at the same time and examine if these changes will satisfy the sum weights larger than 0.1216. Thus, after some attempts, it was concluded that the optimum solution will be: $x_{33} = 1$, $x_{37} = 1$, $x_{310} = 1$, $x_{318} = 0$, $x_{333} = 1$, which means among these variables only x_{318} must be discarded, and the others should be selected to satisfy the constraints of the problem and minimize objective function.

It should be noted that step 4 can be easily done using General Algebraic Modeling System (GAMS) software. Using General Algebraic Modeling System software revealed that the final determined indicators would be the same as those selected through the above step.

As can be summarized from all the followed steps in the above, the final solution for the main problem i.e. the variables to be decided for including in the AWPI, is $X_3 = [1, 1, 1, 1, 1, 1, 1, 1, 0, 1, 1, 0, 0, 1, 1, 1, 1, 0, 1, 0, 0, 1, 1, 1, 0, 0, 0, 1, 1, 1, 1, 0, 1, 1, 1]$. Therefore, R_{11}, R_{12}, R_{13}, R_{21}, R_{22}, R_{23}, A_{11}, A_{12}, A_{14}, A_{21}, U_{11}, U_{12}, C_{11}, C_{12}, C_{21}, C_{24}, C_{25}, C_{26}, C_{210}, C_{211}, C_{212}, C_{31}, E_{11}, E_{12}, E_{13} were chosen as appropriate concrete indicators for measuring agricultural water poverty at the individual level. The final concrete indicators are marked in Table 4 by an asterisk symbol.

5 Conclusion

The increasing stress on water resources brought about by ever-rising demand, inappropriate use of available water resources, and growing pollution worldwide, is of serious concern (Food and Agriculture Organization (FAO) 2007). Growing water scarcity stands as one of the major threats to future agricultural sustainability, especially in rural areas. Within the wider framework of agricultural studies, the sustainability of agricultural water management is of major importance and should be taken into account by translating into practical and appropriate systems of assessment. In this respect, the Agricultural Water Poverty Index was introduced to show continuous update of changes and the state of agricultural water. It can help to assess the situation of agricultural water in a region, to monitor trends in water situations over time, or to diagnose causes of water poverty in a region over a specific time period. Considering the whole picture of water poverty in an agricultural system, the AWPI can be used to classify the farmers' position regarding water scarcity and assess whether they are water poor or not.

Since water poverty concerns both quantity and quality of available water resources, indexes used to assess water situations must embrace different aspects of an agricultural system with specific regard to water. Therefore, the AWPI includes conditions of water availability or water resources, accessibility to available water, water use, capacities to improve water management, and the environmental conditions affecting water poverty. In fact, it encompasses five components: Resources, Access, Use, Capacity and Environment. To measure each component, a set of appropriate and different indicators are needed. Moreover, a rigorous framework for developing its indicators should be considered to better structure them. The various conceptual frameworks including Causal Chain frameworks and the Ecological Hierarchy Network have been introduced. Based on the most crucial factors, and according to different perspectives that researchers keep in mind, a specific framework can be chosen among them (Niemeijer and de Groot 2008).

The AWPI indicators were developed in this paper by compounding diverse subcomponents and indicators related to different dimensions of the agricultural water management with specific regard to an Integrated Causal Network and Ecological Hierarchy Network framework. This framework combines both the Ecological Hierarchy Network and the Causal Network through a hierarchal structure of abstract and concrete indicators. Following the steps proposed in this paper as the conceptual path to develop the AWPI indicators, a various number of indicators were taken into account to measure each component of the AWPI. The five main components of the AWPI, based on the Ecological Hierarchy Network framework, were considered as the first level abstract indicators and by using a Pressure–State–Impact–Response framework, the second level of abstract indicators including subcomponents of the AWPI was determined. Accordingly, each abstract indicator was extended to include concrete indicators which were put into practice directly for accurate measuring of the agricultural water poverty. After that, the concrete indicators were mathematically assigned weights, and an AHP (Analytic Hierarchy Process) process was conducted to evaluate and compare relative importance. Water is a dynamic and complex resource which is difficult to describe with simple indicators, and also data availability limits the application of more sophisticated indicators (Rijsberman 2006). Therefore, criteria were considered for selecting the most proper concrete indicators. However, indicators must be selected and used carefully if they are to be effective. Using a quantitative method i.e. Binary Integer Programming method, 25 indicators were eventually selected for including in an AWPI. The paper concludes by proposing a future research plan aimed at conducting a study to measure the Agricultural Water Poverty by using these developed indicators.

Acknowledgements Part of this paper was completed during a fellowship study at the International Institute for Asian Studies (IIAS), Leiden/Amsterdam, Netherlands. This research was supported in part by the Iran National Drought Research Institute, Shiraz University, Shiraz, Iran under grant No. 333PG.

References

Barry J, Proops J (1999) Seeking sustainability discourses with Q methodology. Ecol Econ 28:337–345

Bell S, Morse S (2003) Measuring sustainability: learning by doing, 1st edn. Earthscan Publications Ltd, London

Borja A, Dauer DM (2008) Assessing the environmental quality status in estuarine and coastal systems: comparing methodologies and indices. Ecol Indic 8:331–337. doi:10.1016/j.ecolind.2007.05.004

Bouleau G, Argillier Ch, Souchon Y, Barthelemye C, Babut M (2009) How ecological indicators construction reveals social changes—the case of lakes and rivers in France. Ecol Indic 9:1198–1205. doi:10.1016/j.ecolind.2009.03.010

Dale VH, Beyeler SC (2001) Challenges in the development and use of ecological indicators. Ecol Indic 1:3–10

Doody DG, Kearney P, Barry J, Moles R, O'Regan B (2009) Evaluation of the Q-method as a method of public participation in the selection of sustainable development indicators. Ecol Indic 9:1129–1137. doi:10.1016/j.ecolind.2008.12.011

Elle M, Dammann S, Lentsch J, Hansen K (2010) Learning from the social construction of environmental indicators: from the retrospective to the pro-active use of SCOT in technology development. Build Environ 45:135–142. doi:10.1016/j.buildenv.2009.05.011

Food and Agriculture Organization (FAO) (2007) Coping with water scarcity, challenge of the twenty-first century. Retrieved from http://www.worldwaterday.org

Forouzani M, Karami E (2010) Agricultural water poverty index and sustainability. Agron Sustain Dev 31(2):415–431. doi:10.1051/agro:2010026

Golusin M, Ivanovic OM (2009) Definition, characteristics and state of the indicators of sustainable development in countries of Southeastern Europe. Agr Ecosyst Environ 130:67–74. doi:10.1016/j.agee.2008.11.018

Ioris AAR, Hunter C, Walker S (2008) The development and application of water management sustainability indicators in Brazil and Scotland. J Environ Manage 88:1190–1201. doi:10.1016/j.jenvman.2007.06.007

Jhorar RK, Smit AAM, Roest CWJ (2009) Assessment of alternative water management options for irrigated agriculture. Agr Water Manage 96:975–981

Kurtz JC, Jackson LE, Fisher WS (2001) Strategies for evaluating indicators based on guidelines from the Environmental Protection Agency's Office of Research and Development. Ecol Indic 1:49–60

Lawrence P, Meigh J, Sullivan C (2002) The water poverty index: an international comparison, vol KERP 2002/19, Keele economics research papers. Centre for Economic Research, Keele University, Keele, pp 1–19, ISSN: 1352–8955

Lin T, Lin JY, Cui Sh, Cameron S (2009) Using a network framework to quantitatively select ecological indicators. Ecol Indic 9:1114–1120. doi:10.1016/j.ecolind.2008.12.009

Malano H, Burton M, Makin I (2004) Benchmarking performance in the irrigation and drainage sector: a tool for change. Irrig Drain 53(2):119–133

Malkina-Pykh IG, Pykh YA (2008) Quality-of-life indicators at different scales: theoretical background. Ecol Indic 8:854–862. doi:10.1016/j.ecolind.2007.01.008

Manoliadis OG (2002) Development of ecological indicators—a methodological framework using compromise programming. Ecol Indic 2:169–176

Muller F, Lenz R (2006) Ecological indicators: theoretical fundamentals of consistent applications in environmental management. Ecol Indic 6:1–5. doi:10.1016/j.ecolind.2005.08.001

Niemeijer D, de Groot RS (2008) A conceptual framework for selecting environmental indicator sets. Ecol Indic 8:14–25. doi:10.1016/j.ecolind.2006.11.012

Pereira LS, Oweis Th, Zairi A (2002) Irrigation management under water scarcity. Agr Water Manage 57:175–206

Perret S, Touchain E (2002) A simulation-based approach to assess the economic viability of smallholding irrigation schemes in South Africa: conceptualization and first implementation, No. 02/02, Research Report to CIRAD-Tera. CIRAD-Tera/University of Pretoria, Pretoria

Rijsberman FR (2006) Water scarcity: fact or fiction? Agr Water Manage 80:5–22. doi:10.1016/j.agwat.2005.07.001

Ronchi E, Federico A, Musmeci F (2002) A system oriented integrated indicator for sustainable development in Italy. Ecol Indic 2:197–210

Sadeghi Niaraki A, Kim K (2009) Ontology based personalized route planning system using a multi-criteria decision making approach. Expert Syst Appl 36:2250–2259. doi:10.1016/j.eswa.2007.12.053

Smeets E, Weterings R (1999) Environmental indicators: typology and overview, vol 25, Technical report. European Environment Agency (EEA), Copenhagen

Sullivan C (2002) Calculating a water poverty index. World Dev 30(7):1195–1211. doi:10.1016/S0305-750X(02)00035-9

Torabi Haghighi A, Keshtkaran P (2008) Method of facing with drought in Fars province-Iran. XXIV conference of the Danubian countries on hydrological forecasting and hydrological bases of water management, Bled, 2–4 June 2008

Turnhout E, Hisschemoller M, Eijsackers H (2007) Ecological indicators: between the two fires of science and policy. Ecol Indic 7:215–228. doi:10.1016/j.ecolind.2005.12.003

Yang H, Zhang X, Zehnder AJB (2003) Water scarcity, pricing mechanism and institutional reform in northern China irrigated agriculture. Agr Water Manage 61:143–161

Yli-Viikari A, Hietala-Koivu R, Hyvönen E, Perälä P, Turtola E (2007) Evaluating agri-environmental indicators (AEIs)—use and limitations of international indicators at national level. Ecol Indic 7:150–163. doi:10.1016/j.ecolind.2005.11.005

Participatory Rural Appraisal to Solve Irrigation Issues

Albert T. Modi

Abstract This report unexpectedly shows that farmers performed better than extension officers for crop management. Local knowledge of small-holder farmers should be considered to enhance technical skills in crop production under irrigation. Small-scale farmers of the Tugela Ferry Irrigation scheme in South Africa own 0.1 ha plots. The plots can be increased by leasing from other plot-holders who fail in crop management. This study was initiated in response to the fact that farmers were constrained by the lack of technical approaches for crop management. The farmers relied on technical advisors, known as extension officers, to improve technical knowledge and skills required for successful production of marketable products. The study used participatory rural appraisal as a tool to identify key technical and institutional constraints to crop production. Matrix and pair-wise ranking were used for data classification and analysis. The key outcomes of the study were (i) identification of 12 and 18 desirable attributes of a good extension officer and a good farmer, respectively and (ii) identification of 18 problems constraining crop management practices on the irrigation scheme, and solutions to these problems. A comparison of farmers and extension officers on key performance areas related to crop management, *inter alia*, skilfulness in use of technology to access water, ability to demonstrate skills to others, achievement of good yields, ability to meet market requirements and gain income, showed that overall, farmers performed better (score = 5.03) compared with extension officers (score = 4.84). The findings therefore demonstrate the usefulness of participatory rural appraisal tools for rural economic development.

Keywords Extension officer • Pairwise ranking • Small-scale farmer

A.T. Modi (✉)
School of Agricultural, Earth and Environmental Sciences, University of KwaZulu-Natal, P/Bag X01, Scottsville 3209, South Africa
e-mail: modiat@ukzn.ac.za

E. Lichtfouse (ed.), *Farming for Food and Water Security*, Sustainable Agriculture Reviews 10, DOI 10.1007/978-94-007-4500-1_7,

1 Introduction

Many circumstances influence the management practice a farmer chooses. Natural circumstances such as climate, soil, pests and diseases impose biological limitations on the cropping system (Fey 2007). On the other hand, many socioeconomic circumstances, such as transportation, capital, markets, labour, farm inputs, credit and technical assistances affect the external environment that conditions the farmers' decision-making (Denison and Manona 2007). By conducting multidisciplinary research and analysing the socioeconomic, technical and ecological constraints, important feedback can be obtained about farm conditions, management practices and farmers' needs. This information can be incorporated into research decisions that are conducive to the development of technologies adapted to farmers' needs and resources (Kundhlande et al. 2004). Descriptions of research methodologies attuned to actual conditions of smallholder farmers in the developing world are available (Milligan 2003; Pretty et al. 1995). These methodologies emerged in response to critics of internationally funded rural development, who charged that in the past, programmes lacked an understanding of the ecological and socioeconomic milieu in which they operated, excluded the small farmers as both collaborators and beneficiaries, and ineptly promoted inappropriate technologies.

Participatory research approaches allow researchers and participants opportunities to investigate issues of concern to resource-poor communities, and to plan, implement and evaluate development activities (Baur and Kradi 2001; Bellon 2001; van Heck 2003; Hart 2007; Wiggins et al. 2010). These approaches challenge prevailing stereotypes about people's knowledge. The methods used range from visualisation to interviewing and group work. The common theme is the promotion of interactive learning, shared knowledge, and flexible, yet structured analysis. These methods have proven valuable in a wide range of sectors and situations. Participatory approaches can also bring together different disciplines, such as agriculture, health and community development, to enable an integrated vision of livelihoods and well-being. They offer opportunities for mobilising local people for joint action (Chambers 1997; Hart 2007). In the late 1970s rapid rural appraisal emerged as a result of dissatisfaction with time consuming surveys as means of gathering information for solving policy-related issues in rural communities. There was also dissatisfaction with superficial visits to rural areas by policy makers and funders, which Chambers (1993) termed "rural development tourism". The approach focused on improving interview techniques by using key informants and semi-structured interviews with checklists and triangulating. It also promoted a spirit of open, cooperative enquiry in exchanging information on new techniques experimented with (Chambers 1997). The focus of participation has shifted in recent years to include, *inter alia*, (1) Emphasis on sub-national, national and international decision making, not just local decision making, (2) Move from projects to policy processes and institutionalisation, (3) Greater recognition of issues of difference and power and (4) Emphasis on assessing the quality and understanding the impact of participation, rather than simply promoting participation (Milligan 2003).

The objective of this study was to use participatory approaches to determine whether farmers and extension officers are able identify problems and solutions to best management practices at the Tugela Ferry Irrigation Scheme. A wide range of similar approaches and methodologies, including Participatory Rural Appraisal, Rapid Rural Appraisal, Participatory Learning Methods, Participatory Action Research, Farming Systems Research, Méthod Active de Recherche et de Planification Participative, and many others, are recognised by researchers who emphasise full participation of people in the processes of learning about their needs and opportunities, and in the action required to address the needs (Anyaegbunam et al. 1998; Milligan 2003; Pretty et al. 1995).

2 Material and Methods

2.1 *The Participatory Rural Appraisal Tools Used in This Study*

Townsley (1996) suggested several tools for participatory research, including secondary data reviews, workshops, and semi-structured interview techniques. Townsley (1996) also discussed techniques for data collation, including ranking and classification, mapping, diagrams and graphics. Accordingly, in this study, workshops were used as tools for participatory identification of problems and solutions. To facilitate the ranking of problems and solutions, matrix ranking was used as a major tool for data classification. Descriptions of the workshop approach and ranking techniques used in this study are presented in Tables 1 and 2.

2.2 *Research Design*

This study was undertaken within the framework of Participatory Rural Appraisal, which entails a creative combination of disciplines to conduct rapid appraisals for generating new technologies. In Participatory Rural Appraisal, a team of researchers visits a region of homogeneous farming systems, attempting to understand them and to identify appropriate technological and institutional improvements. Ideas about improved farm practices emerge from discussions between scientists and farmers. Subsequently, the ideas are tried through on-farm trials by a technology testing team. This study was designed as a planning process to allow the following activities by the participants:

- List problems that limit the farming system,
- Rank, in order of importance, identified problems, including the scope of the problem,
- Analyse the cause of problems for which there is enough evidence,
- Analyse interrelations among problems and causes,
- List solutions to those problems,
- Identify factors for experimentation in order to evaluate proposed solutions (Schoeman and Magongoa 2004).

Table 1 Workshop as a participatory research tool

Definition

A meeting where a series of set tasks are performed and an output produced. In PRA, workshops usually involve the PRA team, but also, if appropriate, local people, officials, technical specialists not taking part in the PRA full-time.

Key features

Everyone involved needs to be encouraged to contribute,

Someone needs to moderate to keep the workshop moving and ensure that the tasks set are performed,

The output of the workshop needs to be recorded and

Some form of media for presenting ideas, findings and reports is required.

Purpose

Preparatory workshop

Assembling the team, introductions and briefing,

Training in RRA techniques (if required),

Discussion and setting of RRA objectives,

Discussion topics for investigation,

Preparation of initial checklist of research topics,

Review of appropriate tools/approaches, and

Planning of PRA.

Periodic recurring workshops

Periodic review of findings of field work,

Monitoring of progress of PRA,

Checking of coverage,

Review of techniques used/discussion of alternatives,

Triangulation – (each topic of research investigated by different team members using different techniques and different sources),

Review of checklist of research topics and

Report updating.

Final workshop

Review of overall findings,

Report preparation,

Discussion of follow-up activities and

Participation of key non-participants (local officials, community leaders).

Note that the definition and key features of a workshop must allow the identification of a purpose for a relevant study objective (Adapted from Townsley (1996))

2.3 *Participants and Sampling Procedure for the Qualitative Study*

The participants in this study were limited to (1) the farmers who practice farming at the Tugela Ferry Irrigation Scheme and (2) extension officers who work with the farmers at the scheme. The numbers of participants in the study were not purposely determined. Invitations were made to all the farmers and extension officers. However, the final sample sizes were determined by the numbers of farmers who attended the workshops. Farmers came from the seven blocks currently under irrigation on the Scheme (Blocks 1, 2, 3, 4, 5, 7A and 7B). At the workshop to identify broad

Table 2 Ranking and classification technique as a participatory research tool

Definition

Tools for encouraging the people being interviewed to divide sets of items or activities into categories and rank them according to different criteria.

Key features

Can be used as a formal exercise or as an aid to interviewing,

Provides focus for discussion,

Can be carried out with individuals or with groups,

Provides a clear, graphic form of presentation of local people's ideas and

Adaptable to local circumstances and can use materials readily understood and manipulated by local people.

Main types relevant to this study

Matrix ranking: Using local classifications, the features or characteristics of groups of items or resources can be ranked according to different criteria such as reliability, seasonal stability, price, income generated, preferences, etc.

Pair-wise ranking: A more detailed ranking can be obtained using pair-wise ranking which compares pairs of items in a group until all are placed in an order of priority according to certain criteria.

Indicative ranking: A notional ranking can be used in many circumstances to provide indications of relative size or importance of particular features, numbers of people involved in activities. Local materials such as stones or beans can be used to quickly indicate proportions or numbers in a more concrete fashion.

Purpose

To understand local people's priorities,

To understand why certain choices are made,

To understand the local environment and people's knowledge about it and

To understand local terminology and classifications.

Note that the definition and key features of the main types relevant to the present study (Adapted from Townsley (1996))

areas for investigation, the farmers were initially grouped according to their blocks, and subsequently they were allowed to mix with members from different blocks. Separation of the farmers was done to compare the blocks, with respect to how the farmers view their problems, and to allow each block to learn about other blocks. During the workshops focusing specifically on linking problems with solutions, farmers from different blocks were mingled into six separate groups of six to eight persons, including extension officers (eight). Data collection was undertaken to determine two specifically enumerated variables: 'problems' and 'solutions'. Categories of each variable were classified by the participants using matrix and pair-wise ranking techniques (Table 2), which were variously undertaken by using materials such as seeds, or by a show of hands (Modi 2003).

2.4 Selection of Farmers for Crop Production Field Trials

A workshop involving the project researcher (author), extension officers and farmers to discuss exploratory agronomic trials was held at Tugela Ferry. After consideration

Table 3 Roles of participants in agronomic trials at Tugela Ferry

Farmer	Extension officer	Project researcher
Donate 0.5–0.1 ha of land and select trial site.	Provide guidance and advice to the farmer on site selection and soil sampling	Provide guidance and advice to the extension officer on site selection; facilitate soil analysis and explanation of recommendations for fertiliser application
Manage trial	Share ideas with the farmer and liaise with Project researcher	Share ideas with extension officer and monitor performance of extension officer and farmer
Provide labour to perform land preparation, planting, plant protection, harvesting and marketing	Facilitate access to inputs by the farmer and liaise with Project researcher	Provide operational capital resources and monitor performance of extension officer and farmer
Monitor and evaluate trial performance	Monitor and evaluate trial performance	Monitor and evaluate trial performance
Report to extension officer, Project researcher and other farmers (privately and publicly at farmers days)	Report to the farmer, Project researcher, other extension officers, other farmers (privately and publicly at farmers days)	Report to extension officer, the farmer, other farmers, Project leader(privately and publicly at farmers days)
Do self-evaluation and propose improvements and future directions to the extension officer	Do self-evaluation and propose improvements and future directions to the Project leader	Recommend future directions and facilitate reaching of consensus

Note the distinction between farmers, extension officers and scientist, who is also the project researcher. All these groups play different complementary roles

of internal operational conditions and farmer preferences eight farmers from each block volunteered to participate in agronomic trials. Farmer volunteerism was based on suggested guidelines that formed three key criteria for participation in the exploratory agronomic trials. These criteria were accepted by consensus (Chavez-Tafur, *et al.* 2007) as (i) the roles of the farmer in the exploratory agronomic trials (ii) the roles of the extension officer and the (iii) roles of the project researcher (Table 3). For all participants (farmers, extension officers, and project researcher) the roles were based on feasibility, practicality and resources as the key factors influencing participation. Hence, it was agreed that the exploratory agronomic trials would be undertaken as a partnership between the three parties (project researcher, extension officers and farmers) (du Plessis et al. 2002). The researcher-extension officer-farmer relationship, guided by the criteria in Table 3.

Farmers planted field trials of maize, potatoes, tomatoes and green pepper. The project researcher and extension officers did not participate in the planting of

Table 4 Criteria for assessment of the performance of extension officers and the farmers in management of agronomic trials

		Performance grades and scores					Maximum score
Farmer	Extension officer	E	VG	G	F	P	
Planning of activities	Planning of activities	1	0.8	0.6	0.4	0.2	1
Efficiency in implementation of advice from extension officers	Evidence and quality of farmer advice						1
Quality of trial	Identification of shortcomings in trial implementation						1
Needs communication	Evidence of response to farmer's needs						1
Ability to explain weaknesses	Identification of relevant farmer's needs						1
Quality of reporting	Quality of reporting						1
Record keeping	Record keeping						1
Dissemination	Dissemination						1
Total Score		**8**	**6.4**	**4.8**	**3.2**	**1.6**	**8**

Note: *E* excellent (more than meets the expectations), *VG* very good (complies with expectations), *G* good (has minor limitations in some areas), *F* Fair (has minor limitations in all areas or significant limitations in some areas), *P* poor (has significant limitations in all areas or shows no evidence of effort to succeed)

trials, and farmers were allowed to demonstrate their skills in making decisions about land preparation, planting densities and plant protection. It was made optional for a farmer or a group of farmers to seek advice from the extension officers, but not from the project researcher. The reason for this approach was to ensure a realistic evaluation of the farmer and extension officer performance during the course of the trials. Information gathered during the trial would be used at workshops to discuss the trials.

Extension officers and the project researcher met regularly (once in 2 weeks) to allow extension officers to report about their interactions with the farmers. The project researcher also met with individual farmers to get their reports, determine internal dissemination within blocks and to assess field trials. At the end of the growing season, a dissemination day was held with farmers and extension officers, where other farmers farming on the scheme, general members of the community and representatives of agricultural companies from the surrounding towns were present. The performance of extension officers and farmers was assessed on criteria based on Table 1. The criteria are shown in Table 4.

3 Results and Discussion

Results of this study were divided into three sections: (i) A brief account of the results from a special activity on training of farmers for participatory research, herein referred to as training for transformation, (ii) A presentation of the farmers' views on problems and solutions and (iii) Performance of farmers and extension officers on participatory agronomic trials.

3.1 The Training for Transformation Workshop

The purpose of the workshop was to create an opportunity for the farmers and extension officers to express their views in a manner that would demonstrate introspection and awareness about each other's expectations from the relevant stakeholders in the business of farming. The key performance areas were (i) What are the attributes of a good farmer? and (ii) What are the attributes of a good extension officer? Originally, these questions were extended to the service providers relevant to best management of small-holder irrigation schemes. However, in the absence of service providers, only the questions relating to the farmers and extension officers were addressed. The questions were discussed by separate groups of farmers, grouped by block, and extension officers. Each group listed the attributes of a farmer or extension officer and ranked them (Tables 5 and 6). Subsequently, the key attributes of a farmer and extension officer, respectively, were consolidated by all groups in a joint exercise. The attributes of a good farmer and a good extension officer are presented as descriptor terms (Tables 5 and 6).

Technical skills (Skilled), organisation of demonstration trials (Demonstrator), ability and willingness to advise farmers (Adviser) and ability to create markets for the farmers (Market) were ranked, by both the farmers and extension officers, as the most important attributes of an extension officer. The extension officers also felt it was crucial that a good extension officer be interested to learn about new developments in agriculture (Curious), understand and articulate relevant government policies (Policy) and have access to their own vehicles (Transport). The farmers differed from the extension officers in their emphasis of an extension officer as a role model, apolitical person, an inspector and a fund-raiser.

The most important attributes of a farmer were viewed by both the farmers and extension officers as diligence, access to income, markets, implements and irrigation water. Organisation of farmers into an association that facilitates cooperation and record keeping were deemed highly important by the extension officers, but most farmers were less aware of that requirement. The introspective view of the extension officers showed a clear awareness about most of the crucial attributes of a government service provider in their field (Table 5). The farmers' self assessment was biased in favour of needs, and less emphasis was placed on performance requirements (Table 6).

Table 5 Attributes of a good extension officer (*EO*) scored (high = 3 points, medium = 2 points or low = 1 point) by separate groups of farmers (n = 65) from seven blocks (Blk1, Blk2, Blk3, Blk4, Blk5, Blk7A and Blk7B) and extension officers (*EO*) (n = 8)

Attribute	Blk1	Blk2	Blk3	Blk4	Blk5	Blk7A	Blk7B	EO	Total	Rank
1. Exemplary	2	3	2	3	3	2	3	1	19	2
2. Apolitical	2	2	2	3	3	2	2	1	17	4
3. Skilled	3	3	3	3	3	3	3	3	24	1
4. Inspector	2	2	2	3	2	3	3	1	18	3
5. Patient	3	3	2	2	2	1	1	1	15	5
6. Demonstrator	3	3	3	3	3	3	3	3	24	1
7. Adviser	3	3	3	3	3	3	3	3	24	1
8. Curious	1	1	1	2	3	1	1	3	13	6
9. Market	3	3	3	3	3	3	3	3	24	1
10. Policy	1	1	1	2	1	1	1	3	11	7
11. Fundraiser	2	1	2	2	3	3	3	1	17	4
12. Transport	2	3	2	1	2	2	2	3	17	4

Table 6 Attributes of a good farmer scored (high = 3 points, medium = 2 points or low = 1 point) by separate groups of farmers (n = 65) from seven blocks (Blk1, Blk2, Blk3, Blk4, Blk5, Blk7A and Blk7B) and extension officers (*EO*) (n = 8)

Attribute	Blk1	Blk2	Blk3	Blk4	Blk5	Blk7A	Blk7B	EO	Total	Rank
1. Diligence	3	3	3	3	3	3	3	3	24	1
2. Income	3	3	3	3	3	3	3	3	24	1
3. Fencing	2	3	3	3	2	2	2	1	18	4
4. Record keeping	1	1	1	2	1	1	2	3	12	8
5. Implements	3	3	3	3	3	3	3	3	24	1
6. Scheduling	1	1	2	2	1	1	2	2	12	8
7. Curiosity	1	2	1	3	1	1	1	2	12	8
8. Soil sampling	2	3	3	2	2	2	3	3	20	3
9. Market access	3	3	3	3	3	3	3	3	24	1
10. Farmers' days	1	2	3	3	2	2	1	3	17	5
11. Skilled	2	1	2	3	1	1	2	2	15	6
12. Conservation	1	1	2	1	2	2	1	2	12	8
13. Media access	1	1	2	3	2	2	3	3	17	5
14. Local knowledge	1	1	2	1	1	2	2	1	11	9
15. Water sufficiency	3	3	3	3	3	3	3	3	24	1
16. Land sufficiency	2	3	3	2	3	3	3	2	21	2
17. Communication	1	2	1	1	2	2	1	2	12	8
18. Transport	1	1	2	2	1	1	1	3	12	8
19. Service access	2	1	2	1	2	2	2	2	14	7
20. Organised	1	2	1	2	3	2	1	3	15	6

Pair-wise ranking of the most important attributes (judging by a total score of 24 points in Tables 5 and 6) was performed to determine their relative importance, and the reasons for their ranking (Tables 7 and 8). To illustrate the importance of organisation and cooperation, a different exercise was given to the farmers and

Table 7 Pair-wise ranking of the four highly ranked attributes of an extension officer

Attributes	S	D	A	M	Score	Reason
Skilled (S)	X	S	S	M	2	Provides competitiveness
Demonstrator (D)		X	D	M	1	Allows sharing of information with farmers, but farmers generally know what to do
Adviser (A)			X	M	0	It is the extension officer's main job description.
Market (M)				X	3	It is ensures income generation

The ranking of each attribute was scored and reasons for preferential ranking were given by the extension officers. Attribute codes are shown in *parentheses*

Table 8 Pair-wise ranking of the five highly ranked attributes of a farmer

	D	I	Im	M	W	Score	Reason
Diligence (D)	X	I	Im	M	W	0	Anyone who is engaged in farming must be doing it because they have a good measure of diligence
Income (I)		X	Im	M	W	0	Easy to achieve once water, implements and markets are sorted
Implements(Im)			X	Im	W	3	Very important for land preparation; tractor is particularly desirable
Market (M)				X	W	2	There is a good local market (hawkers), but a more reliable market would improve livelihoods
Water (W)					X	4	Most crucial to success in crop production and income generation

The ranking of each attribute was scored and reasons for preferential ranking given by the farmers. Attribute codes are shown in *parentheses*

extension officers. The exercise entailed separation of six legume species and a different cultivar of one of the species that were mixed in one bag. Separation was done by selecting one representative seed of each type. Eighty percent of the farmers selected seven individual seeds, 15% selected four types and 5% selected three types. However, none of the participants were able to correctly identify all the species. Fifty percent of the participants identified four types correctly, 30% identified five types correctly and 20% identified two types correctly. The reasons for lack of complete success in identification were cited as: lack of familiarity with the types that are not locally available, such as lupines and soybeans; close resemblances among members of different species, such as green beans and dry beans; significant differences in size and colour among members of the same species, such as cowpeas and green beans.

The second exercise on legume species was used as a basis to get feedback from the farmers about the lessons they learnt from the workshops and how they were relevant to their ability to identify problems and solutions related to best management practices for small-holder irrigation schemes. The lessons were acknowledged as:

- No one knows everything about oneself, so it is important to learn from and with others,
- It is easier to identify the things that are more familiar to oneself than to identify peculiar objects,
- There are generally different aspects of any one issue and
- When many issues are important, some issues are more important than others.

The researcher was satisfied that the training workshops achieved the objective of making the farmers and extension officers conscious about Participatory Rural Appraisal and the importance of introspection in the process of problem identification and seeking of solutions.

3.2 *Workshop on Identification of Problems and Solutions*

Following group discussions and presentation of problems by each group, a glossary of terms was collectively produced and ranked by the farmers. The list of problems identified by the farmers is presented in Table 9. Results of a pair-wise ranking exercise with the farmers are shown in Table 10. The pair-wise rankings of problems (Table 10) were scored and reasons given for the preferential ranking of each problem (Table 11).

For each identified problem, farmers identified a solution. Solutions were purposely identified as a "wish-list", initially, to gain as much information as possible from the farmers. Ranking and scoring of the solutions would provide opportunities for more reasoning. The identification of solutions at Tugela Ferry is presented in Table 12. Matrix ranking was used to score the solutions to the problems listed by the farmers (Table 13).

Comparisons of pair-wise ranking of the problem with the matrix ranking of the solution to each problem are presented in Table 14. The relationship between the problems and solutions were compared using regression analysis.

Indicative ranking showed that the following four problems, in the order presented, were most important for Tugela Ferry farmers:

1. **Water**: particularly the poor state of infrastructure related to its movement from the river to the fields, which had a negative effect on the scheduling of its utilisation by different farmers between and within blocks,
2. **Implements**: a need for more tractors was singled out as being crucial for efficient scheduling of planting dates,
3. **Markets**: lack of standardisation at the local level and absence of any access to the major commercial markets were reported to have a negative impact on the ability of farmers to predict incomes, and

Table 9 List of problems constraining best management at and Tugela Ferry according to the farmers (n=65)

Problem	Code
In-field water scarcity	Wa
Damaged irrigation canals/equipment	Ic
Leaking storage dams	Sd
Expensive water bills	Wb
Poor knowledge of irrigation scheduling	Is
Poor scheduling of planting dates	Ps
Poor access to markets	Ma
Poor farm roads	Fr
Poor record keeping	Rk
Poor farmer organisation	Fo
Poor access to farming implements (Tractor)	Fi
Poor fencing	Fe
Lack of access to outlets for operational resources (fertilisers, herbicides, pesticides and seeds)	Or
Poor access to new technology	Nt
Limited or lack of demonstration trials	Dt
Poor youth involvement	Y
Crime	C
Unfavourable local government/land tenure	Lg

Note that the majority of problems are linked to basic technology requirements. It is also interesting to note that farmers pointed at a government institution as being problematic

4. **Weak institutional arrangements**: particularly at the level of committees representing blocks and the irrigation scheme as a whole.

Training of farmers in irrigation scheduling, crop production, farm management and cooperative arrangements (committees and farmer's associations) were suggested as solutions that could improve the farmer's performance across many areas of their work.

The pair-wise ranking scores produced during the Participatory Rural Appraisal research concurred with the indicative ranking (Tables 5 and 6). Matrix ranking of solutions to the problems also concurred with the indicative and pair-wise rankings by the farmers (Tables 13 and 14). The significant correlation ($r=+0.64$, $P<0.01$) between the problems and solutions confirmed that the ranked problems and solutions were carefully related by the farmers. However, it was shown that, theoretically, only about 40% ($R^2=0.41$) of the suggested solutions could be reasonably explained by the identified problems. Notwithstanding, the regression coefficient was significant ($P<0.01$). Hence, the outcomes of Participatory Rural Appraisal at Tugela Ferry can be taken as a reasonable measure of how the farmers viewed their situation with respect to identifying the problems and solutions to best management practices.

Table 10 Pair-wise ranking of constraints to best management as indicated by Tugela Ferry farmers

	Wa	Ic	Sd	Wb	Is	Ps	Ma	Fr	Rk	Fo	Fi	Fe	Or	Nt	Dt	Y	C	Lg
Wa	X	Wa	Wa	Wa	Wa	Wa	Wa	Wa	Wa	Wa	Wa	Wa	Wa	Wa	Wa	Wa	Wa	Wa
Ic		X	Ic	Ic	Ic	Ic	Ma	Ic	Ic	Ic	Ic	Ic	Ic	Ic	Ic	Ic	Ic	Ic
Sd			X	Wb	Sd	Sd	Ma	Sd	Sd	Sd	Sd	Sd	Sd	Sd	Sd	Sd	Sd	Sd
Wb				X	Wb	Wb	Wb	Wb	Wb	Wb	Fi	Wb	Wb	Wb	Wb	Wb	Wb	Wb
Is					X	Is	Ma	Fr	Is	Fo	Fi	Is	Is	Is	Is	Is	Is	Is
Ps						X	Ma	Ps	Ps	Fo	Fi	Fe	Ps	Ps	Ps	Ps	Ps	Ps
Ma							X	Ma	Ma	Ma	Ma	Ma	Ma	Ma	Ma	Ma	Ma	Ma
Fr								X	Fr	Fo	Fi	Fe	Fr	Fr	Fr	Fr	Fr	Fr
Rk									X	Fo	Fi	Fe	Or	Rk	Dt	Rk	Rk	Rk
Fo										X	Fi	Fe	Or	Fo	Fo	Fo	Fo	Fo
Fi											X	Fi	Fi	Fi	Fi	Fi	Fi	Fi
Fe												X	Fe	Fe	Fe	Fe	C	Lg
Or													X	Or	Or	Or	C	Lg
Nt														X	Dt	Nt	C	Lg
Dt															X	Dt	C	Lg
Y																X	C	Lg
C																	X	Lg
Lg																		X

The ranking of constraints in each column/row identifies the correlation of problems. This is helpful in seeking solutions to problems, because it allows relevant grouping to save money and time

Table 11 Scoring of problems and reasons given by the Tugela Ferry farmers for preferential ranking

Problem	Pair-wise score	Reasons
Wa	17	Crucial for plant growth
Ic	15	Loss of water and money; difficult to fix
Sd	13	Loss of water and money
Wb	14	Thousands of rands are paid to Eskom, but there is no equivalent income
Is	9	Inefficient water distribution among users
Ps	8	Flooding of local market
Ma	13	Reliable market required
Fr	8	Access to cropping land is poor
Rk	4	Uneconomic farm management and frustration to extension officers
Fo	9	Stifles development
Fi	13	Delays land preparation
Fe	8	Animal encroachment
Or	5	Access limited by distance to towns
Nt	1	Generally available for cultivars and agro-chemicals
Dt	3	Only important for new technologies
Y	0	Difficult to attract them to farming; almost given up on them
C	5	Serious, but beyond the farmers influence
Lg	6	Cannot be easily influenced by the farmers; government intervention

Note that the higher the score, the more important is the problem. Farmers generally prefer that the highest ranked problems are solved first. That way, there may be less need to solve less important problems

3.3 Performance of Farmers and Extension Officers in Crop Production Trials

In ranking order, the performance of farmers followed this pattern: Block 5 > Block 2 > Block 3 > Block 7A > Block 4 > Block 7B > Block 1. It is significant that the performance scores for the farmers ranged from 2.6 to 6.8, a difference of ~65% between the best (Block 5) and the poorest farmer (Block 1) (Fig. 1). It is interesting to note that there was a strong correlation ($r = 0.94$) between the performance of farmers and that of extension officers (Fig. 1). However, the farmers generally performed better than the extension officers. At four blocks (Blocks 2, 3, 5 and 7A) out of seven, farmers were found to score better than the extension officers. The extension officers scored better than the farmers at three blocks (Blocks 1, 4 and 7B). The strong correlation between the scores for farmers and those for extension officers suggested that the extension officers and the farmers may have influenced each other in performance. Where the farmer scored high, the extension officer also scored high (Fig. 1).

Findings of this study showed that both the farmers and the extension officers at Tugela Ferry were generally above average in management potential. The mean

Table 12 Matching of problems with solutions by the Tugela Ferry farmers

Problem	Solution
Wa	Repair and maintain irrigation canals and storage dams; schedule water use (**Wa_s**).
Ic	Repair and maintain irrigation canals (**Ic_s**).
Sd	Repair and maintain dams (**Sd_s**).
Wb	Maintain electric pumps; repair and maintain irrigation canals and storage dams; schedule water use through an organised farmer association (**Wb_s**).
Is	Do demonstration and research trials (**Is_s**).
Ps	Do demonstration and research trials (**Ps_s**).
Ma	Organise farmer associations (revive Msinga Vegetable Producers' Organisation – MVEPO) (**Ma_s**).
Fr	Provide government assistance with graders and a programme similar to the Department of Transport's Zibambele for rural roads (**Fr_s**).
Rk	Do demonstration and research trials; training workshops (**Rk_s**).
Fo	Provide training on institutional organisation to create strong block committees and an umbrella farmer association (**Fo_s**).
Fi	Provide government subsidy (Siyavuna?) and encourage farmers to invest in buying and maintenance through a farmer association (**Fi_s**).
Fe	Same as Fi (**Fi_s**).
Or	Revive the defunct Co-op or start a new one (**Or_s**).
Nt	Do Farmer's Days (**Nt_s**).
Dt	Do more on-farm trials with farmer participation (**Dt_s**)
Y	Provide learning opportunities; facilitate youth development programmes (**Y_s**)
C	Provide youth development programmes (**C_s**)
Lg	Sensitise traditional and municipal authorities to the economic significance of the scheme as the major source of food security and livelihood in the area (**Lg_s**).

Solution codes are shown in parenthesis with a subscript 's'

performance score for the farmers was 5.03, compared with 4.84 for the extension officers (SED=0.12). The relationship between the scores for the farmers and the scores for extension officers may have been influenced by the assumptions collectively taken by the project researcher (author), the extension officers and the farmers during pre-initiation, – evaluation interactions that:

1. The role of the extension officer shall be to provide mentorship and enhance the farmers' access to external knowledge and information.
2. The role of the farmer shall be to use local and formal knowledge to manage the trials and labour to channel the agro-ecological resources of the cropping system represented by the trial.

Although the extension officers were expected to do their own record keeping and independently perform dissemination, it was inevitable that the majority of their performances in record keeping and dissemination were influenced by what they gathered from the farmers. In fact, the reports given by the farmers from their farming diaries were significantly more detailed than the reports received from the extension officers. The latter relied too much on verbal reports and had a general tendency to pay less attention to detail.

Table 13 Participatory rural appraisal matrix ranking of solutions produced with Tugela Ferry farmers

	Wa_s	Ic_s	Sd_s	Wb_s	Is_s	Ps_s	Ma_s	Fr_s	Rk_s	Fo_s	Fi_s	Or_s	Nt_s	Dt_s	Y_s	C_s	Lg_s	Total score
Wa	10	10	10	10	10	10	10	1	10	9	8	8	8	8	5	5	7	139
Ic	10	10	10	10	10	5	8	8	6	10	8	4	2	0	0	0	10	111
Sd	10	10	10	10	10	8	8	1	4	6	6	0	5	0	0	0	8	96
Wb	10	10	10	10	10	8	10	3	10	10	1	8	10	6	0	0	9	125
Is	10	10	10	10	10	10	10	7	10	10	5	8	8	8	10	0	0	136
Ps	10	10	9	9	10	10	10	1	10	10	9	10	9	10	10	0	0	137
Ma	10	10	10	10	10	10	10	10	10	10	8	8	5	5	4	9	8	147
Fr	7	9	4	1	6	1	10	10	0	8	10	8	0	5	5	2	10	96
Rk	9	8	7	10	10	10	10	1	10	10	5	8	8	10	10	0	0	126
Fo	10	10	8	10	9	8	10	9	10	10	8	10	6	10	10	8	0	146
Fi/Fe	10	10	10	8	7	7	10	8	7	10	10	9	6	6	10	9	8	145
Or	8	5	5	10	7	10	10	4	10	10	6	10	10	10	10	10	0	135
Nt	8	8	8	8	10	10	10	5	10	10	8	10	10	10	10	0	0	135
Dt	8	5	3	5	10	10	10	5	10	10	10	10	10	10	10	0	0	126
Y	8	8	4	7	10	10	10	2	10	10	6	4	9	10	10	10	8	136
C	5	5	5	5	5	5	5	0	0	10	5	8	5	4	10	10	10	97
Lg	5	5	5	5	0	0	7	10	0	6	1	0	0	0	10	10	10	74
Scores for solutions	148	143	128	138	144	129	156	85	123	159	114	121	111	112	119	68	88	

Note: Score range allowed was 0–10. Note the total scores in the last row/column indicating the weight of each problem in relation to other problems

Table 14 Comparison of problems and solutions ranked according to pair-wise and matrix rankings

Pair-wise ranking scores for problems	Matrix ranking scores for problems	Matrix ranking scores for solutions
17 (Wa)	147 (Ma)	159 (Fo_s)
15 (Ic)	146 (Fo)	158 (Ma_s)
14 (Wb)	145 (Fi)	148 (Wa_s)
13 (Sd)	145 (Fe)	144 (Is_s)
13 (Ma)	139 (Wa)	143 (Ic_s)
13 (Fi)	137 (Ps)	138 (Wb_s)
9 (Is)	136 (Is)	132 (Ps_s)
9 (Fo)	136 (Y)	128 (Sd_s)
8 (Ps)	135 (Or)	127 (Rk_s)
8 (Fr)	135 (Nt)	124 (Y_s)
8 (Fe)	126 (Rk)	123 (Or_s)
5 (Lg)	126 (Dt)	114 (Fi_s)
5 (Or)	125 (Wb)	114 (Fe_s)
5 (C)	111 (Ic)	112 (Dt_s)
4 (Rk)	97 (C)	111 (Nt_s)
3 (Dt)	96 (Sd)	88 (Lg_s)
1 (Nt)	96 (Fr)	85 (Fr_s)
0 (Y)	74 (Lg)	73 (C_s)

Note the emphasis of the need to use different ranking methods to respond to different problems. In-field water scarcity and access to markets are key problems, which can be solved by improvement in farmer organisation. Improved farmer organisation would allow extension officers and technical advisors to reach out to farmers effectively

4 Conclusion

This study showed that workshops were successful Participatory Rural Appraisal tools for identification of problems and solutions related to the technical and institutional issues at the Tugela Ferry irrigation scheme. Through short and focused exercises, Participatory Rural Appraisal workshops also provide opportunities for training of farmers to enhance the farmers' understanding and participation in complicated thought-processes. This study was a significant milestone towards a common understanding between the researcher, farmers and extension officers. The processes used in the study created a better understanding on the part of the researcher that farmers have goals that are not limited to technical issues, and that the fundamental problems at Tugela Ferry are lack of infrastructure and proper institutional arrangements. These problems are viewed as the causes of technical constraints that limit productivity, sustainability, stability and equitability in the management of the Tugela Ferry irrigation scheme.

The findings of this study also showed that smallholder farmers and extension officers from Tugela Ferry have a potential to improve their crop management capabilities. The better management performance by farmers compared with extension

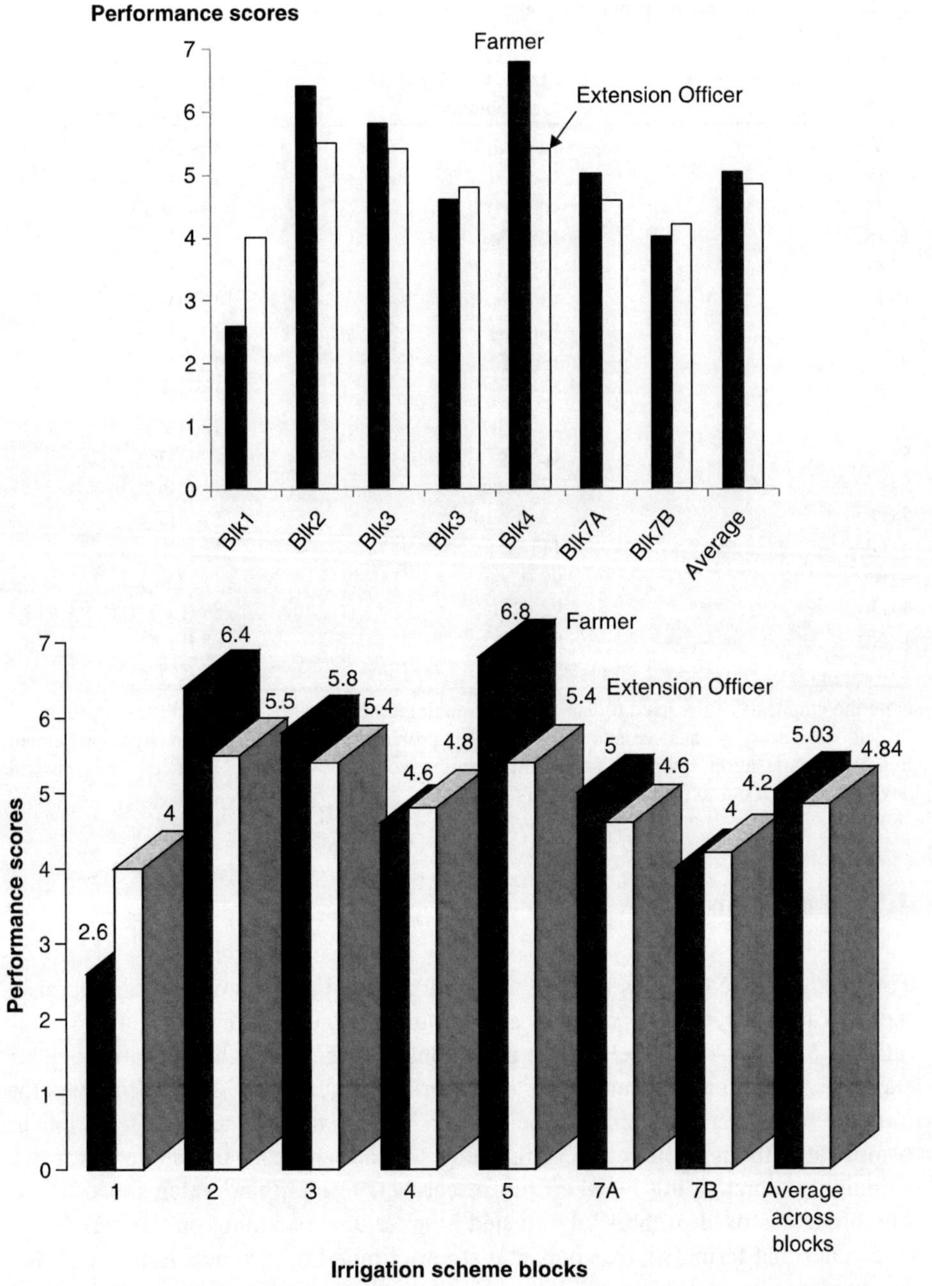

Fig. 1 Comparison of farmers and extension officers for performance in management of crop production trials. The farmers and extension officers were compared on seven production blocks (*Blk1, 2, 3, 4, 5, 7a* and *7B*) of the Tugela Ferry *irrigation scheme*. Note that the average across all blocks showed that the farmers were significantly better than the extension officers ($SED = 0.12$), because the former had local knowledge

officers was encouraging, although a significant improvement is required for both farmers and extension officers. From the crop production perspective, the key lessons learned from this study were that (i) An interactive approach to crop production was important to identify the roles of the farmer, the extension officer and the external resource persons (ii) The farmer needs to be largely self-reliant and innovative concerning decisions about cropping system management (iii) The extension officer is primarily a mentor and he or she must facilitate the transfer of external formal knowledge to the farmer and (iv) Interactive dissemination is a key component of crop production, because it builds self-confidence and it is a socially acceptable way of learning from peer groups.

Acknowledgements The author wishes to thank the Water Research Commission for funding the study as well as the farmers and extension officers of Tugela Ferry for participating.

References

Anyaegbunam C, Mefalopulos P, Moetsabi T (1998) Participatory rural communication appraisal: starting with people. SADC Centre of Communication for Development and FAO Regional Project GCP/RAF/297/ITA, Harare

Baur H, Kradi C (2001) Integrating participatory research methods in a public agricultural research organisation: a partially successful experience in Morocco, vol 109, Network Paper. Agricultural Research & Extension Network, London. ISBN 085003509

Bellon MR (2001) Participatory research methods for technology evaluation: a manual for scientists working with farmers. CIMMYT, Mexico. ISBN 9706480668

Chambers R (1993) Rural development: putting the last first. Longman, Harlow

Chambers R (1997) Whose reality counts? Putting the first last. London Intermediate Technology Publications, London

Chavez-tafur J, Hampson K, Ingeval A, Thijssen R (2007) Learning from experience: a manual for organising and documenting field based information. (ILEIA) Information Centre for Low External Input and Sustainable Agriculture, Amersfoort

Denison J, Manona S (2007) Principles, approaches and guidelines for participatory revitalisation of smallholder irrigation schemes, volume 1 – a rough guide for irrigation development practitioners. Research Report No TT 308/07. Water Research Commission, Pretoria

Du Plessis FJ, van Averbeke W, van der Stoep I (2002) Micro-irrigation for smallholders: guidelines for funders, planners, designers and support staff in South Africa. Research Report No TT 164/01. Water Research Commission, Pretoria

Fey M (2007) Soil science based improvements in subsistence agriculture. In: Ntutela S, Gevers W, Ramaite B (eds) Science-based improvements of rural/subsistence agriculture. Forum proceedings of the committee on science for poverty alleviation, 18 October 2006. Academy of Science of South Africa, Pretoria, pp 30–43

Hart T, Hart T (2007) The socioeconomics of subsistence farmers and the contribution of the social sciences to agricultural development. In: Ntutela S, Gevers W, Ramaite B (eds) Science-based improvements of rural/subsistence agriculture. Forum proceedings of the committee on science for poverty alleviation, 18 October 2006. Academy of Science of South Africa, Pretoria, pp 69–83

Kundhlande G, Groenewald DC, Baiphethi MN, Viljoen MF, Botha JJ, van Rensburg LD, Anderson JJ (2004) Socioeconomic study on water conservation techniques in semi-arid areas. Research Report No. 1267/1/04. Water Research Commission, Pretoria

Milligan A (2003) Participatory learning and action notes 47: general issue. IIED, London

Modi AT (2003) What do subsistence farmers know about indigenous crops and organic farming? preliminary experience in KwaZulu-Natal. Dev S Afr 20:675–684
Pretty JN, Guijt I, Scoones I, Thompson J (1995) A trainer's guide for participatory learning and action. IIED, London
Schoeman G, Magongoa B (2004) Community identified performance indicators for measuring water services. Research Report No. TT 228/04. Water Research Commission, Pretoria
Townsley P (1996) Rapid rural appraisal, participatory rural appraisal and aquaculture. FAO Fisheries Technical Paper 358. Food and Agriculture Organisation of the United Nations, Rome
Van Heck B (2003) Participatory development: guidelines on beneficiary participation in agricultural and rural development. Food and Agriculture Organization of the United Nations, Rome
Wiggins S, Kirsten J, Llambi L (2010) The future of small farms. World Dev 38:1341–1348

Bioavailability of Soil P for Plant Nutrition

R.S. Yadav, S.C. Meena, S.I. Patel, K.I. Patel, Mohd. Sayeed Akhtar, B.K. Yadav, and Jitendra Panwar

Abstract The poor mobility of soil inorganic phosphorus is due to the large reactivity of phosphate ions with numerous chemical, mineralogical and biological soil constituents. Adsorption-desorption and precipitation-dissolution equilibria control the concentration of P in the soil solution and P chemical mobility. P bioavailability depends on soil pH, concentrations of anions that compete with P ions for ligand exchange and metals that can coprecipitate with P ions. The mineralization and hydrolysis of organic phosphorus also maintain the phosphorus solution concentration.

R.S. Yadav (✉)
Directorate of Groundnut Research, Indian Council of Agricultural Research, 362001, P.O. No. 5, Junagadh, Gujarat, India
e-mail: yadavrs2002@gmail.com

S.I. Patel • K.I. Patel
Centre of Excellence for Research on Wheat, S. D. Agricultural University, 382870 Vijapur, Mehsana, India

S.C. Meena
Krishi Vigyan Kendra, Maharana Pratap University of Agriculture and Technology, 326001 Jhalawar, Rajasthan, India
e-mail: subhmeena@yahoo.com

M.S. Akhtar (✉)
Department of Biology, College of Natural Sciences, Jimma University, 378 Jimma, Ethiopia
e-mail: sayeedbot@yahoo.co.in

B.K. Yadav
Department of Agricultural Chemistry and Soil Sciences, Rajasthan College of Agriculture, MPUAT, 313001 Udaipur, Rajasthan, India
e-mail: bkyadav74@yahoo.co.in

J. Panwar (✉)
Department of Biological Sciences, Centre for Biotechnology, Birla Institute of Technology and Science, 333031 Pilani, Rajasthan, India
e-mail: drjitendrapanwar@yahoo.co.in

E. Lichtfouse (ed.), *Farming for Food and Water Security*, Sustainable Agriculture Reviews 10, DOI 10.1007/978-94-007-4500-1_8,

Thus, only a marginal proportion of soil phosphorus is present as P ions in the soil solution. Due to lack of appropriate methods for studying its speciation and biogeochemical behavior, the mobility of inorganic P in most soils is still poorly understood and hardly predictable. This is even worse when considering the problem of the bioavailability of P to plants.

The characterisation and quantification of soil phosphorus and the factors that control the availability of P to plants are of utmost importance to define the bioavailability of soil P. Thus, the measurement of intensity and capacity factors together can describe P supply with considerable precision. Phosphorus flux to the root system is mainly controled by diffusion. Soil water content and diffusion coefficient, including the soil buffer power are key factors in the diffusion of P to plant roots. Root systems have the ability to increase the bioavailable pool of P due to their influence on soil chemistry either directly by the activity of plant roots or indirectly by enhancing the activity of rhizospheric microflora. Phosphorus uptake by roots, effect of soil pH, anion/cation balance, gaseous exchanges and release of root exudates are major rhizosphere processes controling the bioavailability of soil P. Soil P bioavailability can also vary with plant species or genotypes, plant nutritional status and ambient soil conditions.

Various extractions procedures are now widely used to assess P bioavailability based on the knowledge of phosphorus dynamics in soils and P mobilization by plants. Furthermore, physiological processes occurring in the rhizosphere also provide a better understanding of P availability. ^{31}P-NMR technique clearly indicates the distribution of inorganic phosphorus in living cell. Kinetics have elucidated the functional characteristics of plasma membrane and tonoplast inorganic phosphate transporters. Molecular studies have confirmed the presence of multiple genes encoding phosphate transporters. Modeling the P bioavailability is a logical approach to understand the complexity of the P nutrition. Though, the authenticity of the models will largely depends on the accuracy and quality of input data that are very subjective to varying soil conditions. Inclusion of bioavailability of P as a parameter in crop modeling will help in solving the large black box of processes and mechanisms of P uptake from soil that has perplexing and hitherto not well understood. The aim of this chapter is to give an overview of P uptake processes and mechanisms involved in plants and the chemical processes that are directly induced by plant roots which can affect the concentration of P in soil solution and, ultimately, the bioavailability of soil phosphorus to plants.

Keywords Phosphorus bioavailability • P fixation • Rhizosphere processes • P characterization • Modelling

1 Introduction

A model of bioavailability of nutrients has been proposed by Barber (1995). It can be defined as the mobility of soil nutrients in relation to their availability to plants which mainly depends on quantity/intensity ratio of nutrient availability altered by the soil and plant dynamics. The release of nutrients from their solid phase to solution phase in the soil; the movement of nutrients through soil solution to the plant and the absorption of nutrients by plant root-mycorrhizal system are the important process which control the availability of soil nutrients (Comerford 1998). Release of nutrients from the solid phase to soil solution is controlled by the physico-chemical processes of adsorption and/or dissolution as well as biochemical process of mineralization. All the essential plant nutrients absorbed by the root system are always in specific ionic form and reach to the surface of roots by the process of diffusion or mass flow, and could be controlled by the interactions between soil and plant roots. The plant roots take up nutrients from the soil solution by an ion exchange mechanism which mainly depends on the amount of root-mycorrhizal surface and its uptake characteristics. Therefore, the concentrations of mineral nutrients in the soil solution as well as the buffer capacity of the soil are of utmost importance for mineral nutrient supply to growing plants.

Phosphorus (P) is one of the major plant growth limiting nutrients despite being abundant in soils both in organic and inorganic forms. However, many soils throughout the world are P deficient as out of total phosphorus concentration present in soil, only 1–3% is available to the plants. These low levels of plant available P are due to the high reactivity of soluble P with calcium, iron and aluminum that leads to its precipitation in soil. Further, due to ever reduced natural P deposits with increased P fertilizer cost strongly demands a major shift in P fertiliser use (Cordell et al. 2009). A better understanding of the processes governing the dynamics and availability of P in soil-plant system is thus needed. Soil phosphorus takes part in many soil P reactions, which occur between the liquid and solid phases. Sorption and desorption are the two important processes for movement of P in the soil depending upon the intensity and capacity factors. The movement of soil P to plant roots is mainly governed by rate of diffusion which is directly related to the degree of saturation of the P adsorption capacity which divided soil P into labile and non-labile fractions. The addition or removal of P or any changes in the P adsorption capacity by the soil system altered the diffusion rate. Thus, the behaviour of P in soil can be summarized in the following equilibrium reaction

Solution P $\leftrightarrow$ Labile P $\leftrightarrow$ Non-labile P

The reactions between solution P and labile P are rapid, but that between labile and non-labile P are slow. This chapter will stress on the processes which control the release and movement of P in soil, and explains the important factors influencing the acquision of soil phosphorus by plants.

2 Phosphorus Cycle in Soils

The phosphorus cycle in soil is a dynamic process involving soil, plants and microorganisms. The major processes include uptake of soil P by plants, recycling through return of plant and animal residues, biological turnover through mineralization-immobilization, fixation reactions at clay and oxide surfaces, and solubilization of mineral phosphates through the activities of microorganisms. In the natural state,

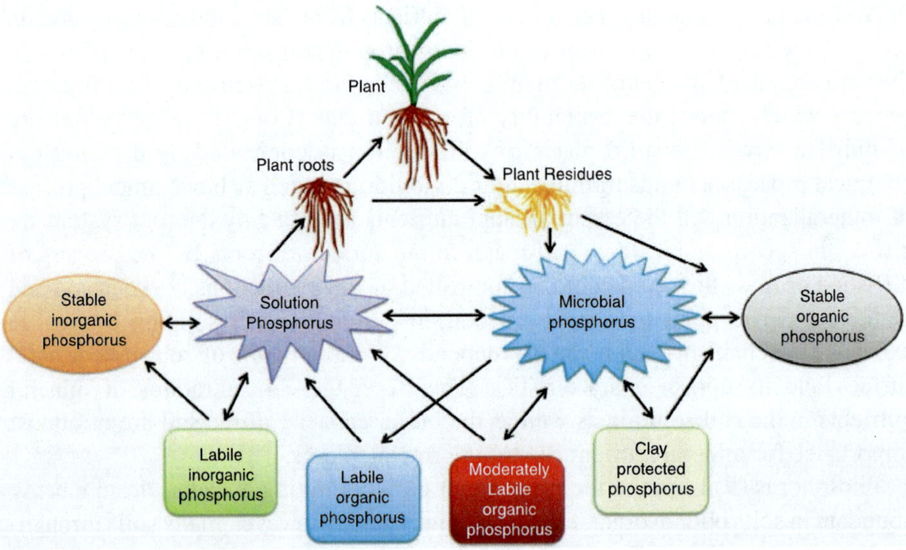

Fig. 1 Schematic representation of the P cycle in soil showing organic and inorganic pool based on availability to plant. The clay protected phosphorus (both inorganic and organic) and the stable organic phosphorus fractions are generally not available to plants at short term basis. The rest inorganic as well as organic fractions are directly or indirectly available to plants

essentially the entire P consumed by plants is returned back to the soil as plant and animal residues; under cultivation conditions, some P is removed in the harvest and only a part of it is returned back. Losses of soil P occur through leaching and erosion. The P cycle in soil depends on the biological transformation, soil-plant interrelationship and phosphate reactions. In the plant-soil microbial interrelationships, P is partitioned in two major pools based on the availability of various organic and inorganic forms to plants (Fig. 1). Soil solution P is shown to be in equilibrium with a given quantity of labile P such that as P is taken up by the plants, or immobilized by the microorganisms, additional inorganic P is solubilized. Chemical and biochemical aspects of the P cycle have been reviewed from several stand points, including fluxes of P on a global scale (Richey 1983; Oelkers and Valsami-Jones 2008), pedogenesis (Pierzynski et al. 2005), plant nutrition, inorganic forms and fixation reactions (Shen et al. 2003) and soil organic P and associated transformations (Anderson 1980; Condron et al. 2005).

Phosphorus is constantly cycling in agricultural fields, mediated by the biological P requirements of all living organisms. When soil is harnessed for agricultural production the natural cycling get disturbed by fertiliser additions and removal of nutrients with crop harvesting. The effects of human activity on soil P have been studied for quite some time. In agricultural fields which receive chemical P fertilization, the total P in the surface soil tends to increase as compared to soils of barren fields (Perrot et al. 1990). On the other hand, phosphorus fertilization given through manures has been found to increase the concentration of organic phosphorus (Sharpley et al. 2004) as well as inorganic phosphorus (Uusitalo et al. 2007). Precisely, it can be concluded that application of any type of phosphorus fertilizers in agricultural soils alter the native soil P pools. In addition, the phosphorus cycle in agricultural fields is also subjected to seasonal changes, due to temperature fluctuations or owing to agronomic management practices like tillage and cropping systems (Styles and Coxon 2007; Muukkonen et al. 2007).

Therefore, phosphorus cycle in any agroecosystem is a dynamic process transformed through soil, plants and microorganisms including phosphorus uptake by plants, phosphorus fixation reactions, mineralization and immobilization in rhizosphere. Generally, the addition of phosphatic fertilizers in the soils tends to increase the phosphorus unavailability either in inorganic or organic forms on long term basis.

3 Nature of Soil Phosphorus and Its Availability

Phosphorus occurs in soil in several inorganic and organic forms. Inorganic phosphorus includes apatitic minerals, secondary precipitates formed with Ca, Fe and Al and free phosphate ions ($H_2PO_4^-$, HPO_4^{2-}, and PO_4^{3-}) attached to sorption surfaces or dissolved in the soil water. Organic phosphorus includes a group of organic molecules having P as a part of their structure. Orthophosphate esters are Po compounds having an ester linkage joining PO_4-P to organic moiety and they are further divided into mono- and di-esters according to the number of ester groups attached to each PO_4-P. The largest group of organic phosphorus in most soils is the monoesters including sugar phosphates, phosphoproteins, mononucleotides and inositol phosphates (Turner et al. 2002). Diesters including nucleic acids, phospholipids, teichoic acid and aromatic compounds generally occur in smaller concentrations in soil than monoesters. Phosphonates form a special group of organic phosphorus that contains a C-P bond. Polyphosphates complicate the simple division of P into organic and inorganic pools, since most of them are inorganic but few are of biological origin. However, polyphosphates such as ADP and ATP are, chemically, organic compounds.

The availability of soil P in most contemporary studies still relies on relatively simple acid extractions, despite the passage of nearly 150 years of Hilgard's analytical approaches. Even still, most research and management studies that evaluate bioavailable P continue to characterize P availability using only extractants designed to correlate with yields of P-fertilized agricultural crops grown in the same year as that of the soil sampling and analysis (Bray and Kurtz 1945; Mehlich 1978; Olsen et al. 1954). These correlative methods support a short-term dose–response approach to intensive soil management (Richter and Markewitz 2001), and seemingly ignore slowly cycling fractions of soil P. The most direct way of determining nutrient availability in soil is to measure the growth responses of plants by means of field plot fertilizer trials. This is a time consuming procedure and the results are not easily extrapolated from one location to another. In contrast, chemical soil analysis is a comparatively rapid and inexpensive procedure for obtaining information about nutrient availability in soil as a basis for recommending fertilizer application, which has been practiced for many years with relative success particularly in agriculture and horticulture. Therefore, the characterization of soil phosphorus using conventional techniques like *ex-situ* soil analysis, fertilizer response field experiments, isotopic dilution techniques etc. are unable to characterize the phosphorus bioavailability as well as the root induced changes in the rhizosphere. The recent analytical techniques like use of ion exchange resigns, sensors, chips, plant growth models would be able to partially characterize the nature and behaviour of bioavailable phosphorus in the soil.

4 Factors Affecting Phosphorus Availability in Soil

Plants absorbed the inorganic phosphate either in the form of HPO_4^{2-} or $H_2PO_4^-$ forms from the soil solution (Hansen et al. 2004; Turner et al. 2007) which has been replaced through movement of inorganic phosphorus in the soil, resulting in a depletion profile of inorganic phosphorus near the root surface (Jungk 2001). As a result, diffusion of inorganic phosphorus down its concentration gradient is the most important transport process by which Pi moves through the soil to reach roots (Barber 1995). The gradients generated by uptake deplete the nutrients in the soil solution which may be buffered by re-equilibration with nutrients held in other form. It is also known as "nutrient kinetics re-equilibrium". However, the kinetics of re-equilibration is often poorly understood and most quantitative models of root uptake have ignored this factor particularly when the available nutrient pool is adequate to fulfil the plant requirement. Further, another factor which is not simply the passive sink of nutrients but also exert an enormous effect on the soil is plant roots, which may affect soil pH, redox, soil biology, soil chelation chemistry etc. Some of these changes are discussed below.

4.1 Solubilization by Organic Anion Excretion

Excretion of nutrient-solubilizing organic anions from roots is thought to be an important mechanism by which plants obtain insoluble nutrients. The important processes are: the diffusion of the organic anion away from the root, its decomposition by soil microbes, its reaction with the soil in solubilizing the nutrients, and diffusion of the nutrients away from the solubilization zone both towards the root and away from it. Kirk (1999) has developed a model for these processes and corroborated it by comparing predictions with measured concentration profiles of P in the rhizosphere of rice plants in aerobic soils. A sensitive analysis showed the particular importance of rooting geometry and density. Root density affects both the accumulation of solubilizing agents in soil and the interception of solubilized P by neighbouring roots and resulted in complex interactions between solubilization and root geometry (Gahoonia and Nielsen 2004; Lynch and Brown 2008).

4.2 Root Mediated pH Changes

Plant roots acquire most of their essential mineral nutrients either in cationic or in anionic forms. The simultaneous uptake of several charged ion species by roots disturb the electro potential of plant tissues, consequently the roots must compensate this charge imbalance by exertion of charged compounds back into the soil to preserve electro-neutrality in their tissues. Now it is evident that these compensating ions are H^+ in case of excess cation uptake and OH^- for excess anion uptake (Hinsinger 2001). The excretion is also stoichiometrical equivalent to the charge imbalance (van Beuschichem et al. 1988; Tang and Rengel 2003). As the rate of H^+ or OH^- flux depends on the cation/anion imbalance, therefore conceptually one would predict that the flux would itself change temporally as the development of depletion zone for each cation and anion changed their concentrations, and hence uptake at the root surface. This temporal dynamics would be translated into a spatial effect with changes at the root apex being either reinforced or reserved depending on the uptake history of a root segment. Such spatial pH differences have been observed frequently between different regions of the root (Marschner et al. 1991; Shen et al. 2003).

4.3 Root Morphology

Plants growing in P-deficient soil allocate a greater proportion of assimilates to root growth and tend to have fine roots of a small diameter and therefore a large surface area. P-efficient barley (Gahoonia et al. 2011) and cowpea cultivars (Krasilnikoff et al. 2003) have longer root hairs allowing them to take up more P in comparison with P-inefficient genotypes. P-deficient *Lupinus angustifolius* increased the primary root elongation and developed a large number of the cluster-like first-order lateral roots with dense root hairs, thus allowing efficient P acquisition under low P supply (Wang et al. 2008). Fine roots and especially root hairs effectively scavenge P from soils because of a large surface area of contact with the soil. The shoot P status may regulate the formation of cluster roots, as specialised structures of selected plant species for thorough exploration of the soil volume. However, the form of P in soils may also regulate cluster root formation (Shu et al. 2007). In addition, the development of cluster roots can respond to presence of organic matter adjacent to the root (Hodge 2004).

4.4 Microbial Dynamics in the Rhizosphere

Darrah et al. (2006) have developed a model for the dynamics and activity of microbial population growing on root exudates. Important aspects allowed in this model were the time course of substrate excretion, three dimensional diffusion through the soil, re-sorption of certain substrates by the root, and predation of different microbes on each other. The model showed that the population dynamics were surprisingly complex. The width of zone of increased activity becomes progressively narrower after its initial growth as the population close to the root consumes an increasing proportion of substrates. The time course of excretion is found to be important. For example a diurnally varying excretion, as has been observed for the excretion of

Fe-solubilizing siderophores in some plants, could greatly enhance the longevity of the substrate. Such models are necessarily a great simplification of what actually happens, but they are nonetheless a great aid in unravelling what could otherwise be an endless sink for empirical observation.

4.5 Arbuscular Mycorrhizal Fungi

The arbuscular mycorrhizal fungi have the biochemical and physiological capabilities for increasing the supply of available P or other immobile nutrients. These mechanisms may involve rhizosphere acidification, and increases in root phosphatase activity or excretion of chelating agents (Gianinazzi et al. 2010). Recent gene expression study shows that plants induce a common set of mycorrhiza specific genes but there is also variability, indicating that there exists functional diversity in arbuscular mycorrhizal symbioses (Feddermann et al. 2010). The differential expression of symbiosis associated genes has been related to the fungal species, plant genotypes and environmental factors. In a study reported by Tarafdar and Marschner (1994), 48–59% of the P taken up by wheat was a result of mobilization of phosphorus from organic phosphorus sources hydrolysed by arbuscular mycorrhizal fungi produced phosphatases. However, Habte and Fox (1993) found that effective mycorrhizal associations did not occur in soils with high organic matter content without addition of inorganic phosphorus suggesting that mobilization of inorganic phosphorus from organic sources was insufficient.

Thus, buffering capacity and rhizosphere of the soil are the two important factors to determine the phosphorus bioavailability in the soils through indirect effect on soil pH, redox potential, soil biology, soil chelation chemistry etc. All these processes altered the excretion of organic anions, root mediated pH changes, root morphology, microbial dynamics in the rhizosphere, colonization of arbuscular mycorrhizal fungi and ultimately the bioavailability of the soil phosphorus for plant nutrition.

5 Solubilization and Mineralization of Soil Phosphorus

Several reports have suggested the ability of different microbial species to solubilize insoluble inorganic phosphate compounds, such as tricalcium phosphate, dicalcium phosphate, hydroxyapatite and rock phosphate (Goldstein 1986; Khan et al. 2007). The solubilizations of different types of insoluble phosphates vary with the type of microorganisms, media conditions and available carbon and phosphate sources (Harvey et al. 2009; Khan et al. 2010; Zaidi et al. 2009). The organic acids secreted by plants or microbes also increase the available P content breaking up the Ca, Fe and Al phosphate and by decreasing adsorption of P compounds in soil (Bolan et al. 1990) (Fig. 2). Organic acids are the low molecular weight compounds, which are characterized by the possession of one or more carboxyl groups. The study of low

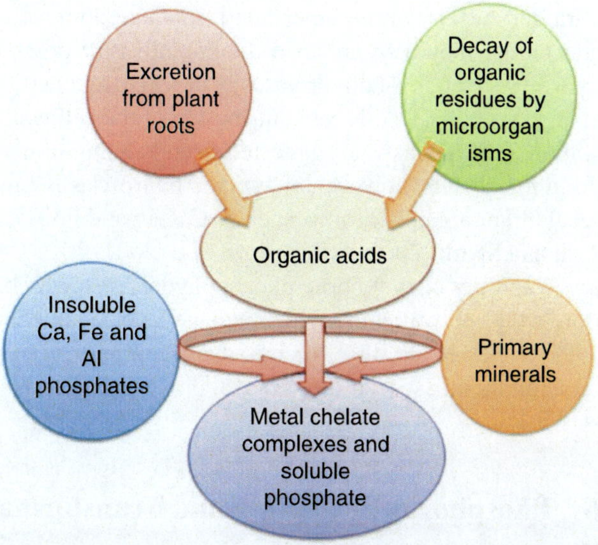

Fig. 2 Release of insoluble phosphates to soluble form through the action of organic acids and other naturally occurring chelates

molecular weight organic acids and the ability of their carboxylate functional groups to interact with soil, by occupying anion adsorption sites and competing with phosphate, have an expressive importance to increase available P in soils (Guppy et al. 2005). The change in soil pH in rhizosphere by presence of organic acids increase the phosphorus solubility by complexing Fe and Al oxides, besides the competition for adsorption sites promoted by carboxylate groups (Jones 1998; Ehlers et al. 2010).

Organic phosphate solubilization is also called as mineralization of organic phosphorus and it occurs in soil at the expense of plant and animal remains, which contain a large amount of organic phosphorus compounds. The decomposition of organic matter in soil is carried out by the action of numerous saprophytes, which produce the release of radical orthophosphate from the carbon structure of the molecule. The organo-phosphonates can equally suffer a process of mineralization when they are victims of biodegradation (McGrath et al. 1995; Cheng 2009). The degradability of organic phosphorus compounds depend mainly on the physicochemical and biochemical properties of their molecules among which the nucleic acids, phospholipids and sugar phosphates are easily broken down, but phytic acid, polyphosphates, and phosphonates are decomposed at a slower rate (McGrath et al. 1995, 1998). Phosphorus can be released from organic compounds in soil by three groups of enzymes: *Nonspecific phosphatases*, which perform dephosphorylation of phospho-ester or phospho-anhydride bonds in organic matter; *Phytases*, which specifically cause P release from phytic acid; *Phosphonatases* and *C-P Lyases*, enzymes that perform C-P cleavage in organophosphonates.

The main activity apparently corresponds to the work of acid phosphatases and phytases because of the predominant presence of their substrates in soil (Yadav and Tarafdar 2003). Under P-deficient conditions, enhanced secretion of phosphohydrolases by plant roots and also by rhizosphere microorganisms can enhance the availability of Pi in the rhizosphere by hydrolysis of soluble organic P esters (Yadav and

Tarafdar 2001). On the other hand, root excretion of carboxylates such as oxalate and citrate may also enhance the solubility of organic P forms available for the hydrolysis by phosphohydrolases in the rhizosphere (Otani and Ae 1999; Marschner et al. 2005, 2006). A direct contribution of the cell wall in mobilization of sparingly soluble P form was also suggested from P-fixing minerals (Ae and Otani 1997).

In many soils, however, enzymatic hydrolysis is limited by the low solubility of recalcitrant organic P forms such as Ca/Mg and Fe/Al phytates (Luengo et al. 2006; Arai and Sparks 2007; Richardson et al. 2001, 2005). Another limiting factor is the low solubility of root borne phosphohydrolases, which is restricted by immobilization at the cell wall and in the mucilage layer of apical root zones (Dinkelaker et al. 1997; Jones et al. 2009) or by adsorption and inactivation on clay minerals and organomineral associations (Rao et al. 1996; Devau et al. 2010).

6 Phosphorus Speciation and Transformations

The reactions controlling the P cycle in soil include solubilisation, precipitation, sorption, desorption, leaching, immobilisation and mineralisation. Apatitic P, the primary sources of P in soils, is solubilised through weathering processes. The ionic form of solubilised P depends on the pH of the solution, the predominating species in slightly acidic soils being $H_2PO_4^-$ and in soils having a pH over 7 being HPO_4^{2-} (Fig. 3). Solubilised P can leached from the soil, sorbed onto Fe and Al oxides and

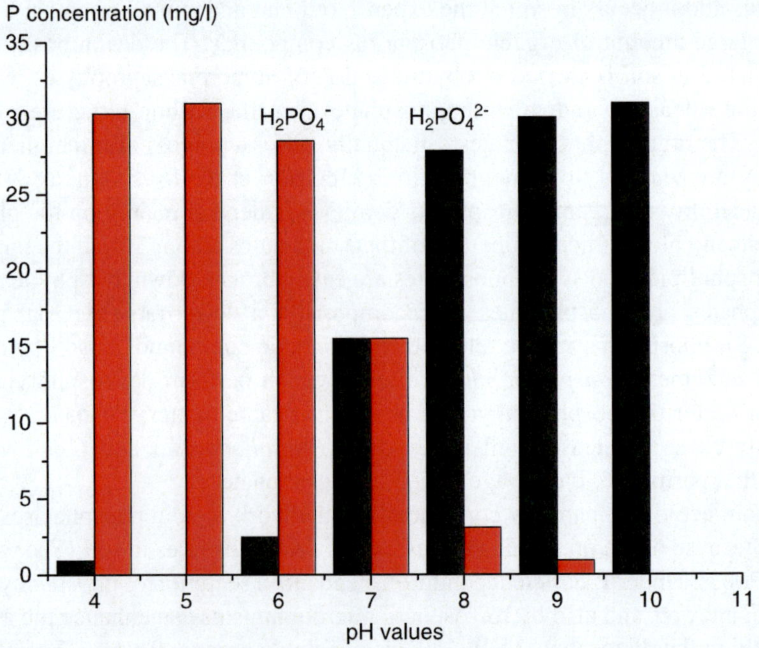

Fig. 3 Effect of pH on ionic forms of dissolved phosphate

mineral edges of clay particles, precipitate as secondary Ca, Fe or Al-minerals or be taken up by living organisms. Only a small proportion of soluble P is dissolved in soil water. The negatively charged PO_4-P can bound to positively charged surfaces through electrostatic forces. However, the importance of this mechanism is minimal in soil environments compared to ligand exchange.

Ligand exchange, originally discovered in by Hingston et al. (1967) allows fairly strong sorption of P and is considered as the main retention mechanism for inorganic phosphorus in non-calcareous soils. In this reaction, PO_4-P is chemisorbed onto variably-charged sorption components. Ligand exchange refers to the formation of an inner-sphere complex where PO_4^- group displaces through its oxygen-bearing group on OH^- or H_2O group from Al or Fe in a hydrous oxide or on the edge surface of a clay mineral. The reaction can take place on a positively, negatively or uncharged surface. However, the H_2O ligand, predominant in acidic conditions, increases the positive charge on a sorption surface and is easier to replace with a PO_4 group (Vance et al. 2003; Hinsinger et al. 2005), thus enhancing the sorption of P at lower pH. The PO_4-P in soil solution tries to reach equilibrium with PO_4^-P sorbed onto soil surfaces. Therefore, an increase in PO_4-P concentration in soil solution enhances the P sorption. At lower concentrations, the formation of binuclear or bidentate complexes, where PO_4 group displaces two coordinated OH^- or H_2O groups from a hydrated surface, is favoured, but as the level of adsorption increases the formation of monodentate structures is preferred and the sorption strength is decreased (White 1979). The fast adsorption of PO_4-P onto variably-charged soil oxides and hydroxides is followed by slower diffusion of P into the hydroxide (Barrow 1983). This diffusion proceeds with time, decreasing the plant availability of P (Barrow 1979, 1983; Hinsinger 2001).

Desorption of P is enhanced in conditions opposite to those favouring sorption. With increasing pH, desorption is enhanced and similarly, a decrease in the PO_4-P in the soil solution enhances P desorption from the sorption sites. Further, sorption and desorption of P is affected by P concentration and ionic strength in the soil solution and temperature (Barrow 1983). An increase in salt concentration increases the sorption of P (Ryden and Syers 1975) and decreases the desorptions (Luengo et al. 2006; Arai and Sparks 2007). The rate of sorption and desorption of P increases with temperature (Hinsinger 2001). Precipitation can be seen as a reverse reaction to mineral dissolution and can be defined as formation of discrete, insoluble compounds in soil (Pierzynski et al. 2005). In alkaline soils where Ca is the predominant cation, the soluble PO_4-P can precipitate, forming Ca phosphates. In these soils, the dicalcium phosphate dihydrate ($CaHPO_4 \cdot 2H_2O$) which forms readily after the addition of P to soil, can in the long term turn into more stable forms of calcium phosphates (Freeman and Rowell 1981; Pierzynski et al. 2005). In acidic soils, Ca phosphates are unstable and in favourable conditions P can form precipitates with Al and Fe instead of Ca. Separate crystals of Al and Fe phosphates have been detected (Martin et al. 1988). The evidence of the surface precipitation of P has also been reported (Ler and Stanforth 2003). However, some studies support the idea that the slow diffusion of P into the pores of amorphous oxides is the reaction mechanism rather than precipitation (Magid and De Arambarri 1985), even though sorption is thought to proceed to precipitation in the long run (Pierzynski et al. 2005).

Inorganic P is transformed and incorporated into the organic pool through the biological cycle. Plants and microbes take up P required for their growth and after the death of living organisms, the immobilised P returns to the soil through mineralisation by microbes. The microbiological decomposition process results in a release of both inorganic and organic phophorus. The esters of organic phosphorus compounds can be adsorbed onto soil sorption surfaces through the PO_4 group by a similar mechanism to that for inorganic phosphorus. Diesters have only one ionisable proton per PO_4 group compared to two ionisable protons in monoester PO_4 groups and, thus, are less strongly sorbed onto soil surfaces. Lacking the sorption-induced protection, the diesters are more easily mineralised than monoesters (Celi and Barberis 2005). Guan et al. (2006) reported myo-inositol hexaphosphates, monoesters which contain six PO_4 groups to adsorb on Al-hydroxides with three PO_4 groups and Celi et al. (1999) reported them to absorb on goethite with four of its six groups. The strong sorption protects inositol hexaphosphates from mineralisation and leads to their accumulation in soil (Celi et al. 1999). Further, inositol hexaphosphates have been reported to displace inorganic PO_4-P from sorption sites (Berg and Joern 2006). In addition to sorption to Fe and Al oxides and mineral edges of clay particles, inorganic and organic phosphorus can be bound to organic molecules through cationic bridges (Gerke and Hermann 1992).

The P present in soil solution is negatively charged, either monovalent or divalent orthophosphate ions. The major P sorbents are those soil constituents that bear positive charges. These comprise various variable charge compounds that contain hydroxyl (Fe and Al oxides), carboxyl (organic matter) or silanol (clays) groups. In addition, they occur mostly as small crystals and more or less poorly ordered minerals that have a considerable specific surface area, and hence a strong reactivity as sorbents (Schwertmann and Taylor 1989). Thus, they play a prominent role in the adsorption of P ions in most soils; not only in ferralsols from tropical regions that are known for their properties being largely influenced by Fe and Al oxides, but also in soils in the alkaline pH range such as calcareous soils (Rahmatullah and Torrent 2000). Being variable charge minerals means that their capacity to adsorb anions such as P ions will increase with decreasing pH, because of the increase in positive charge of such minerals as a consequence of their larger protonation at low pH (Strauss et al. 1997). Therefore, when considering the sole process of adsorption of P ions onto Fe and Al oxides, decreasing the pH should result in a stronger retention and hence in a decreased mobility of inorganic P.

7 Mechanisms of Phosphate Solubilization and Acquisition by Plants

7.1 Release of Nutrients from the Soil Solid Phase

The key solid phase properties are the quantity of labile nutrients, and the manner of nutrient release to the soil solution. The term labile, in this context, is defined as nutrients in a plant-accessible form during a given time period. Soil solution concentration is important because (1) the quantity of nutrients in solution is low relative to a plants demand and, therefore, must be replenished continuously; (2) practically all nutrient uptake occurs from the soil solution; and (3) nutrients must be in solution to move effectively to the plant's root-mycorrhizal surface, thereby giving the plant a larger soil volume from which it can extract nutrients. The solid phase maintains the soil solution nutrient concentration. Release from the solid phase to soil solution occurs biochemically through mineralization and immobilization or physiochemically through adsorption and desorption, and precipitation and dissolution or it is added externally as fertilizers.

The greater the sorptive capacity of the soil, the greater should be the potential for the mineralized ion to be sorbed onto the mineral surface. Therefore, the role of mineralization should be (1) to control the soil solution concentration directly when the nutrient does not sorb onto the mineral soil, or (2) contribute to the soil solution concentration and to the nutrient concentration on the soil's mineral surface when soil's sorption capacity is greater than zero. In the latter case, nutrient distribution between the solution and the mineral surface should be related to nutrient affinity for the mineral surface. While it can be shown that mycorrhizal hyphae can intercept mineralized P before other microorganisms, or before sorption to some clays (Joner and Jakobsen 1994), there is very little relevant literature directly targeting this topic, even though the interaction between mineralization and sorption is an integral element of modeling bioavailability.

Adsorption-precipitation and desorption–dissolution reactions regulate the removal of nutrients from, or release into, the soil solution. Adsorption-desorption reactions are classified as either outer sphere for cations or inner sphere for anions. The later one is notably for phosphate and sulfate. Anions participate in outer sphere reactions only when the soil has a positive charge. Except for specific circumstances, however, this does not add much to our understanding of their bioavailability. An outer sphere reaction is an electro-negative attraction of charged nutrients to the exchange complex of soil surface. Cation exchange is the sorption process when cations are involved, and anion exchange is the process when anions are concerned. Outer sphere reactions are rapid, have comparatively weak bonding between the nutrient and soil surface, and do not have kinetic limitations for equilibrium. An inner sphere reaction occurs when the solute becomes part of the soil mineral surface, such as during ligand exchange. In this case, the bond is covalent, can exhibit strong hysteresis between sorption and desorption, and is considered not to have kinetic limitations to equilibrium, unless the time frame is on the order of minutes to hours.

7.2 Nutrient Movement Through the Soil Solution

Nutrients move to roots through the solution phase. Movement in soil occurs by mass flow and/or diffusion. These processes are soil state-dependent (Table 1). Mass flow is defined as the quantity of nutrient flowing to the root system due to transpiration stream. Therefore, the most obvious controls of mass flow is the water regime of soil, or how the soil water content and its energy status change over time, and the

Table 1 Soil processes and the soil characteristics defining each process, along with a qualitative evaluation of the plant's influence on the process

Soil process	Defining soil characteristic	Potential plant influence on process
Sorption	Temperature (T)	Low, shading
	Kd of adsorption, pH	High, root exudates
	Solution ionic strength	None to low
Desorption/dissolution	Temperature	Low, shading
	Kd of desorption,	High, root exudates
	Solution ionic strength	None to low
Mineralization/ immobilization	Volumetric water content (*q*)	High, transpiration
	Temperature	High, shading
	Organic matter quality	High
	Enzyme concentration	High, phosphatase exudation
Mass flow	Hydraulic conductivity, *q*	High, transpiration effect on *q*
	Bulk density (*r*)	Low to medium, root action
	Pore-size distribution	Low to medium, root action
	Solution concentration (*Cl*)	Low to medium, root exudates
Diffusion	Impedance	High, transpiration effect on *q*, root action on soil structure
	Kd	Low to medium, root exudates
	Hydraulic conductivity	High, transpiration
	Pore-size distribution	Low to medium, root action
	Solution concentration	Low to medium, root exudates

(Source: Comerford 2005)

desorption-mineralization processes that determine the soil solution concentration. This is followed by the pore-size distribution of soil, which determines hydraulic conductivity changes with soil water content. The best conditions for mass flow dominance are when (1) the soil sorption capacity for a nutrient is in the diminishing returns phase, and (2) water content does not seriously limit water flow. Under these conditions, nutrient delivery by mass flow is maximized. Note that the plant plays a key role in determining the importance of mass flow through its effect on the water potential gradient driving mass flow, and its effect on the soil solution concentration.

When mass flow does not meet the plant's nutrient demand, i.e., amount arriving at the root system is less than that required amount by plant, the nutrient concentration in rhizosphere get reduced as compared to the soil solution present beyond the rhizosphere, a nutrient gradient develops in the soil. Fick's law states that diffusive flux is the function of this concentration gradient. This law illustrates that nutrient deficiencies or the soil's inability to support luxury consumption promote the onset of diffusion. A plant experiencing a nutrient deficiency must have diffusive flow supplying a portion of that nutrient. The rate of nutrient flux by diffusion is retarded by any sorptive interaction that the nutrient has with soil. This interaction is contained in the buffer power term of diffusive flow. The greater interaction the nutrient has with soil surface, the greater will be the *b* and *Kd*, and slower will be the nutrient delivery to plant root system via diffusion. When the *Kd* is near 0, as with nitrate in most soils, the buffer power is equivalent to the soil water content.

Soil water content and soil water regime are other critical components of bioavailability. Soil water content regulates diffusion through determining (1) the cross-sectional area available for diffusion, (2) determining the path length of diffusion by controlling diffusive impedance (*f*) and, (3) contributing to the retardation of nutrient movement, as indicated by the soil buffer power (*b*). The impedance factor is the ratio of the length of straight-line path of nutrient movement to the actual path. The impedance factor is a function of bulk density and water content (Bhadoria et al. 1991; Kirk et al. 2003), so it is manageable through irrigation, physical soil management, and the action of plant roots and soil biota, such as earthworms etc.

7.3 Nutrient Uptake by the Root System

Phosphate must pass through the cell wall before entering a plant cell. Between pH 3 and 7, the predominant form of inorganic phosphorus is $H_2PO_4^-$ (Sentenac and Grignon 1985). Due to its negative charge, inorganic phosphorus passes readily through the spaces between cellulose microfibrils and is repelled by negatively charged surfaces in the pectin network (Schindler 1998). The presence of enzyme that are capable of hydrolyzing organic phosphorus, such as acid phosphatases, have been reported in cell walls (Hurley et al. 2010). Also, cell walls of peanut (*Arachis hypogaea* L.) roots have been reported to solubilize $FePO_4$ (Ae and Otani 1997). The plasma membrane (PM) with its lipid bilayer is a barrier to the passage of charged ions, such as $H_2PO_4^-$ (Marschner 1995). Intrinsic proteins span the lipid bilayer and are responsible for transport of ions across biological membranes. Since transport systems are via membrane-bound proteins, it is not surprising that uptake processes can be described by Michaelis-Menten enzyme kinetics (Glass 1989). The Michaelis-Menten constant (Km) indicates the affinity of the enzyme for substrate; smaller the value, greater the affinity (Fig. 4). Primary pumps use the energy gained from the breakdown of ATP to ADP to drive the movement of H^+ or other ions across the membrane (Marschner 1995). Primary pumps and secondary transport systems are considered to be active uptake, because ions can be moved against their electro-chemical potential gradient.

Cytoplasmic homeostasis is defined as a set point or desired level of an ion that is maintained in the cytoplasm (Mimura 1999). During changes in external Pi, cytoplasmic Pi concentration in roots remained fairly constant, whereas, vacuolar Pi concentration varied widely (Lee and Ratcliffe 1993; Pratt et al. 2009). Dunlop and Phung (1998) have reported the presence of inward and outward rectifying vacuolar channels that allow movement of Pi ions into or out of the vacuole in sugar beet (Fig. 5). These channels may be involved in maintaining cytoplasmic homeostasis of inorganic phosphorus.

Radial movement of ions and solutes from the exterior of the root to the xylem occurs in either the apoplasm or apoplast (via cell walls) or the symplasm or symplast

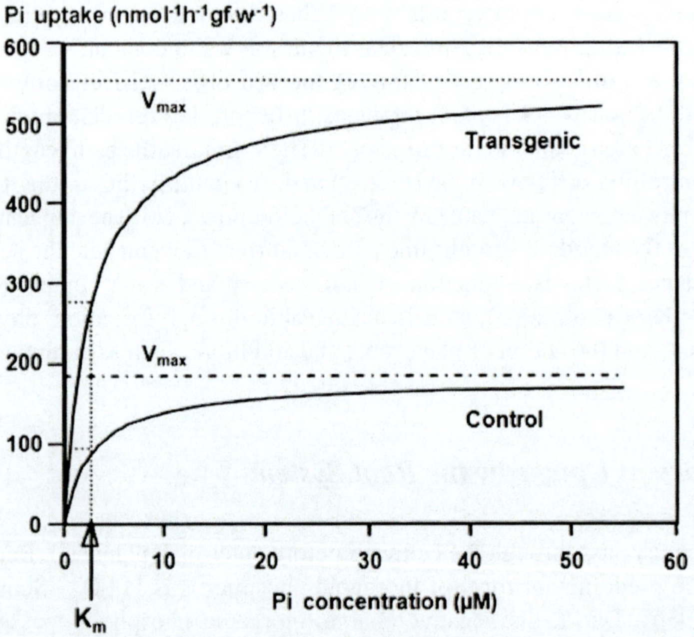

Fig. 4 Michaelis-menten kinetics of tobacco suspension cultured cells with (*Transgenic*) and without (*Control*) insertion and overexpression of *Arabidopsis thaliana* gene, PHT1, encoding for a putative high affinity P transporter (Mitsukawa et al. 1997)

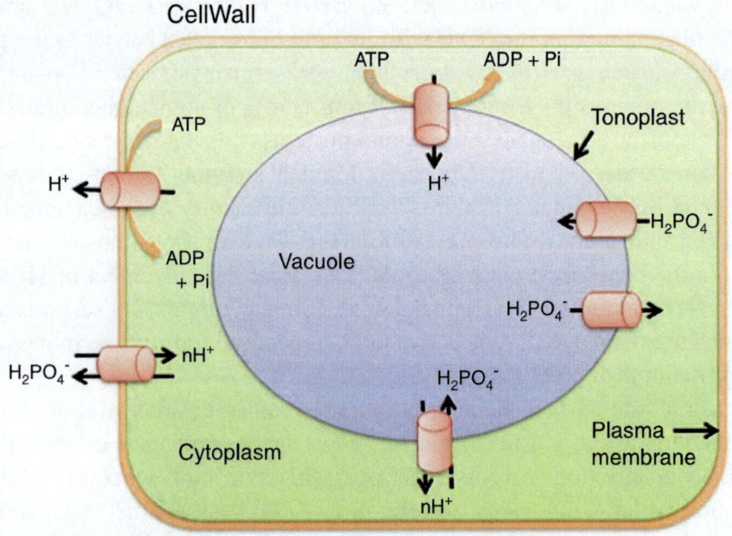

Fig. 5 Transport systems at plasma membrane (H^+-ATPase and a $nH^+/H_2PO_4^-$ symport) and tonoplast (H^+ -ATPase, *outward* and *inward* rectifying P channels, and putative nH^+/H_2PO_4 + antiport) (Miyasaka and Habte 2001)

(via plasmodesmatal connenctions between cells) (Marschner 1995). Ions can absorb by epidermal cells and then moved either (a) via the symplastic pathway or (b) via the apoplastic pathway upto the endodermal cells and then enter the symplastic pathway due to the presence of casparian strips on the endodermal cells. Recent advances in molecular biology have allowed cell-specific localization of high affinity Pi transporters in roots using *in situ* hybridization (Liu et al. 1998; Doerner 2008; Lin et al. 2009; Rouached et al. 2010). Under P sufficient conditions, root hairs and epidermal cells absorb most of the inorganic phosphate which then moves via the symplastic pathway to vascular tissue. Under P deprivation, however, more cells in the root are involved in inorganic phosphorus uptake, including young stelar tissue.

Apoplastic transport via the xylem allows the movement of both water and ions from roots to shoots. In the xylem sap of castor bean, nitrate was the predominant inorganic anion, followed by $H_2PO_4^-$ (Jeschke et al. 1997). Proton pumps were localized in the endodermis and living stelar parenchyma cells, using a polyclonal antibody to ATPase (Parets-Soler et al. 1990; Hamburger et al. 2002; Stefanovic et al. 2007). As a result of the proton pump generated by H^+-ATPases, secondary transport systems or channels located in xylem parenchyma cells can move ions into the xylem. The whole interior of sieve tubes is considered part of the symplasm, because phloem conducting as elements retains their PM at maturity (Versaw and Harrison 2002). In the phloem of castor bean, $H_2PO_4^-$ was found to be the predominant inorganic anion and approximately 17% of total P was present as organic phosphorus (Jeschke et al. 1997). Phosphorus is highly mobile in the phloem, and net export of P from older leaves has been observed (Jeschke et al. 1997).

Therefore, phosphate solubilization and acquisition by plants is a multiphase phenomenon including release from solid phase, movement to root surface and uptake by the roots. The phosphate release from soil solid phase i.e. the capacity factor is important to replenish phosphorus in soil solution for plant nutrition which mainly occur biochemically, physico-chemically or added externally as fertilizers. However, the replenishment from the soil solid phase depends upon various processes like sorption, adsorption/dissolution, and mineralization/immobilization. Further this phosphorus flux through the soil solution to plant root surfaces by mass flow, diffusion or both, where the mass flow is due transpiration and diffusion is due to phosphorus depletion gradient near root surface. Soil water content and hydraulic coefficient are the important parameters for these processes. As phosphorus is an immobile nutrient in the soil therefore mostly diffusion is the major process for its movement in soil system. Lastly, the phosphorus uptake by root system is mainly by primary pumps and secondary transport as the ions can be moved against their electro-chemical potential gradient. The movement of ions or solutes from root surface to xylem occurs in either through apoplast or symplast against electro-chemical potential gradient or specific to localized cells of high affinity inorganic phosphate transporters in roots.

8 Modeling Phosphorus Bioavailability

There are several soil processes which influence nutrient bioavailability (Table 1). The most important tend to be root amount, mineralization regimes, soil water regimes, and soil partition coefficients-buffer powers. The complexity of nutrient bioavailability lends itself to computer modeling. The problem is multi-dimensional in both the number of variables and in space. Thus, typical mechanistic models have upward of ten input variables, and both soil depth and horizontal variability add to this complexity. These models are based on the mechanistic approach comsidering-nutrient release from the solid phase, movement to the root by mass flow-diffusion, and uptake at the root system surface.

A bioavailability evaluation is more complex than the simpler process used to evaluate soil fertility in the past, i.e., soil testing correlated with plant growth, nutrient uptake, or fertilizer response. The soil testing approach was key to the green revolution. It continues to provide nutrient management recommendations, and has fueled agriculture for decades. Today, large-scale, multi-site field research is becoming excessively expensive, and nutrient management practices are requiring minimum nutrient inputs for maximum return on investment with minimum offsite effects. A more useful approach under these limiting circumstances is to use field testing to verify or debunk selected model predictions.

With regard to soil parameters, predicting desorption from soil characteristics and its hysteresis with sorption need to be better understood (Barros Filho et al. 2005). The control of the soil solution concentration by mineralization and desorption needs to be clarified. Spatial variability of *Kd* and labile nutrient pools, along with methods to predict this variability, will improve model accuracy and precision. With regard to the plant itself, a better understanding of the roots Michaelis-Menten parameters and of the dynamics of parameter change in the short and long term is necessary. For severe nutrient deficiencies, a much better understanding of *C*min will be fruitful. More accurate measurement and prediction of root system dynamics are required.

Even with these limitations, the utility of nutrient bioavailability models has been shown under both greenhouse and field conditions, particularly when time frames are relatively short (Barber 1995). Further development of mechanistic models based on bioavailability concepts will be useful to improve nutrient management, to gain a better understanding of nutrient cycling, and to predict stress responses of extensively managed ecosystems. Their incorporation into plant growth models, such as TREGRO (Weinstein and Yanai 1994), or into P cycling component models such as ANIMO, GLEAMS, and DAYCENT (Lewis and McGechan 2002) shines light into the "black box" that has often characterized soil processes.

Modeling bioavailability is a logical approach to solve the complexity of this topic. Mostly a set of input parameters control nutrient uptake which requires accurate and quality input data. However, these input data sets may change with soil and species conditions. Inspite of limitations like inaccuracy and poor quality inputs, bioavailability models have been shown to be useful. Their incorporation into plant growth models has been successful, and has taken soil processes out of the black box in which the soil is so often confined.

9 Conclusion

The phosphorus concentration in soil solution is determined by a number of processes of soil-plant continuum and their interactions with soil phosphorus. All these parameters depends on various common factors like pH, concentrations of metal cations such as Ca, Fe and Al as well as on the concentrations of competing inorganic (especially bicarbonate, and possibly sulphate) and organic ligands (such as carboxylic anions). These factors could be modified in rhizosphere and the plant roots can indeed shift the chemical equilibria that determine mobility and bioavailability of soil inorganic P. However, many aspects of overall P dynamics in the soil-plant continuum are not thoroughly understood like regulation of phosphorus acquisition and phosphorus starvation rescue mechanisms in plants; the complex coordination of root morphology, physiological and biological responses under varying phosphorus availability and plant sensing of heterogenous phosphorus supply in the soil. All these processes will possibly result in a build-up of P concentration in the soil solution and, hence in an increased bioavailability of P to plants.

Future research should thus try to examine the relative contributions of the various processes through which plant roots can alter the bioavailability of soil inorganic P. Besides mathematical modelling, the development of new tools provided by plant molecular biology and soil geochemistry will also help ascertaining our current knowledge of such plant–soil interactions. These rhizosphere processes are even further complicated by many additional interactions that rhizosphere microflora can bring about. It is therefore not surprising that the quest for a universal soil testing procedure for adequately predicting the bioavailability of soil P to a range of plant species in a large range of soil types is unsuccessful till date and will remain vain for long. Many of the above mentioned processes can be affected by the nutritional status of the plants. In addition, they vary considerably between plant species and even among the various genotypes of a given species. This should be further studied in order to better understand the adaptation of plants to low P soils and their consequences for the dynamic competition within plant communities in both natural ecosystems and agro-ecosystems. In addition, this should also be better exploited in breeding programs aiming at finding out more P-efficient genotypes. Many of above processes are indeed involved in the efficiency of P acquisition of a plant, although this also depends on numerous other traits like root growth and architecture, development of root hairs and symbiotic mycorrhizal associations that are crucial for an efficient acquisition of poorly mobile phosphorus nutrient.

References

Ae N, Otani T (1997) The role of cell wall components from groundnut roots in solubilizing sparingly soluble phosphorus in low fertility soils. Plant Soil 196:265–270

Anderson G (1980) Assessing organic phosphorus in soils. In: Khasawneh FE, Sample EC, Kamprath EJ (eds) The role of phosphorus in agriculture. American Society of Agronomy, Madison, pp 411–432

Arai Y, Sparks DL (2007) Phosphate reaction dynamics in soils and soil minerals: a multiscale approach. Adv Agron 94:135–179

Barber SA (1995) Soil nutrient bioavailability: a mechanistic approach, 2nd edn. Wiley, New York

Barros Filho NF, Comerford NB, Barros NF (2005) Phosphorus sorption, desorption and resorption in soils of the Brazilian Cerrado supporting eucalyptus. Biomass Bioener 28:229–236

Barrow NJ (1979) Three effects of temperature on the reactions between inorganic phosphate and soil. J Soil Sci 30:271–279

Barrow NJ (1983) A mechanistic model for describing the sorption and desorption of phosphate by soil. J Soil Sci 34:733–750

Berg AS, Joern BC (2006) Sorption dynamics of organic and inorganic phosphorus compounds in soil. J Environ Qual 35:1855–1862

Bhadoria PBS, Kaselowsky J, Claassen N, Jungk A (1991) Impedance factor for chloride diffusion in soil as affected by bulk-density and water-content. Z Pflanzenernahr Bodenkd 154:69–72

Bolan NS, White RE, Hedley MJ (1990) A review of the use of phosphate rocks as fertilizers for direct application in Australia and New Zealand. Aust J Exp Agric 30:297–313

Bray RH, Kurtz LT (1945) Determination of total, organic, and available forms of phosphorus in soils. Soil Sci 59:39–45

Celi L, Barberis E (2005) Abiotic stabilisation of organic phosphorus in the environment. In: Turner BL, Frossard E, Baldwin DS (eds) Phosphorus in the environment. CABI Publishing, Wallingford, Oxfordshire, UK, pp 113–132

Celi L, Lamacchia S, Marsan FA, Barberis E (1999) Interaction of inositol hexaphosphate on clays: adsorption and charging phenomena. Soil Sci 164:574–585

Cheng W (2009) Rhizosphere priming effect: its functional relationships with microbial turnover, evapotranspiration, and C-N budgets. Soil Biol Biochem 41:1795–1801

Comerford NB (1998) Soil P Bioavailability. In: Lynch JP, Deikman J (eds) Phosphorus in plant biology: Regulatory roles in molecular, cellular, organismic and ecosystem processes. American Society of Plant Physiologist, Maryland, USA, pp 136–147

Comerford NB (2005) Soil factors affecting nutrient bioavailability. In: BassiriRad H (ed) Nutrient acquisition by plants: an ecological perspective. Springer, Berlin/Heidelberg

Condron LM, Turner BL, Cade-Menun BJ (2005) Chemistry and dynamics of soil organic phosphorus. In: Sims JT, Sharpley AN (eds) Phosphorus: agriculture and the environment. American Society of Agronomy/Crop Science Society of America/Soil Science Society of America, Inc, Madison, pp 87–121

Cordell D, Drangert JO, White S (2009) The story of phosphorus: global food security and food for thought. Glob Environ Chang 19:292–305

Darrah PR, Jones DL, Kirk GJD, Roose T (2006) Modelling the rhizosphere: a review of methods for 'upscaling' to the whole-plant scale. Eur J Soil Sci 57:13–25

Devau N, Le Cadre E, Hinsinger P, Gerard F (2010) A mechanistic model for understanding root-induced chemical changes controlling phosphorus availability. Ann Bot 105:1183–1197

Dinkelaker B, Hengeler C, Marschner H (1997) Root exudates and mobilization of nutrients. In: Rennenberg H, Eschrich W, Ziegler H (eds) Trees contribution to modern tree physiology. Backhuys, Leiden, pp 441–452

Doerner P (2008) Phosphate starvation signaling: a threesome controls systemic Pi homeostasis. Curr Opin Plant Biol 11:536–540

Dunlop J, Phung T (1998) Phosphate and slow vacuolar channels in *Beta vulgaris*. Aust J Plant Physiol 25:709–718

Ehlers K, Bakken LR, Frostegard A, Frossard E, Bunemann EK (2010) Phosphorus limitation in a ferralsol: impact on microbial activity and cell internal P pools. Soil Biol Biochem 42:558–566
Feddermann N, Finlay R, Boller T, Elfstrand M (2010) Functional diversity in arbuscular mycorrhizal-the role of gene expression, phosphorus nutrition and symbiotic efficiency. Fungal Ecol 3:1–8
Freeman JS, Rowell DL (1981) The adsorption and precipitation of phosphate onto calcite. J Soil Sci 32:75–84
Gahoonia TS, Nielsen NE (2004) Root traits as tools for creating phosphorus efficient crop varieties. Plant Soil 260:47–57
Gahoonia TS, Nielsen NE, Joshi PA, Jahoor A (2011) A root hairless barley mutant for elucidating genetic of root hairs and phosphorus uptake. Plant Soil 235:211–219
Gerke J, Hermann R (1992) Adsorption of orthophosphate to humic-Fe-complexes and to amorphous Fe-oxide. Z Pflanzenernahr Bodenkd 155:233–236
Gianinazzi S, Gollotte A, Binet MN, van Tuinen D, Redecker D, Wipf D (2010) Agroecology: the key role of arbuscular mycorrhizas in ecosystem services. Mycorrhiza 20:519–530
Glass ADM (1989) Plant nutrition: an introduction to current concepts. Jones and Bartlett Publishers, Boston
Goldstein AH (1986) Bacterial solubilization of mineral phosphates: historical perspective and future prospects. Am J Altern Agric 1:51–57
Guan XH, Shang C, Zhu J, Chen GH (2006) ATR-FTIR investigation on the complexation of myo-inositol hexaphosphate with aluminum hydroxide. J Colloid Inter Sci 293:296–302
Guppy CN, Menzies NW, Moody PW, Blamey FPC (2005) Competitive sorption reactions between phosphorus and organic matter in soil: a review. Aust J Soil Res 43:189–202
Habte M, Fox RL (1993) Effectiveness of VAM fungi in nonsterile soils before and after optimization of P in soil solution. Plant Soil 151:219–226
Hamburger D, Rezzonico E, Petetot JMC, Somerville C, Poirier Y (2002) Identification and characterization of the Arabidopsis *PHO1* gene involved in phosphate loading to the xylem. Plant Cell 14:889–902
Hansen JC, Cade-Menun BJ, Strawn DG (2004) Phosphorus speciation in manure amended alkaline soils. J Environ Qual 33:1521–1527
Harvey PR, Warren RA, Wakelin S (2009) Potential to improve root access to phosphorus: the role of non-symbiotic microbial inoculants in the rhizosphere. Crop Pasture Sci 60:144–151
Hingston FJ, Atkinson RJ, Posner AM, Quirk JP (1967) Specific adsorption of anions. Nature 215:1459–1461
Hinsinger P (2001) Bioavailability of soil inorganic P in the rhizosphere as affected by root-induced chemical changes: a review. Plant Soil 237:173–195
Hinsinger P, Gobran GR, Gregory PJ, Wenzel WW (2005) Rhizosphere geometry and heterogeneity arising from root mediated physical and chemical processes. New Phytol 168:293–303
Hodge A (2004) The plastic plant: root responses to heterogeneous supplies of nutrients. New Phytol 162:9–24
Hurley BA, Tran HT, Marty NJ, Park J, Snedden WA, Mullen RT, Plaxton WC (2010) The dual-targeted purple acid phosphatase isozyme AtPAP26 is essential for efficient acclimation of Arabidopsis to nutritional phosphate deprivation. Plant Physiol 153:1112–1122
Jeschke WD, Kirkby EA, Peuke AD, Pate JS, Hartung W (1997) Effects of P deficiency on assimilation and transport of nitrate and phosphate in intact plants of castor bean (*Ricinus communis* L.). J Exp Bot 48:75–91
Joner EJ, Jakobsen I (1994) Contribution by two arbuscular mycorrhizal fungi to P uptake by cucumber (*Cucumis sativus* L.) from ^{32}P-labelled organic matter during mineralization in soil. Plant Soil 163:203–209
Jones DL (1998) Organic acids in the rhizosphere: a critical review. Plant Soil 205:25–44
Jones DL, Nguyen C, Finlay RD (2009) Carbon flow in the rhizosphere: carbon trading at the soil root interface. Plant Soil 321:5–33

Jungk A (2001) Root hairs and the acquisition of plant nutrients from soil. J Plant Nutr Soil Sci 164:121–129

Khan MS, Zaidi A, Wani PA (2007) Role of phosphate-solubilizing microorganisms in sustainable agriculture – a review. Agron Sustain Develop 27:29–43

Khan MS, Zaidi A, Ahemad M, Oves M, Wani PA (2010) Plant growth promotion by phosphate solubilizing fungi – current perspective. Arch Agron Soil Sci 56:73–98

Kirk GJD (1999) A model of phosphate solubilization by organic anion excretion from plant roots. Eur J Soil Sci 50:369–378

Kirk GJD, Solivas JL, Alberto MC (2003) Effects of flooding and redox conditions on solute diffusion in soil. Eur J Soil Sci 54:617–624

Krasilnikoff G, Gahoonia T, Nielsen NE (2003) Variation in phosphorus uptake efficiency by genotypes of cowpea (*Vigna unguiculata*) due to differences in root and root hair length and induced rhizosphere processes. Plant Soil 251:83–91

Lee RG, Ratcliffe RG (1993) Subcellular distribution of inorganic phosphate and levels of nucleoside triphosphate, in mature maize roots at low external phosphate concentrations: measurements with ^{31}P-NMR. J Exp Bot 44:587–598

Ler A, Stanforth R (2003) Evidence for surface precipitation of phosphate on goethite. Environ Sci Technol 37:2694–2700

Lewis DR, McGechan MB (2002) SW-soil and water: a review of field scale phosphorus dynamics models. Biosyst Eng 82:359–380

Lin WY, Lin SI, Chiou TJ (2009) Molecular regulators of phosphate homeostasis in plants. J Exp Bot 60:1427–1438

Liu C, Muchhal US, Uthappa M, Kononowicz AK, Raghothama KG (1998) Tomato phosphate transporter genes are differentially regulated in plant tissues by phosphorus. Plant Physiol 116:91–99

Luengo C, Brigante M, Antelo J, Avena M (2006) Kinetics of phosphate adsorption on goethite: comparing batch adsorption and ATR-IR measurements. J Colloid Interf Sci 300:511–518

Lynch JP, Brown KM (2008) Root strategies for phosphorus acquisition. In: White PJ, Hammond JP (eds) The ecophysiology of plant-phosphorus interactions. Springer, Dordrecht, pp 83–116

Magid L, De Arambarri P (1985) Adsorption of phosphate by two iron oxides in relation to their porosity. J Soil Sci 36:523–530

Marschner H (1995) Mineral nutrition of higher plants. Academic, New York, pp 447–476

Marschner H, Haussling M, George E (1991) Ammonium and nitrate uptake rates and rhizosphere pH in non-mycorrhizal roots of Norway spruce [*Picea abies* (L.) *Karst.*]. Trees 5:14–21

Marschner P, Solaiman Z, Rengel Z (2005) Growth, phosphorus uptake, and rhizosphere microbial-community composition of a phosphorus-efficient wheat cultivar in soils differing in pH. J Plant Nutr Soil Sci 168:343–351

Marschner P, Solaiman Z, Rengel Z (2006) Rhizosphere properties of *Poaceae* genotypes under P-limiting conditions. Plant Soil 283:11–24

Martin RR, Tazaki K, Smart RStC (1988) Direct observation of phosphate precipitation in the goethite/phosphate system. Soil Sci Soc Am J 52:1492–1500

McGrath JW, Wisdom GB, McMullan G, Larkin MJ, Quinn JP (1995) The purification and properties of phosphonoacetate hydrolase, a novel carbon-phosphorus bond-cleaving enzyme from *Pseudomonas fluorescens* 23F. Eur J Biochem 234:225–230

McGrath JW, Hammerschmidt F, Quinn JP (1998) Biodegradation of phosphonomycin by *Rhizobium huakuii* PMY1. Appl Environ Microbiol 64:356–358

Mehlich A (1978) New extractant for soil test evaluation of phosphorus, potassium, magnesium, calcium, sodium, manganese, and zinc. Commun Soil Sci Plant Anal 9:477–492

Mimura T (1999) Regulation of phosphate transport and homeostasis in plant cells. Intl Rev Cytol 191:149–200

Mitsukawa N, Okumura S, Shirano Y, Sato S, Kato T, Harashima S, Shibata D (1997) Overexpression of an *Arabidopsis thaliana* high-affinity phosphate transporter gene in tobacco cultured cells enhances cell growth under phosphate-limited conditions. Proc Natl Acad Sci 94:7098–7102

Miyasaka SC, Habte M (2001) Plant mechanisms and mycorrhizal symbiosis to increase phosphorus uptake efficiency. Commun Soil Sci Plan 32:1101–1147

Muukkonen P, Hartikainen H, Lahti K, Sarkela A, Puustinen M, Alakukku L (2007) Influence of no-tillage on the distribution and lability of phosphorus in Finnish clay soils. Agric Ecosys Environ 120:299–306

Oelkers EH, Valsami-Jones E (2008) Phosphate mineral reactivity and global sustainability. Elements 4:83–87

Olsen SR, Cole CV, Watanabe FS, Dean LA (1954) Estimation of available phosphorus in soils by extraction with sodium bicarbonate. United States Department of Agriculture Circular 939. United States Government Printing OYce, Washington, DC

Otani T, Ae N (1999) Extraction of organic phosphorus in andosols by various methods. Soil Sci Plant Nutr 45:151–161

Parets-Soler A, Pardo JM, Serrano R (1990) Immunocytolocalization of plasma membrane H^+-ATPase. Plant Physiol 93:1654–1658

Perrot KW, Sarathchandra SU, Waller JE (1990) Seasonal storage and release of phosphorus and potassium by organic matter and microbial biomass in a high producing pastoral soil. Aust J Soil Res 28:593–608

Pierzynski GM, McDowell RW, Sims JT (2005) Chemistry, cycling, and potential moment of inorganic phosphorus in soils. In: Sims JT, Sharpley AN (eds) Phosphorus: agriculture and the environment. American Society of Agronomy/Crop Science Society of America/Soil Science Society of America, Inc, Madison, pp 53–86

Pratt J, Boisson A-M, Gout E, Bligny R, Douce R, Aubert S (2009) Phosphate (Pi) starvation effect on the cytosolic Pi concentration and Pi exchanges across the tonoplast in plant cells: an in vivo ^{31}P-nuclear magnetic resonance study using methylphosphonate as a Pi analog. Plant Physiol 151:1646–1657

Rahmatullah MA, Torrent J (2000) Phosphorus dynamics and uptake by wheat in a model calcite-ferrihydrite system. Soil Sci 165:803–812

Rao MA, Gianfreda L, Palmiero F, Violante A (1996) Interactions of acid phosphatase with clays, organic molecules and organo-mineral complexes. Soil Sci 161:751–760

Richardson AE, Hadobes PA, Hayes JE (2001) Extracellular secretion of *Aspergillus* phytase from *Aradiopsis* roots enables plant to obtain phosphorus from phytase. Plant J 25:641–6496

Richardson AE, George TS, Hens M, Simpson RJ (2005) Utilization of soil organic phosphorus by higher plants. In: Turner BL, Frossard E, Baldwin DS (eds) Organic phosphorus in the environment. CABI, Wallingford, pp 165–184

Richey JC (1983) The phosphorus cycle. In: Bolin B, Cook RB (eds) The major biogeochemical cycles and their interactions. Wiley, New York, pp 51–56

Richter DD, Markewitz D (2001) Understanding soil change. Cambridge University Press, Cambridge

Rouached H, Arpat AB, Poirier Y (2010) Regulation of phosphate starvation responses in plants: signaling players and cross-talks. Mol Plant 3:288–299

Ryden JC, Syers JK (1975) Rationalization of ionic strength and cation effects on phosphate sorption by soils. J Soil Sci 26:395–406

Schindler TM (1998) The new view of the primary cell wall. Z Planzenernahr Bodenk 161:499–508

Schwertmann U, Taylor RM (1989) Iron oxides. In: Dixon JB, Weed SB (eds) Minerals in soil environment, 2nd edn. Soil Science Society of America, Madison, pp 379–438

Sentenac H, Grignon C (1985) Effect of pH on orthophosphate uptake by corn roots. Plant Physiol 77:136–141

Sharpley AN, McDowell RW, Kleinman PJA (2004) Amounts, forms, and solubility of phosphorus in soils receiving manure. Soil Sci Soc Am J 68:2048–2057

Shen J, Rengel Z, Tang C, Zhang F (2003) Role of phosphorus nutrition in development of cluster roots and release of carboxylates in soil-grown *Lupinus albus*. Plant Soil 248:199–206

Shu L, Shen J, Rengel Z, Tang C, Zhang F, Cawthray GR (2007) Formation of cluster roots and citrate exudation by *Lupinus albus* in response to localized application of different phosphorus sources. Plant Sci 172:1017–1024

Stefanovic A, Ribot C, Rouached H, Wang Y, Chong J, Belbahri L, Delessert S, Poirier Y (2007) Members of the *PHO1* gene family show limited functional redundancy in phosphate transfer to the shoot, and are regulated by phosphate deficiency via distinct pathways. Plant J 50:982–994
Strauss R, Brummer GW, Barrow NJ (1997) Effects of cristallinity of goethite: II. Rates of sorption and desorption of phosphate. Eur J Soil Sci 48:101–114
Styles D, Coxon C (2007) Meteorological and management influences on seasonal variation in phosphorus fractions extracted from soils in western Ireland. Geoderma 142:152–164
Tang C, Rengel Z (2003) Role of plant cation/anion uptake ratio in soil acidification. In: Rengel Z (ed) Handbook of soil acidity. Marcel Dekker, New York, pp 57–81
Tarafdar JC, Marschner H (1994) Phosphatase activity in the rhizosphere and hydrosphere of VA mycorrhizal wheat supplied with inorganic and organic phosphorus. Soil Biol Biochem 26:387–395
Turner BL, Paphazy M, Haygarth PM, McKelvie ID (2002) Inositol phosphates in the environment. Phil Trans R Society Lond B 357:449–469
Turner BL, Richardson AE, Mullaney EJ (2007) Inositol phosphates: linking agriculture and the environment. CAB International, Wallingford
Uusitalo R, Ylivainio K, Turtola E, Kangas A (2007) Accumulation and translocation of sparingly soluble manure phosphorus in different types of soils after long-term excessive inputs. Agric Food Sci 16:317–331
van Beuschichem ML, Nelemans JA, Bienfait HF (1988) Interrelationship between trans-plasma membrane electron/proton transfer stoichiometry, organic acid metabolism, and nitrate reduction in dwarf beans (*Phaseolus vulgaris*). Plant Physiol 86:269–273
Vance CP, Uhde-Stone C, Allan DL (2003) Phosphorus acquisition and use: critical adaptations by plants for securing a nonrenewable resource. New Phytol 157:423–447
Versaw WK, Harrison MJ (2002) A chloroplast phosphate transporter, PHT2;1, influences allocation of phosphate within the plant and phosphate-starvation responses. Plant Cell 14:1751–1766
Wang B, Shen J, Tang C, Rengel Z (2008) Root morphology, proton release, and carboxylate exudation in lupin in response to phosphorus deficiency. J Plant Nutri 31:557–570
Weinstein DA, Yanai RD (1994) Integrating the effects of simultaneous multiple stresses on plants using the simulation model TREGRO. J Environ Qual 23:418–428
White RE (1979) Retention and release of phosphate by soil and soil constituents. In: Tinker PB (ed) Soils and agriculture, vol II, Critical reports on applied chemistry. Blackwell Scientific Publications, Oxford, pp 71–114
Yadav RS, Tarafdar JC (2001) Influence of organic and inorganic phosphorus supply on the maximum secretion of acid phosphatase by plants. Biol Fertil Soils 34:140–143
Yadav RS, Tarafdar JC (2003) Phytase and phosphatase producing fungi in arid and semi-arid soils and their efficiency in hydrolyzing different organic P compounds. Soil Biol Biochem 35:745–751
Zaidi A, Khan MS, Ahemad M, Oves M (2009) Plant growth promotion by phosphate solubilizing bacteria. Acta Microbiol Immunol Hungar 56:263–284

Animal Manure for Smallholder Agriculture in South Africa

Alfred Oghode Misaiti Okorogbona and Lasisi Olaosebikan Adebisi

Abstract In South Africa, the annual quantity of animal manure is sufficient to meet 13.3%, 9.9% and 27.6% of the country N, P and K soil requirements, respectively. While 25% of the estimated 3 million tons of animal manure is used as fertilizers, most of the remaining 75% is wasted, with a small portion used as energy in heating. Animal manure is thus underutilized. Indeed, most resource-poor farmers have little technical know-how of manure use. This chapter reviews smallholder agriculture in the South African context. We discussed the meaning, uses and role of African leafy vegetables in the lives of resource-poor farmers. Animal manure is an organic source of nutrients for leafy vegetable production among resource poor farmers. Cattle, goat and chicken manures are the common types of animal manure used as nutrient sources. We discussed the classification of manure, the quality of manure and the effect of application rates of animal manure on crop growth and yield.

In South Africa, black farmers dominate the smallholder farming system. The goal of smallholders is to increase their agricultural production for commercialization and in turn having a better livelihood. Unfortunately, this goal is limited by the availability of resources and socio-economic factors. African leafy vegetables are consumed mainly by black people who are the main producers of the leafy vegetables. They use primarily animal manure for soil amendment source in fertilizing their cropped land. Rural dwellers also harvest wild vegetables. Among organic sources of nutrients, ruminant manure and chicken manure are of major concern. The leafy vegetables provide economic, nutritional and medicinal advantages to the rural resource poor blacks.

This review highlights that animal manure contains essential nutrients required by plant. Animal manure is heterogeneous in nature and its quality as fertilizer is affected by many factors. Ruminant manure differ from chicken manure. Chicken manure

---

A.O.M. Okorogbona (✉) • L.O. Adebisi
Department of Crop Sciences, Center for Organic and Smallholder Agriculture,
Tshwane University of Technology, P Bag X680, Pretoria 0001, South Africa
e-mail: alfredoghode@yahoo.com

E. Lichtfouse (ed.), *Farming for Food and Water Security*, Sustainable Agriculture Reviews 10, DOI 10.1007/978-94-007-4500-1_9,

contains higher concentration of nutrient than ruminant manure, particularly nitrogen. Manures have positive effects on crop growth and yield, with chicken manure applied at rate lower than those of ruminants for optimum yield. Crops differ in nutrient requirements and the optimum application rate of animal manure is crop and manure specific. Amaranth thrived well and had considerable yield while the crop was yet to attain its peak biomass production at application rates of 9.22 t poultry manure ha^{-1}, 10 tha^{-1} sheep manure, 10.30 t goat manure ha^{-1} and 11.7 cattle manure ha^{-1}. Chinese cabbage was yet to reach peak biomass production at the application rates of 3.42 t layer chicken manure ha^{-1}, 11.9 t goat kraal manure ha^{-1} and 23.8 t cattle manure ha^{-1}. Biomass production of nightshade continued to increase at 68.25 tha^{-1} of goat manure, 170 tha^{-1} cattle manure and peaked at 17 t chicken manure ha^{-1}. Pumpkin reaches peak biomass production at 8.53 t chicken manure ha^{-1} while the peak biomass production was yet to be attained at 170 t cattle kraal manure ha^{-1} and 68.25 t goat kraal manure ha^{-1}.

Keywords Smallholder agriculture • African leafy vegetables • Animal manure • Manure quality • Manure nutrients • Manure application

1 Introduction

The use of animal manure is a wide-spread ancient practice for maintaining soil productivity, which continue to gain popularity, especially in cost-effective integrated smallholder farming systems of developing countries. Prior to the widespread use of chemical fertilizer in South Africa, animal manure was used throughout the country. As reported by Malherbe (1964), about 1.5 million tons of farm manure used to be applied to soils annually. The district of Malmesbury in the Western Cape, South Africa, had the highest manure utilization (Malherbe 1964). Animal manure continues to be available as fertilizer. According to Fertilizer Society of South Africa (1989), in 1983, about 3 million tons of animal manure was available from feedlots- cattle, broiler and pigs, in South Africa. Based on the N, P and K contents, it was estimated that the monetary value amounted to about R29,700,000. It was also estimated that the quantity of animal manure produced annually in South Africa was sufficient to meet 13.3%, 9.9% and 27.6% of the country's N, P and K soil requirements respectively. While 25% of the estimated 3 million tons were used as fertilizers (FSSA 1989), the bulk of the remaining 75% was mostly wasted, with a small portion used as energy in heating (Mkhabela and Materechera 2003; Mkhabela 2004). As postulated by Mkhabela (2004), the numbers are not expected to have changed since then.

According to Van Averbeke (2008), report on livestock population provided by Department of Agriculture in 1995, indicated that there were 4,651,946 million cattle, 4,087,435 million goats and 3,302,659 million sheep held by South African smallholder sector. Muller et al. (2006) reported that a live cow in Holstein herd (replacement cattle) in South Africa weighed 533 ± 71 kg. Wilkerson et al. (1997)

Table 1 Estimates of quantity of manure produced from livestock raised by smallholder farmers in South Africa

Parameter	Animal manure species		
	Cattle	Goats	Sheep
Year	1995	1995	1995
Population of animals (million)	4,651,946	4,087,435	3,302,659
Live weight of animal (kg)	350.0–500.0	30.0	45.4
Manure excreted per animal (tons)	1.30	0.11	0.32
Manure produced by total population (tons)	6,047,530	899,236	985,843

Sources: (Peacock 1996; Wilkerson et al. 1997; Muller et al. 2006; Van Averbeke 2008; Shoenian 2010; Synman 2010; Gwaze et al. 2010; Almeida 2011)

NB Indications from Table 1 shows that manure from cattle was the most available, followed by manure from sheep and the least was manure from goats despite the estimated population of goats being higher than sheep population. Although, information obtained indicates that sheep excreted more quantity of manure than goats

pointed out that a growing and replacement cattle excrete approximately 2.63 t of manure on a dry matter basis yearly. Since cattle are left to graze during the day and only kept in the kraal at night, only manure produced in kraal could be accumulated for use. Based on calculations, 50% of the daily excretion per cow could be estimated as 1.315 kg of manure, which meant that by practical implication, in 1995, South African smallholder cattle sector produced approximately 6,047,530 tons of cattle manure.

A live goat in South Africa weighed between 33.96 and 38.60 kg (Synman 2010; Gwaze et al. 2010). Peacock (1996) pointed out that goat weighing 30 kg excretes approximately 0.22 tons of manure on a dry matter basis depending on the feed intake and digestibility of the diet during the year. Since manure excreted by goats reared in South African smallholder sector is mainly obtainable from the kraals while the animals remain in kraal, basically at night. Based on calculations, 50% of the daily excretion per goat could be estimated as 0.603 kg of manure, meaning by practical implication, in 1995, South African smallholder goat sector produced approximately 899,236 tons of goat manure.

According to Almeida (2011), the average weight of Damara sheep (fat-tailed sheep), considered to be the most internationalized breed of sheep in South Africa and other parts of Southern region as a whole, had average weight of 46.6 kg. With reference to the postulation of Shoenian (2010) who pointed out that a 45.4 kg sheep excretes 0.597 tons of manure annually. Since manure excreted by sheep reared in South African smallholder sector is obtained similarly as in the case of goats, while the animals remain basically in enclosures at night. Based on calculations, 50% of the daily excretion per sheep could be estimated as 1.64 kg of manure, meaning by practical implication, in 1995, approximately 985,843 tons of sheep manure was produced from South African smallholder sheep sector. Estimates of the quantity of manure produced from macro-livestock species kept by smallholders in South Africa in 1995 are presented in Table 1. These data are used in this review based on the assumption that the numbers estimated are not expected to have changed much since 1995.

In 2005, 624 million broilers with a live weight of 1.94 kg and a carcass weight of 1.32 kg were slaughtered (Germishuis 2006). According to Shortall and Liebhardt (1975), a broiler typically produces 1.225 kg of manure on a dry matter basis during its life cycle. Similar results were reported by Van Averbeke (2008), who found that broilers had produced 1.512 kg of dry manure by the time they had reached the live weight of 2.3 kg after 7 weeks, which was a higher live weight than the industry standard of about 2 kg. By implication, in 2005, the South African poultry sector produced approximately 0.75 million metric tons of poultry manure on a dry matter basis. Despite this huge quantity of animal manure availability, agricultural production continue to be on a nose diving trend in Africa.

Manure from animal origin play a significant role in sustainable crop production (Dakora and Keya 1997; Mkhabela and Materechera 2003; Bationo et al. 2006; Van Averbeke 2008; Materechera 2010). Animal manure is a useful organic nutrient source obtained from livestock and poultry production. The use of chemical fertilizers increases crop growth and yield but their high costs (Khaliq et al. 2004; Kimetu et al. 2004; Duada et al. 2008; Fening et al. 2010; Mkhabela and Materechera 2003; Ogbona and Umar-Shaaba 2011) and the inaccessibility of these inorganic nutrient sources by the marginal low input farmers, in addition to the present attention given to organic food products which necessitate the need to revive organic farming that is presently taking a worldwide dimension, has reawaken the use of solely organic nutrient sources by large and small scale crop producers. These factors continue to make the use of animal manure to be the most tenable and affordable organic soil amendment source among African resource poor farmers. In Britain, Wilkinson (1979) reported that about 42% of N, 29% of P and 57% of K in fertilizer could be replaced by manure if all the potentially available manure produced by animals housed in confinements were spread on cropland. Despite this enormous quantity and the potential of these organic materials, animal manure remain underutilized while Africa continue to be in distress by a widening gap between food production and human population growth coupled with a declining agricultural production as aforementioned, leading to food inadequacy and unsustainable development. A school of thought opined that the under-utilization could be due to the misgivings and complexities associated with the use of animal manure while other scientists postulated that the under-utilization is associated with lack of technical know-how of the scenery which encompasses manure phenomena by majority of resource poor African farmers. Hence review on some of the research works that dealt with various aspects of animal manure becomes imperative as the application of manure from different types of animal requires thorough understanding of the relationship between crop responses and availability of nutrients in the soil following animal manure incorporation.

Most studies on the use of animal manure in Africa have awarded limited information to a canopy of researches comprising the main users of these organic nutrient sources in developing countries, the classification of manure, types of manure commonly used in African smallholder crop production, importance of animal manure in crop production, quality of animal manure as a fertilizer, effects of application rates of animal manure on crop growth and yield. Instead, most studies on the use of animal manure for crop production in Africa have focused mainly on the addition of these organic nutrient sources to soil for improved crop yield and as such, these research works are blanket to most African agrarian setting, with the assumption that a smallholder farmer is basically the anonym of a commercial farmer; manure from animal origin can be classified as the same; all types of animal manure have the same quality; animal manure can be used for soil amendment at the same application rates; nutrient availability and nutrient release pattern are the same, for all types of manure. It is in line with the foregoing that we have reviewed a few research works to unveil a few complexities surrounding the use of animal manure, to contribute to past review works carried out on the use of animal manure in African smallholder crop production through the alignment of the elucidation from various researchers to a development trajectory on the aforementioned misconceptions about animal manure in general. This review focused on the context of the concepts that encompasses animal manure and its use for crop production. These include nutrition of African leafy vegetables with special reference to the use of animal manure as fertilizer, which is the common practice among smallholders in Africa; meaning and types of animal manure used in African smallholder crop production with particular reference to South African smallholder; characteristics and quality of manure accumulating in kraals; effects of animal manure application rate on crop growth and yield; nutrient release from animal manure and forms of availability of nutrients in animal manure.

2 Smallholder Farming in South Africa

The smallholder farming sector in South Africa is very diverse and therefore difficult to define. As a standalone concept, a smallholding was broadly defined 'as a farm enterprise that is small relative to one or other local or universal norm' (Kirsten and van Zyl 1998). However, in South Africa, the term smallholder is commonly used to refer to a black farmer, probably because the majority of black farmers operate enterprises that are small in size relative to those of white farmers (Machete et al. 2004). The link between size of holding and race was also pointed out by Pote (2008) who described the smallholder sector as being dominated by black households which produced on relatively small plots of land with limited resources for the purpose of household subsistence or sale. Pote's (2008) description added two new elements to the South African concept of smallholder, namely, the prevalence of resource limitations and the inclusion of subsistence as one of the purposes of farming. Consequently, Kirsten and Van Zyl (1998) defined South African smallholder as a black farmer operating on a relatively small piece of land with limited resources for subsistence or commercial purposes or both.

For social and economic reasons, which have political roots, South African smallholders are constrained by technical and institutional factors, such as insufficient farm inputs and farm equipment, inappropriate transportation infrastructure and high transaction costs (Pote 2008). In Southern Africa, smallholders often practise mixed farming, which involves the cultivation of crops and the rearing of animals (Nyamangara et al. 2005; Materechera and Modiakotla 2006). South African case studies in KwaZulu-Natal Province (Adey 2006), the Eastern Cape Province (Yoganathan et al. 1998; Van Averbeke et al. 2008) and North West Province

(Materechera and Modiakotla 2006) showed that cattle farming typically dominated the livestock component of this mixed farming system and maize, the crop component. Besides the production and consumption of grain crops, black smallholder communities in South Africa also consume vegetables (Mkhabela and Materechera 2003; Hebinck and Monde 2007; Faber et al. 2010).

In recent years, following the end of the apartheid era in South Africa, three categories of smallholders were identified (Department of Agriculture 2001; Van Averbeke and Mohammed 2006). Studies carried out in South Africa by Van Averbeke and Mohammed (2006) indicated that smallholders form a large and diverse group of farmers operating with strategies that are in line with their farming objectives. Estimates of the human population that engaged in smallholder farming activities, as reported by Niewoudt (2000) indicated that about 2.1 million black small-scale farmers were in South Africa in the year 1999.

The review divulged that smallholder farming system in South Africa is predominantly dominated by blacks whose aspirations are geared towards increasing the commercialization of their agricultural production activities, which are hindered by limited availability of resources and a number of socio economic factors including low level of income, poverty and lack of good educational background.

3 Leafy Vegetables

South Africa is endowed with wide varieties of biodiversity among which are wild vegetable species of remarkable nutritious qualities (Reinten and Coetzee 2002; Lewu and Mavengahama 2010). These vegetables include those growing in the wild. Mnzava (1995) defined vegetables “as all categories of plants of which the leaves, fruits or roots are acceptable and used as food by rural and urban communities”. Among the different vegetables consumed by black communities in South Africa, leafy vegetables have been shown to be of considerable importance (Mnzava 1995; Maundu 1995; Jansen van Rensburg et al. 2007; Faber and Wenhold 2007; Vorster et al. 2008). Leafy vegetables are plant species of which the leafy parts, which may include young, succulent stems, flowers and very young fruit, are used as a vegetable (Jansen van Rensburg et al. 2007).

In South Africa, people obtain leafy vegetables in different ways. Most commonly, the vegetables are harvested from the wild or from crop lands where these vegetables grow as weeds but some leafy vegetable species are obtained exclusively through cultivation (Jansen van Rensburg et al. 2007). These leafy vegetables include indigenous, indigenized and introduced plant species (Jansen van Rensburg et al. 2007). Maundu (1995) defined an indigenous crop as a crop whose natural home is known to be in a specified region. Within the context of Africa, Schippers (2002) and Laker (2007) defined indigenized species as species that originated in other continents, especially Asia, South and Central America, but have become part and parcel of traditional African food culture and Agriculture. Maundu (1995) described introduced species as species whose origin is known to be outside a specified region. Therefore, an indigenous or introduced species which due long use has become part of the culture of a community is termed traditional (Africa in our case) African leafy vegetable (Asfaw 1995; Maundu 1995). Faber et al. (2010) and Jansen van Rensburg et al. (2007) defined African leafy vegetables as "the collective of leafy vegetable species that form part of the culinary repertoire of particular contemporary African communities". This definition captured the meaning of the dynamic concept of "imifino" (isiZulu and isiXhosa) or "*merogo*" (Sesotho and Setswana), used by black people in South Africa to refer to the green leafy vegetables they consumed. Own production of African leafy vegetables was advocated as part of a food security strategy aimed at combating micronutrient deficiencies among poor African people residing in the rural areas (Faber et al. 2007; Jansen van Rensburg et al. 2007; Maunder and Meaker 2007). As a result, different agencies with an interest in rural development, including the South African Department of Agriculture and the Water Research Commission (Backeberg and Sanewe 2010), have been promoting the production of African leafy vegetables as an important step towards alleviating nutritional deficiencies among affected communities in South Africa.

In South Africa, there has been growing interest in the development of African leafy vegetables into crops for use by local smallholders. One of the reasons is that producing these plants could improve the diet of members of smallholder households, particularly children, by providing a rich source of micronutrients, such as vitamin A, iron and zinc (Faber and Wenhold 2007). Another reason is that these plants, specifically the indigenous and indigenized species, appear to be well adapted to local growing environments and are thought to have lower input requirements than exotic leafy vegetables (Mnzava 1995; Eyzaguirre 1995; Van den Heever 1995; Diouf et al. 2007; Faber and Wenhold 2007). A third reason is that the production of these plants creates new economic activity and sources of income for rural people (Ramrao et al. 2006; Law-Ogbomo and Ajayi 2009).

With specific reference to South Africa, Jansen van Rensburg et al. (2007) reported that eight leafy vegetable species were of particular importance, namely *Amaranthus cruentus* (*Amaranthus cruentus* L.); spider flower (*Cleome gynandra* L.); Jew's mallow (*Corchorus olitorius* L.); pumpkin (*Cucurbita maxima* Duchesne); cowpea (*Vigna unguiculata* (L.) Walp.); non-heading Chinese cabbage (*Brassica rapa* L. subsp. *chinensis*); nightshade (*Solanum retroflexum* Dun.) and tsamma melon (*Citrillus lanatus* Thunb.). Overall, *Amaranthus cruentus* was shown to be the most commonly consumed African leafy vegetable in South Africa, Faber et al. (2010) but *Amaranthus cruentus* was rarely cultivated. Spider flower and jute were two other popular species that were rarely cultivated (Jansen van Rensburg et al. 2007). Commonly cultivated species included Chinese cabbage, nightshade, cowpea, pumpkin and to a lesser extent tsamma melon (Jansen van Rensburg et al. 2007).

3.1 Leafy Vegetable in the Diets of Low or Marginal Income Bracket of Human Population

Hunger and under-nutrition remains a global problem. The latest estimate indicates that a total of 925 million people globally are undernourished, most of them living in Asia, the Pacific and sub-Saharan Africa (FAO 2010). Acute under-nutrition causes wasting and chronic under-nutrition results in stunting. It is the latter type of under-nutrition that is prevalent in South Africa (Wenhold and Faber 2008). Stunting tends to be associated with hidden hunger, which refers to chronic micronutrient deficiencies in the human diet. Globally, the most important micronutrient deficiency disorders are those of vitamin A, iron, iodine and zinc (Wenhold and Faber 2008). Hidden hunger also affects a substantial proportion of the South African human population. For example, in 1994, one-third of all South African children between the ages of 6 months and 71 months old suffered from vitamin A deficiency (Wenhold and Faber 2008). The rate of hidden hunger is highest in rural areas and is linked to poverty (Faber and Wenhold 2007). Maunder and Meaker (2007) argued that the prevalence of hidden hunger amongst poor rural people was the result of diets that consisted mainly of starchy plant foods and lacked dietary diversity.

Foods of animal origin, including dairy products, liver and egg yolk are the richest sources of vitamin A and bio-available iron but for reasons of affordability, such foods are often beyond the reach of poor rural dwellers in Africa (Faber and Wenhold 2007). Consumption of home-grown yellow-or orange-fleshed non-citrus vegetables and dark-green leafy vegetables is one of the ways in which this population group can meet its vitamin A requirements and at least part of its iron requirements (Faber and Wenhold 2007). Generally, dark-green leafy vegetables are known to be rich sources of micronutrients, such as beta-carotene, thiamine, vitamin C, calcium, non-haem iron, folic acid and riboflavin (Singh et al. 2001; Faber and Wenhold 2007; Odhav et al. 2007).

3.2 *General Significance of African Leafy Vegetable in South Africa*

In South Africa, leaves of majority of these vegetables are consumed as a cooked vegetable (Jansen van Rensburg et al. 2007; Faber et al. 2010). According to Fox and Norwood Young (1982), the leaves of the vegetables may be dried and stored for use in winter. Apart from their uses as leafy vegetables, some of the vegetables serve as herbal remedy for people with hypertension and cardiovascular disease (Martirosyan et al. 2007). The leaves could also be used to treat anaemia and diabetes (Ogar and Asiegbu 2005; Jansen van Rensburg et al. 2007). Some of the plants are dried and burnt into ash for use as potassium fertilizer (Grubben et al. 2004). Berries of some of the vegetable plants are also used as dye (Poczai and Hyvonen 2011).

This review indicated that a wide variety of vegetables including the indigenous, indigenized and traditional vegetables found in South Africa provide advantages to the rural resource poor farmers by providing income for women for improved livelihood, adding nutritional value to the diets of the resource poor people.

4 Nutrient Supply in Crop Production

Nutrients used for crop production activities are mainly supplied to plants through soil reserves, atmospheric nitrogen from biological fixation, aerial deposition caused by rain and wind, mineral fertilizers and organic input sources. Soil contains natural reserves of plant nutrients in quantities that depend on soil composition and stage of weathering. Soil reserves are often in forms unavailable to plants and only a minor portion is released each year through biological activity or chemical processes. Although, it was pointed by Dakora and Keya (1997) that biological nitrogen fixation is certainly the cheapest and the most effective tool for maintaining sustainable yields in African agriculture but according to FAO (1998), the quantity of nutrient from soil reserves and biological fixation are far less than the amount that could compensate for agricultural production to meet up with the increasing demand for food due to reasons associated with their slow release pattern. In addition, the ability to fix nitrogen is not common to all crops but a factor associated majorly to leguminous crops (Stacey et al. 1992; Tikhonovich et al. 1995 and Postgate 1998). However, in order to supply agricultural production activities with nutrients to produce yield that would meet up with the increasing demand for food, farmers have to supply nutrients through mineral fertilizer and organic nutrient sources (Okwu and Ukanwa 2007).

According to FAO (1998), mineral fertilizers are inorganic substances applied to soil, irrigation water or a hydroponic medium to supply plants with nutrients. These are also referred to as chemical fertilizers. Dakora and Keya (1997) pointed out that the increased use of N from chemical fertilizers has resulted in significant increase in food production world-wide. Low soil fertility is a primary constraint in African crop production (Dakora and Keya 1997; Kimetu et al. 2004; Sanchez and Swaminathan 2005; Okorogbona 2011). Africa's consumption of mineral fertilizers per unit cropped land is the lowest in the world, ranging from 2.2% to 3.9% of the global expenditure compared to 3.6–7.3% for South America, 30.4–43.1% for Asia, and 31.4–42.4% for Europe from 1982 to 1990 (Dakora and Keya 1997). High cost and poor marketing and distribution infrastructure are factors that have contributed to the low use of chemical fertilizers in African agriculture (Rautaray et al. 2003; Okwu and Ukanwa 2007). All these factors have collectively limited N fertilizer use by subsistence and smallholder farmers throughout Africa (Dakora and Keya 1997). All these have made animal manure the most prominent organic source of nutrients among resource poor farmers in developing countries. This was confirmed by the postulation of Chang et al. (2008) who reported that globally, animal manure used to be the most important organic nutrient source employed by farmers to increase the fertility of crop land. This review revealed that nutrients supplied to arable soils through natural means are not enough for crop production activities that would meet up with the increasing demand for food, due to the world increasing human population. This resulted to the use of organic and inorganic fertilisers for increased agricultural production but the high cost and low accessibility of the latter hinder its use among the marginal low input farmers throughout the globe.

5 Animal Manure in Smallholder Vegetable Production

African smallholders are aware of the need to augment the fertility of their cropped soils and use local resources, usually organic materials, for this purpose. Many African smallholders use animal manure as fertilizers. In the South African smallholder sector, which at present is largely responsible for the production of African leafy vegetables and also constitutes the target group for recommendations on the production of these vegetable species, animal manure has long been the primary way in which plant nutrients are returned to cultivated soils (Yoganathan et al. 1998; Materechera and Modiakotla 2006; Mkhabela 2006; Van Averbeke 2008; Okorogbona et al. 2011). Animal manure is classified as organic fertilizer. Kuun et al. (1997) described organic fertilizers as natural materials consisting of organic matter usually from plants or plant residues, which have decomposed or have been digested and excreted as wastes by animals.

5.1 *Classification of Animal Manure*

Animal manure can be classified based on whether it contains only animal excreta, excreta and bedding materials, or excreta and water. Accordingly, animal manure is classified as animal excreta, farmyard manure and slurry manure.

5.2 *Animal Excreta*

Bowman (2009) defined animal excreta as "the excrement that has been naturally produced by animals as a routine part of their digestion, passed out of their body and deposited purely in their holding area without mixing with any other material." Examples include manure from goats, cattle, sheep, layer chickens, turkeys and pigs deposited on floors of animal pens in the absence of bedding materials and running water.

5.3 *Farmyard Manure*

FAO (1998) defined farmyard manure as a mixture of animal excreta, often including urine and litter used for bedding. Examples include manure obtained from dairy cattle and broiler chickens.

5.4 *Slurry Manure*

Johnston et al. (2002) defined slurry manure as "animal waste consisting of water, typically 90–95% by volume, feed particles, feces and urine." Apart from the aforementioned constituents, it may also contain one or more products that were added to limit odour. According to Johnston et al. (2002), most of these products are found in one of the following categories: acidifying agents such as acids, base precipitating salts and substrates that induce acid production, absorbents including clinoptilolite or peat. These might also include saponins, e.g. yucca [*Yucca* spp.] plant extract, masking agent. Also, mixture of aromatic oils, digestive additives such as selected microbial strains and enzymes. Among these products are disinfectants in the form of surfactants, chlorine, orthodichlorobenzene and chemical oxidising agent, e.g. H_2O_2, O_3, $KMnO_4$.

5.5 *Animal Manure as a Resource in Crop Production*

Animal manure play a significant role in crop growth and yield, both from the perspective of crop cultivation and even crops growing in the wild, for example Van Averbeke et al. (2007a, b) reported that vegetables including nightshade were found growing under trees where bird droppings were common in the wild.

5.6 *Importance of Animal Manure*

Vitosh et al. (1973) pointed out that the benefits of animal manure applications in crop farming are well known. The main positive effect of applying animal manure is that it improves growth and yield of crops as a result of plant uptake of the nutrients contained in the manure (Maerere et al. 2001; Gosh et al. 2004; Kihanda et al. 2004). Also important is the effect of raising organic matter content of cropped soils, which improves their production capacity (Fatunbi and Ncube 2009). Manure has a greater effect on soil organic matter content and related soil properties than chemical fertilizer when applied at the same elemental nutrient rate because manure is an exogenous source of organic matter (Edmeades 2003). Typically, soils amended with animal manure have lower bulk densities and higher porosities, hydraulic conductivities and aggregate stabilities than soils amended with chemical fertilizers (Edmeades 2003; Gilley and Eghball 2002; Esminger 1971; Mahimairaja et al. 1995). Soils amended with animal manure also have larger populations of microfauna than chemically fertilized soils and are more enriched in phosphorus (P), potassium (K), calcium (Ca) and magnesium (Mg) in the topsoil and nitrate (NO_3)-nitrogen (N), Calcium (Ca) and Magnesium (Mg) in the subsoil. Animal manure can be beneficial in alleviating micronutrient deficiencies in soils because they contain these nutrient elements or by generating chelating agents, which are known to assist in the solubilization of insoluble micro-nutrients present in the soil, thereby rendering them more available to plants (Wilkinson 1979). Van Averbeke and Yoganathan (1997) pointed out that even though animal manure contains the different nutrients needed by plants; these nutrients are rarely present in the proportions desired by plants.

5.7 *Types of Animal Manure Used in South African Smallholder Crop Production*

The available evidence indicates that South African smallholders mainly use animal manure to maintain the fertility of their crop land. The two most commonly used types of animal manure appeared to be ruminant kraal manure and chicken manure (Harris 2002; Materechera and Mkhabela 2002; Mkhabela and Materechera 2003; Mphaphuli 2003; FAO 2005; Materechera and Modiakotla 2006; Mkhabela 2006; Van Averbeke 2008; Materechera 2010). According to Mkhabela (2006) and Van Averbeke (2008), cattle and goats produce the bulk of the manure accumulating in animal kraals belonging to South African smallholders. In some parts of the country, such as the Eastern Cape Province, manure from sheep is also important (Mkile 2001). Mkhabela (2006) and Van Averbeke (2008) reported increased use of chicken manure by smallholders, who sourced this type of manure from large- and small-scale chicken production enterprises operating in their neighbourhood or through commercial distribution networks.

Kraal manure is an organic material that is composed mainly of the excretions of domestic animals, such as feces, urine and plant residues that have accumulated on the floor of a livestock enclosure (Archer 1988; Van Averbeke and Yoganathan 1997). Kraal manure is usually categorized on the basis of the livestock species contained in the kraal. Accordingly, reference is made to cattle, sheep and goat kraal manure.

The term chicken manure is used to refer to the excrements that are produced by chickens (*Gallus domesticus*) (Gilmour et al. 2004). Chicken manure is a form of poultry manure. Lu and Edwards (1994) differentiated between two main types of chicken manure, namely broiler litter and layer manure. Lopez-Mosquera et al. (2008) and Tyson and Cabrera (1993) defined broiler litter as a heterogeneous organic material that consists of the combination of excrements, feathers, waste feeds and any bedding material that was used during the process of rearing the chickens. Bedding materials used in broiler production include wood shavings, peanut hulls, soybean pods, sunflower husks and seed coats (Lu and Edwards 1994). According to Lu and Edwards (1994), broiler litter typically has a moisture content of approximately 20%.

Layer manure or more specifically layer cage manure was defined by Lu and Edwards (1994) as the excreta of layer chickens that are kept in cages. Layer cage manure is free of bedding materials and typically contains about 75% moisture (Lu and Edwards 1994). In some cases, layers are produced using deep litter systems. In these systems, layers roam freely on the floor of the poultry house, which is covered with bedding material. Excrements and bedding material are allowed to accumulate over a fairly long period of time explaining the term deep litter. Promis®, chicken manure, commonly used in South Africa is an example of manure obtained from layers reared by means of a deep litter system.

5.8 *Mineral Composition in Animal Manures*

The mineral composition of animal manure varies from one animal to the other. Variation in constituents of manure occur within animals of the same species and even animal raised in different geographical locations with varying vegetations. For example in the case of poultry manure where the compositions of layer manure varies from broiler litter due to production system. Evidence of this was pointed out by Mkhabela (2004) who reported that on an average, broiler litter contained 29 kg N ton^{-1}, 12.9 kg P ton^{-1} and 12.5 kg K ton^{-1} and layer manure contained 30.5 kg N ton^{-1}, 9.6 kg P ton^{-1}, and 11.3 kg K ton^{-1} on a wet basis. Wild (1988) reported that dry farmyard manure in Britain normally contains 1.7–2.5%N, 0.2–0.5% P and 1.3–2.0%K while Van Averbeke (2008) reported that on an average, dry kraal manures of cattle, goats and sheep found in the rural areas of South Africa contained 1.64% nitrogen (N), 0.36% phosphorus (P) and 1.58% potassium (K). The slight differences in the reports provided on the nutrient contents of manure from Britain and South Africa, indicated that manure also varies with geographical location perhaps pointing out a possible relationship between vegetation type in a particular region and manure quality (Mkile 2001). Table 2 shows the differences in the nutrient composition of manure obtained from cattle, poultry and goats on dry weight basis. This review pointed out that the main soil nutrient source used by smallholders is predominantly manure from animal origin. Animal excreta in its natural state, manure from farm yard and manure mixed with water used in animal pens include the various classes of organic nutrient materials used for soil fertility by farmers. The main advantage of animal manure is that farmers use the nutrient source to improve soil for increased crop growth and yield. It was also revealed that manure contain all essential nutrients required by plant even though the nutrients are not in same quantity as those available in synthesized fertilizers.

6 Quality of Animal Manure for Use as Fertilizer

Materechera and Modiakotla (2006) used the concept "manure quality", which was defined as 'the value of manure in improving soil properties and enhancing crop yields', to integrate the various factors that affected the worth of manure for use as soil amendment or fertilizer. A wide range of factors that affect manure quality have been identified. Many of these factors are related to one another, because they result from the animal husbandry system that is used and the methods adopted during the collection and storage of manure.

Considering that type of livestock is fundamental to any livestock production system, it made sense to relate the review of different factors that affect manure quality to livestock type. Since the current review was primarily concerned with the manure of ruminants and chicken, factors affecting the quality of the manure obtained from these two types of livestock are discussed accordingly.

Table 2 Nutrient composition of three types of animal manure

Property	Unit	Poultry manure	Cattle kraal manure	Goat kraal manure
pH[a]		6.9	9.2	9.7
Moisture[b]	%	12.1	7.3	9.0
Solids	%	88.0	92.7	91.0
Mineral content	%	15.5	55.0	44.3
Organic matter content	%	72.4	37.7	46.7
Total organic carbon	%	35.5	17.5	25.0
Mineral content (dry)	%	17.7	59.3	48.7
Total nitrogen	%	3.7	1.7	2.2
Carbon/Nitrogen ratio		9.6	10.3	11.3
Calcium (Ca)	%	2.5	1.9	2.3
Magnesium (Mg)	%	0.7	0.9	1.0
Phosphorus (P)	%	1.5	0.5	0.4
Potassium (K)	%	1.5	1.7	4.0
Sodium (Na)	%	0.5	0.3	0.4
Aluminium (Al)	%	0.1	1.1	1.1
Iron (Fe)	%	0.2	1.1	1.0
Copper (Cu)	mg kg^{-1}	99.1	34.8	39.6
Manganese (Mn)	mg kg^{-1}	759.5	336.8	310.0
Zinc (Zn)	mg kg^{-1}	545.3	99.3	94.3

After Okorogbona et al. (2011)

NB Data from Table 2 indicates that N was most available in chicken manure, followed by goat manure and manure from cattle had the least amount of N. P was highest in chicken manure, followed by the P in cattle manure and the least amount of P was available in goat manure. Goat manure had the highest amount of K, followed by cattle manure and chicken had the least amount of K

[a]The sample to water ratio used for pH measurement was1 g manure: 5 ml water

[b]With the exception of mineral content, all values in the table are reported on an air dry basis and are based on the dried and milled samples provided to the Institute of Soil, Climate and Water – Agricultural Research Council, South Africa, 2008

6.1 Quality of Ruminant Manure Accumulating in Kraals or Animal Enclosure

Kraal is a terminology used for an enclosure where livestock, particularly ruminants including cattle, sheep and goats are kept in the Southern African context. The word 'Kraal' originated from the language called Afrikaans, which is one of the languages predominantly spoken by South African whites. Different types of kraals common among South African smallholder farmers are presented in Fig. 1.

Type of livestock, age of the animal, livestock diet, mineral soil particle content, weed seed content, nutrient content, lignin and polyphenol contents, C:N and C:P ratios of the manure, the conditions under which manure accumulates and managed during storage and composting have all been identified as factors that affect the quality of manure that accumulates in ruminant kraals. Wide variability in quality indicators is characteristic of the manure found in the ruminant kraals of smallholders, even when only one type of ruminant is being reared and the production system is fairly uniform. This was demonstrated by the results of a survey of the quality of

Fig. 1 Structure of ruminant kraals: (**a**) a kraal used for keeping free ranging animals at night (kraal belong to a cattle farmer at Tweefontein: a black settlement in Mpumalanga Province, South Africa); (**b**) a kraal used by a smallholder farmer for keeping cattle in confinement during the day and in the night (kraal belonging to a black farmer at KwaMhlanga in Mpumalanga Province); (**c**) a combination of sheep and goat (kraal belonging to a smallholder black farmer in Moloto village at the outskirt of Pretoria metropolis); (**d**) a goat held by young herd keepers inside kraal used for housing sheep and goats

manure stored on 300 small-scale dairy farms in the central Kenyan highlands (Lekasi et al. 2003). All 300 farmers had dairy cattle of the Friesland or Ayrshire breed or crosses between these breeds and reared their dairy cattle using a zero grazing system (Lekasi et al. 2003). Yet, the manure quality indicators that were determined were highly variable. For example, P contents ranged between 0.06% and 0.75%, N contents between 0.33% and 1.91%, K contents between 0.43% and 7.00%, organic carbon content between 6.5% and 49.2% and the C:N ratio between 5.3 and 81.0 (Lekasi et al. 2003). Wide variability in quality indicators of manure stored in the ruminant kraals of South African smallholders were also reported by Yoganathan et al. (1998), Mkile (2001) and Materechera and Modiakotla (2006). Another characteristic of ruminant manure stored on smallholdings is the general absence of statistically significant associations between individual causal factors and quality indicators, emphasizing that the quality of manure produced and stored on smallholdings is the result of a complexity of factors (Lekasi et al. 2003).

The general effect of type of ruminant on the nutrient content of the manure accumulating in smallholder kraals in South Africa was estimated by Van Averbeke (2008), using various local studies that reported on this factor. The findings which

Table 3 Average nutrient content of manure found in ruminant kraals of smallholders in South Africa

	Nutrient content (dry matter basis) (%)		
Type of ruminant	Nitrogen	Phosphorus	Potassium
Cattle	1.45	0.40	1.28
Goats	1.87	0.30	1.75
Sheep	1.62	0.29	1.92
Average	**1.65**	**0.33**	**1.65**

After Van Averbeke (2008)
NB Of all these three types of manure from ruminant kraals-cattle manure had the most P, least N and K; goat manure had the most N, more P and K; sheep manure had the most K, more N and least P

are summarized in Table 3, indicates that among the three types of ruminant manure, cattle kraal manure tended to have the lowest N and K contents but the highest P content. Goat kraal manure tended to have the highest N content, whilst sheep kraal manure tended to have the highest K content. However, a study of 140 manure samples collected in the border region of the Eastern Cape Province of South Africa failed to identify any statistically significant association between type of ruminant and nutrient content of manure stored in kraals (Yoganathan et al. 1998) indicating that the effect of type of ruminant on the nutrient content of kraal manure could be concealed by other factors.

Age of ruminants is known to affect the nutrient content of their excreta. Young animals reportedly use more nitrogen, phosphorus and calcium for building their bodies than mature animals, causing the content of the dung of young animals to be lower in these nutrients than in the case of mature animals (Malherbe 1964). However, in smallholder agriculture, animals of different ages share the same enclosure resulting in the mixing of their excreta and the obscuring of possible animal age effects on the nutrient content of excreta.

Diet of ruminants has been shown to affect the nitrogen, phosphorus, lignin and polyphenol contents of their excreta. The proportion of nutrients excreted by animals tends to increase as the concentration of nutrients in the feed ingested is increased. For example, Lekasi et al. (2003) found that supplementing the Napier grass diet of dairy cattle with concentrates and licks significantly raised the P content of manure stored in the kraals of smallholders in the central Kenyan highlands. Malherbe (1964) stated that the bulk of the N excreted by ruminants occurred in the urine but studies have shown that type of feed ingested plays a role in the partitioning of excreted N into urine-N and feces-N. For example, comparing the N content of the feces and urine of free ranging sheep diet with that of sheep on a diet containing cowpea leaves, Harris (2002) found that the N content of the feces of free ranging sheep was higher than that of sheep that were fed with cowpea leaves but the N content of the urine of the free ranging sheep was lower than that of sheep on the cowpea leaf diet. Rufino et al. (2006), who reviewed studies that determined the partitioning of dietary N into milk, meat and excreta in ruminants, concluded that diet affected both the recovery of dietary N in the excreta and the partitioning of excreted N into urine-N and fecal-N. Relationships between excreted fecal fiber and consumed forage fiber have been reported for small ruminants and for dairy cattle. According to Powell et al. (2009), N contained in ruminant feces consists of endogenous N and undigested fiber-N. Powell et al. (2009) found that when ruminant manure was applied to the soil, the endogenous N mineralized quickly, whilst fiber-N degraded more slowly.

Mineral soil particle content, which dilutes the concentration of nutrients in manure, was identified as one of the main factors explaining variability in the nutrient content of manure stored in the ruminant kraal of South African smallholders (Van Averbeke and Yoganathan 1997). For example, Mkile (2001) reported that the mineral soil particle content of manure stored in cattle, goat and sheep kraals in the Transkei region of the Eastern Cape Province in South Africa ranged between 21% and 68%. In North West Province of South Africa, Materechera and Modiakotla (2006) recorded mineral soil particle contents ranging between 15% and 50% in cattle kraal manure. The origin of soil in manure has been ascribed to soil particles sticking to the hooves of livestock being deposited in the kraals where they get mixed with the manure (Materechera and Modiakotla 2006)

Weed seed content of animal manure has an adverse effect on manure quality because it increases labour requirements in crop production. Materechera and Modiakotla (2006) reported that weed populations in soils amended with cattle kraal manure were on average 15 times higher than in soil that received no kraal manure addition.

The forms in which nitrogen, phosphorus and potassium occur in ruminant manure affect manure quality. The key nitrogenous constituent of urine in ruminants of high protein diet is usually urea but some ammonia, allotoin, creatinine and creatine is also present (Wilkinson 1979). According to Wilkinson (1979), most of the nitrogen in the urine of ruminants is readily available because much of it occurs in inorganic, highly soluble forms, such as ammonia and urea, which are about as available for plant uptake, as the N present in chemical fertilizers. Much of the N excreted in the feces requires mineralization before it is available for crop use. Wilkinson (1979) reported that 45–65% of the total N excreted in the feces of sheep and cattle formed part of alpha-amino acids. The bulk of phosphorus excreted by ruminants is contained in the feces (Malherbe 1964). Wilkinson (1979) pointed out that the P content in the feces of ruminants tended to increase as the P content of the feed was increased and this was confirmed by Kimani and Lekasi (2004) who pointed out similar findings. According to Wilkinson (1979), between 36% and 58% of the P in ruminant manure is plant available with the inorganic phosphorus being present as calcium hydrogen phosphate ($Ca\ (HPO_4)_2 2H_2O$). Malherbe (1964) stated that much of the potassium in ruminant manure was water soluble and plant available. This was confirmed by Wilkinson (1979) who reported that most of the potassium excreted by ruminants was contained in the urine. Wilkinson (1979) estimated that between 75% and 97% of the total potassium contained in ruminant manure was immediately available for plant use.

Storage conditions affect the quality of ruminant manure. These conditions include whether the housing units used for the animals are covered with roofing or not and whether the manure is stored without intervention or turned regularly. According to Harris (2002), covering livestock pens reduces losses in the organic

Table 4 Effect of composting on the components concentration and ratios of fresh and composted cattle manure

Properties	Fresh manure	Composted manure
Total carbon (kg t^{-1})	314	161
Total nitrogen (kg t^{-1})	16.1	14.0
Total phosphorus (kg t^{-1})	4.6	5.2
Carbon-nitrogen ratio	19.7	11.3
Nitrogen-phosphorus ratio	3.5	2.7

After Larney et al. (2006)
NB Composting had the greatest effect on carbon, N content was reduced slightly and phosphorus content increased

Table 5 Physicochemical properties of fresh and composted manure

Properties	Fresh cattle manure	Composted
pH	7.70–8.94	8.86–8.07
Total nitrogen (g kg^{-1})	23.9±9	22.0±0.3
Available P (mg kg^{-1})	211±6	342±22
Total carbon (g kg^{-1})	399.2±2.8	384.9±2.7
Carbon-nitrogen ratio	17.0±0.74	17.5±0.33

After Lazcano et al. 2008
NB Composting reduced carbon and nitrogen contents and increased phosphorus content

content of manure compared to manure accumulating in open kraal where animals and their excreta are exposed to direct sunlight and rainfall. Lekasi et al. (2003) reported that turning manure increased its mineral N concentration and lowered its C:N ratio compared to stored manure that was not turned.

The age of ruminant manure affects its quality because the composition of ruminant manure changes during composting (Malherbe 1964). This is illustrated in Table 4, which compares the composition of fresh manure with that after composting the same manure over a period of 135 days (Larney et al. 2006). Table 4 shows that composting had the greatest effect on carbon, which was reduced to half its original concentration. Nitrogen content was reduced slightly and phosphorus content was increased. Similar trends were evident in an experiment that involved fresh cattle manure and same cattle manure that had a 15-day composting period as shown in Table 5, containing results provided by Lazcano et al. (2008). The results in Table 5 again show that composting reduced carbon and nitrogen contents and increased phosphorus content.

6.2 *Quality of Chicken Manure Obtained in Intensive Production Systems*

Figure 2 represents a typical smallholder intensive poultry system where chicken manure accumulation is collected for soil amendment purpose to grow vegetables in

Fig. 2 Structure of a smallholder poultry house where vegetable garden farmers collects chicken manure for soil amendments: (**a**) external view of the poultry house; (**b**) chicken birds inside the poultry house

South African rural vegetable garden setting. The quality of chicken manure is also variable. Type of chicken, diet, presence or absence of bedding material, conditions under which the manure is stored and age of the manure are among the factors that affect the quality of chicken manure (Wilkinson 1979; Chardwick et al. 2000; Mkhabela 2004; Amanullah et al. 2010; Hossain et al. 2010).

Type of chicken, diet and the presence or absence of bedding material are often linked because they form part of the production system. In intensive production systems, broiler chickens, which are reared for meat, are usually produced on the floor of the poultry house, which is covered with bedding material and the birds are given diets in which the protein content is high, ranging between 18% and 22%, depending on the growth stage of the broilers (Aviagen 2002). The layer cage system is commonly used in intensive egg production (Shini 2003). In this system, the excreta accumulate on the floor below the cages without bedding material being applied (Amanullah et al. 2010). Layer diets typically contain 16% protein (Koelkebeck et al. 2006; Koelkebeck and Anderson 2007), which is lower than the protein content of broiler diets, but layer diets contain more calcium than broiler diets (Jacobus 2004). In some cases, layers, particularly those kept for reproduction purposes, are kept on floors covered with a layer of bedding material of about 10–15 cm deep. In this system, which is referred to as the deep litter system, the excreta accumulate over a longer period than in the case of broiler production (Amanullah et al. 2010). Effects of production system on the nutrient content of chicken manure are shown in Table 6. From Table 6, layer cage manure is richer in most nutrients because of the absence of bedding material. Broiler litter tends to be higher in N than deep litter manure. This could be due to the high protein content of broiler diets relative to layer diets or to differences in the proportion of bedding material contained in the manure. The calcium content of cage layer manure and deep litter manure was similar to that of broiler litter, suggesting that much of the additional calcium contained in layer diets was used in the production of egg shells (Jacobus 2004).

Table 6 Average nutrient content of chicken manure obtained by means of three different production systems

	Chicken production system		
Properties	Deep litter	Broiler	Layer cage
Total N (%)	1.70–2.20	2.40–3.60	3.68–5.30
C:N ratio	9.5–11.5	9.4–11.2	5.8–7.6
Total P (%)	0.62–0.79	0.68–1.22	0.67–1.26
Total K (%)	0.77–1.08	1.16–1.92	2.08–2.41
Ca (%)	0.90–1.10	0.86–1.11	0.80–1.02
Mg (%)	0.45–0.68	0.42–0.65	0.40–0.56
Zn (ppm)	90–308	160–315	290–460

After Amanullah (2007) and Amanullah et al. (2010)
NB Data shows that layer cage manure is richer in most nutrients. Broiler litter tends to be higher in N than deep litter manure

Table 7 Effect of composting period on the nitrogen, phosphorus and potassium contents of chicken manure

Composting period (days)	Nitrogen (%)	Phosphorus (%)	Potassium (%)
0	2.25	2.45	2.23
30	2.05	2.43	2.00
60	1.50	1.94	1.54
90	1.22	1.75	1.15
120	0.83	1.62	0.80
135	0.50	0.94	0.75

After Hossain et al. (2010)
NB Nitrogen, phosphorus and potassium contained in the chicken manure declined during storage and composting processes

The nutrient content of chicken manure also varies considerably depending on the conditions under which the manure is stored and processed. Amanullah et al. (2010) reported that conservation of N was better under anaerobic storage condition than under aerobic condition because less N was lost through volatilisation due to exposure to air. The nutrient content of chicken manure and the availability of the N contained in chicken manure tend to decline during the storage and composting process (Amanullah et al. 2010). Evidence of the loss of nutrients during composting was provided by Hossain et al. (2010) who investigated the effect of composting period on the N, P and K contents of chicken manure. Their findings are presented in Table 7. Wilkinson (1979) reported that 61% of the nitrogen excreted by poultry was in the form of uric acid, which he considered to be plant available and 94% of the P in broiler litter and 88% of the P in cage layer manure was available for plant use. Wilkinson (1979) also indicated that the proportion of potassium contained in chicken manure that was available for plant use was more or less the same as in the case of ruminant manure, that is 75–97%.

This review revealed that the variability in the quality of manure obtained from animals could be due to various factors which are dependent on animal type, age, diet, materials used for bedding in animal houses, content of weed seeds consumed during grazing and browsing, nutrient content of the manure, lignin and polyphenol contents, C:N and C:P ratios of the manure, the conditions under which manure accumulates and managed during storage and composting.

7 Effects of Animal Manure Application Rate on Crop Growth and Yield

Effective and efficient utilisation of animal manure in the production of crops requires thorough understanding of crop response to application rate of animal manure, because application of animal manure can have both positive and negative effects on plant growth and yield depending on the application rate used (Kotze and Joubert 1992; Maerere et al. 2001; Azeez et al. 2010). Positive crop responses to increases in the application rate of animal manure are obtained as a result of increased quantities of nutrients being available for plant uptake (Maerere et al. 2001; Gosh et al. 2004; Kihanda et al. 2004). As was pointed out earlier, the various types of animal manure differ in terms of nutrient content and the immediate plant availability of the nutrients contained in these nutrient sources. Release of the nutrients that are not immediately available in manure depends on the microbial processes of organic matter decomposition, mineralization and immobilization (Wild 1988; Ma et al. 1999; Silva et al. 2006; Fageria 2009). The process of decomposition which leads to the release of nutrients from complex insoluble substances such as animal manure is a function of edaphic and biotic factors and microorganisms play a significant role in these processes (Esse et al. 2001; Okorogbona et al. 2011). Mineralization of N refers to the transformation of organic N to mineral N while immobilization of N refers to the incorporation of N into the tissues of the decomposers (Okorogbona et al. 2011). Nitrogen release from animal manure appears to be the critical factor determining crop response to application rate. Indications are that when the addition of animal manure results in sufficient plant available N being released, adequate plant availability of phosphorus and potassium is also ensured (Mkhabela et al. 2001; Sharifi and Taghizadeh 2009). The rate of N-mineralization of manure is a function of particle size, C/N ratio and in some cases the lignin and polyphenol/N ratio of the animal manure (Chardwick et al. 2000; Agehara and Warcnke 2005; Abbasi et al. 2007). According to Sanchez et al. (2001), animal manure with low N and high concentrations of lignin and polyphenol decomposes and releases N slowly while those rich in N and low in lignin and polyphenols decompose rapidly. Particle size of animal manure and plant residue affect the surface area of the N source exposed to micro organisms (Harris 1988; Agehara and Warcnke 2005).

The negative effect of applying animal manure on crop growth and yield mostly arises from applying manure at rates that are too high. Using application rates that are excessive can cause the building up of soil salinity (Lu and Edwards 1994) and soil alkalinity (De Campos et al. 2004). In the case of poultry manure, it appears that fatty acids present in the manure could have a toxic effect on germinating seeds and young seedlings (Fujiwara et al. 2009). Applying animal manure at excessive rates above plant absorbable level is also known to cause environmental pollution, such as ground water contamination (Sims and Bitzer 1988).

7.1 *Effects of Rate of Application of Ruminant Manure on Crop Growth and Yield*

The available evidence suggests that growth and yield of crops are progressively improved as the rate of ruminant manure is considerably raised to plant optimum requirement level (Reddy et al. 2000; Mhlonto et al. 2007; Azeez et al. 2010). In the case of single applications and their effect on the subsequent crop, the positive crop response to increases in application rate appears to extend to fairly high levels of at least 50 $t\,ha^{-1}$ and as high as 170 $t\,ha^{-1}$, as was demonstrated by Azeez et al. (2010) in a pot experiment using cattle kraal manure applied to a low fertility soil. When ruminant manure is applied annually, it would appear that an application rate of about 20–25 $t\,ha^{-1}$ is optimal (Vitosh et al. 1973). Over a 9 year period, the use of rates higher than this apparent optimum of up to 67 $t\,ha^{-1}$ $annum^{-1}$ did not have a negative effect on crop yield but resulted in net additions to the soil of all nutrients contained in the manure, which was wasteful and posed an environmental hazard (Vitosh et al. 1973). Fraser et al. (2006) referred to studies where negative effects of single applications of ruminant manure on the growth and yield of the subsequent crop were recorded. Fraser et al. (2006) indicated that such effects only occurred when the application rate exceeded 200 $t\,ha^{-1}$ but pointed at one experiment in which the crop response to application rate of cattle manure was positive up to the rate of 312 $t\,ha^{-1}$. One of the problems with studies that deal with application rate effects is that it is not always clear whether the rates used refer to manure as in its wet form or to the dry substance.

7.2 *Effects of Rate of Application of Chicken Manure on Crop Growth and Yield*

As in the case of ruminant manure, the growth and yield of crops are progressively improved as the rate of chicken manure is increased (Goh and Vityakon 1983; Ogbonna and Obi 2007; Azeez et al. 2010). In the case of single applications and their effect on the subsequent crop, the positive crop response to increases in application rate appears to extend to about 8 $t\,ha^{-1}$ and as high as 55 $t\,ha^{-1}$ for certain crops. Azeez et al. (2010) reported that application rate of 8.5 t chicken manure ha^{-1}

(316 kg N ha^{-1}, 126 kg P ha^{-1} and 153 kg K ha^{-1}) was optimal for the production of pumpkin (*Cucurbita maxima* Duchesne). Duada et al. (2008) found the rate of 9.9 t chicken manure ha^{-1} (369 kg N ha^{-1}, 147 kg P ha^{-1} and 179 kg K ha^{-1}) to be optimal for the production of water melon. Farren et al. (1993) reported that the application rate of 10.8 t chicken manure ha^{-1} (403 kg N ha^{-1}, 160 kg P ha^{-1} and 195 kg K ha^{-1}) was optimal for strawberry production and 12.8 t ha^{-1} (478 kg N ha^{-1}, 190 kg P ha^{-1} and 231 kg K ha^{-1}) for grape production. Law-Ogbomo and Ajayi (2009) identified the application rate of 12 t chicken manure ha^{-1} as the optimum rate for the production of amaranth. Azeez et al. (2010) reported the application rate of 17 t chicken manure (633 kg N ha^{-1}, 252 kg P ha^{-1} and 307 kg K ha^{-1}) to be optimal for the production of nightshade (*Solanum retroflexum* Dun.). Lu and Edwards (1994) reported the application rate of 26 t chicken manure ha^{-1} (969 kg N ha^{-1}, 386 kg P ha^{-1} and 469 kg K ha^{-1}) to be optimal for the production of collard seedlings and Shortall and Liebhardt (1975) found that the application of 55 t chicken manure ha^{-1} produced the highest maize yield in the first year of application. Information on the long term effects of application rate of chicken manure on crop production involving annual application is limited but in a field experiment with maize, Shortall and Liebhardt (1975) recorded a reduction in yield when chicken manure was applied at the rate of 55 t ha^{-1} compared to the application rate of 22 t ha^{-1} from the third season onwards.

Negative effects on plant growth resulting from applying chicken manure at rates that are too high are particularly common. Indications are that in the case of first time application, these negative effects started to occur at rates of about 10 t chicken manure ha^{-1} onwards, depending on the crop and probably other factors, such as soil type and manure properties. Azeez et al. (2010) reported a decline in the biomass production of pumpkin when the application rate of chicken manure was increased from 8.5 to 17 t ha^{-1} and a decline in the biomass production of nightshade (*Solanum retroflexum* Dun.) when the application rate of chicken manure was increased from 17 to 34 t ha^{-1}. Lu and Edwards (1994) reported that collard seedlings were severely damaged within 7 days after transplanting when the application rate of poultry manure exceeded 58 t ha^{-1} (2,163 kg N ha^{-1}, 862 kg P ha^{-1} and 1,046 kg K ha^{-1}). Philip et al. (2009) reported a reduction in the yield of fluted pumpkin when the application rate of chicken manure was increased from 40 to 80 t ha^{-1}. Shortall and Liebhardt (1975) recorded a reduction in the grain yield of maize when the rate of chicken manure was increased from 55 to 90 t ha^{-1}. The negative crop response to the addition of ruminant manure at high rates has been attributed to soil salinity (Shortall and Liebhardt 1975), ammonium toxicity (Goh and Vityakon 1983) and in the case of poultry manure, it appears that fatty acids present in the manure could have a toxic effect on germinating seeds and young seedlings (Fujiwara et al. 2009).

This review indicated that the use of animal manure for soil amendment had positive and negative effect to plant growth. The positive effect was that it resulted in increased in crop yield when used adequately and the negative effect was that over application to soil resulted in the hindrance of seed germination, reduced crop growth and crop yield. Also, among the negative effects was the contamination of soil and water when applied above plants critical point or optimum level. For soils amended with manure from ruminants, the application rates could range between 10 and 300 $t\,ha^{-1}$. Revelation from this review showed that chicken manure could be applied to soil used for crop growth at rates ranging between 5.5 and 22 $t\,ha^{-1}$. These application rates are highly dependent on crop type, crop nutrient absorbable level and time of planting following manure incorporation into the soil.

8 Selected African Leafy Vegetables, Uses, Nutrient Requirements and Effect of Animal Manure Application Rate on Their Yield

Ensuring adequate availability of nitrogen, phosphorus and potassium is the key concern of soil fertility management practices in most cropping systems (Prasad and Power 1997; Tshikalange 2007). Nitrogen, phosphorus and potassium are the three most important macro nutrients required for plant growth and development (Jang et al. 2008). In Africa, the low N and P contents of soils limit plant production (Sanchez and Swaminathan 2005). Crop species and even varieties of the same crop species are known to differ in terms of their nutrient requirements (Grattan and Grieve 1999; Marten and Abdoellah 1988; Pospisil et al. 2006).

For logistic reasons, the review was limited to four of the eight African leafy vegetable species commonly grown by smallholders in South Africa. In the selection of these crops, the primary criterion was whether the crop was being cultivated or not. For this reason, Chinese cabbage, nightshade and pumpkin were included in the selection. The decision to select *Amaranthus cruentus* instead of cowpea or tsamma melon was motivated by the apparent dominance of *Amaranthus cruentus* among the various leafy vegetables consumed in South Africa, the indication that cowpea was mainly cultivated for its grain and the lesser importance of tsamma melon as a cultivated species (Vorster et al. 2008).

8.1 Amaranth (Amaranthus cruentus L.)

The *Amaranthus* species that featured in this review is presented in Fig. 3. Several *Amaranthus* species are used as leafy vegetables including *A. thunbergii* (L), *A. cruentus* (L), *A. greazicans* (L), *A. spinosus* (L), *A. deflexus* (L), *A. hypochondriacus* (L), *A. viridus* (L), *A. hybridus* (L) and *A. caudatus* (Van den Heever and Coertze 1996; Jansen van Rensburg et al. 2007) but in South Africa, *A. cruentus* is the species most commonly used for the harvest of its leaves (Jansen van Rensburg et al. 2007). *A. cruentus* is known as *Amaranthus cruentus*, cockscomb and hell's curse in English (Jansen van Rensburg et al. 2007).

Amaranthus cruentus belongs to the Amaranthaceae and is an erect to spreading annual herbaceous plant (Grubben et al. 2004; Van Wyk 2005; Jansen van Rensburg et al. 2007). The height of *Amaranthus cruentus* ranged between 30 and 38 cm,

Fig. 3 Amaranth (*Amarathus cruentus*)

5–7 weeks after sowing (Makinde et al. 2010; Omolayo et al. 2011). The inflorescences of *Amaranthus cruentus* occur terminally and auxiliary (Jansen van Rensburg et al. 2007). *Amaranthus cruentus* has small seeds that are usually shiny and dark brown to black (Jansen van Rensburg et al. 2007).

Amaranthus cruentus has a C4 cycle of photosynthesis and grows optimally under warm conditions with day temperatures above 25°C and night temperature not lower than 15°C, bright light and adequate availability of plant nutrients (Schippers 2002; Jansen van Rensburg et al. 2007; Faber et al. 2010). *Amaranthus cruentus* is said to be tolerant to adverse climatic condition and drought but sensitive to photoperiod and starts to flower as soon as day length shortens (Jansen van Rensburg et al. 2007). Prolonged dry spells induce flowering and reduce leaf production of the plant (Jansen van Rensburg et al. 2007). According to Place et al. (2008), *Amaranthus* species have well developed root systems that have the ability to penetrate through compacted soil layers.

Among the weeds that feature as leafy vegetables in South Africa, *Amaranthus cruentus* probably has the best potential to be developed into a cultivated crop (Jansen van Rensburg et al. 2007). This potential was confirmed by recent case studies conducted by Faber et al. (2010) in the Limpopo and KwaZulu-Natal Provinces, South Africa which showed that *Amaranthus cruentus*, was persistently the most widely consumed African leafy vegetable in the country. The extent of cultivation of *Amaranthus cruentus* in South Africa have been somehow controversial. Fox and Norwood Young (1982), Van den Heever (1995) Jansen van Rensburg et al. (2007) and Faber et al. (2010) all pointed out that *Amaranthus cruentus* was rarely cultivated in South Africa. However, Jansen van Rensburg et al. (2007) indicated that in the Bushbuckridge area of the Limpopo and Mpumalanga Provinces of South Africa, women harvested and stored the seeds for broadcasting when a decline in the *Amaranthus cruentus* population growing as weeds in fields was observed, suggesting that cultivation of the crop did take place in South African smallholder communities, albeit men unsophisticated way. Grubben et al. (2004), on the other hand, stated that *Amaranthus cruentus* was grown commercially in South Africa for the purpose of fresh produce sales and canning. There is no doubt that in other parts of Africa, particularly West Africa, *Amaranthus cruentus* is a popular leafy vegetable that is cultivated all year round for both home consumption and commercial purposes by women in rural and urban areas (Law-Ogbomo and Ajayi 2009).

In South Africa, leaves, plant tips and whole seedlings of *Amaranthus cruentus* are consumed as a cooked vegetable (Jansen van Rensburg et al. 2007; Faber et al. 2010). According to Fox and Norwood Young (1982), the leaves are also dried and stored for use in winter. Apart from its use as a leafy vegetable, *Amaranthus cruentus* also serves as an herbal remedy for people with hypertension and cardiovascular disease (Martirosyan et al. 2007) and in the Republic of Benin, *Amaranthus cruentus* plants are dried and burnt into ash for use as potassium fertilizer (Grubben et al. 2004).

Harvesting of the leafy parts of *Amaranthus cruentus* is carried out at different growth stages, which extend from the young seedling stage to the late vegetative stage (Modi 2007). The harvesting of the consumable parts of this vegetable involves different practices, which include uprooting the entire young plants, cutting back established plants to encourage lateral growth and picking of the top part of stem and branches close to the growing point, a practice referred to as tipping (Schippers 2002; Jansen van Rensburg et al. 2007).

The fresh leaf yield of *Amaranthus cruentus* appears to vary widely from as low as 7 to up to 77 $t\,ha^{-1}$. At Rodeplaat, near Pretoria in South Africa, Van den Heever and Coertze (1996) recorded average fresh leaf yields of 20–25 $t\,ha^{-1}$, when different leafy amaranth species were harvested once and 30–60 $t\,ha^{-1}$, when harvested continuously. These fresh leaf yields were of the same order as the 40 $t\,ha^{-1}$ recorded in Tanzania and the 30 $t\,ha^{-1}$ obtained from 4-week old shoots of *Amaranthus cruentus* in the Benin Republic (Law-Ogbomo and Ajayi 2009). Using tipping to harvest leaves of *Amaranthus cruentus*, Chabalala and Van Averbeke (2011) reported fresh leaf yields ranging between 15 and 20 $t\,ha^{-1}$ under conditions of adequate water and plant nutrient availability in the Vhembe District (Limpopo Province, South Africa). In the USA, fresh leaf yields of *Amaranthus cruentus* as high as 77.27 $t\,ha^{-1}$ have been obtained (Law-Ogbomo and Ajayi 2009) but the world average is only 14.27 $t\,ha^{-1}$, whilst in Nigeria, the average fresh leaf yield obtained by farmers was reported to be 7.6 $t\,ha^{-1}$ (Law-Ogbomo and Ajayi 2009). Law-Ogbomo and Ajayi (2009) pointed out that for optimum growth and yield, *Amaranthus cruentus* required the application of fertilizers and needed to be planted at high planting densities.

Little research attention has been given to the response of amaranth to the application of animal manure. According to Schippers (2002), improved yield of amaranth can be expected from the application of organic fertilizers at rates that range between 20 and 40 t ha^{-1}. Schippers (2002) also indicated that amaranth preferred organic fertilizers that had been composted prior to application and warned against the application of poultry manure at high rates because this caused scorching of the crop. Schippers (2002) suggested that chicken manure should be mixed well with soil or diluted with water and be applied well before planting to avoid damage to the crop. According to Schippers (2002), the most appropriate use of organic fertilizers in the production of amaranth was the application of compost. This was identified as one of the main reasons why amaranth was so popular in urban and peri-urban areas where the use of compost was particularly common.

According to Maerere et al. (2001), biomass production of *Amaranthus cruentus* increased by applying three types of manure up to the highest application rates of 9.22 t poultry manure ha^{-1}, 10.30 t goat manure ha^{-1} and 11.7 cattle manure ha^{-1} that were included in the study. In a field experiment, Mhlonto et al. (2007) also reported steady increases in biomass production of an unclassified *Amaranthus* accession as the rate of application of sheep kraal manure was increased up to the highest rate treatment of 10 t ha^{-1} used in the experiment.

8.2 Chinese Cabbage (Brassica rapa L. subsp. chinensis)

Figure 4 represents the Chinese cabbage (*Brassica rapa* L. subsp. *chinensis*) that featured in this review. *Brassica rapa* L. subsp. *chinensis* is known as Chinese cabbage, rape or Chinese mustard cabbage in English and *mutshina* in Sepedi and Tshivenda, the local languages spoken where the vegetable is predominantly grown in South Africa (Jansen van Rensburg et al. 2007). Chinese cabbage belongs to the Cruciferae. *Brassica campestris*, the progenitor form of Chinese cabbage, is believed to have evolved in the Mediterranean area but the crop was developed by farmers in Eastern Asia (Jansen van Rensburg et al. 2007). Asian farmers developed two main types of Chinese cabbage, namely the heading type (*Brassica rapa* L. subsp. *pekinensis*) and the non-heading Chinese cabbage (*Brassica rapa* L. subsp. *chinensis*). Chinese cabbage is believed to have been introduced to Africa by Asian traders (Van Averbeke et al. 2007a, b). Non-heading Chinese cabbage is very popular among black people in the Limpopo and Mpumalanga Provinces of South Africa (Van Averbeke et al. 2007a, b). Van Averbeke et al. (2007a, b) reported that the Vhembe District in the north of Limpopo Province was the center of cultivation of non-heading Chinese cabbage and farmers in the District maintained an informal seed multiplication and distribution system. Non-heading Chinese cabbage is one of the few African leafy vegetables that have been developed into a fresh produce commodity in South Africa (Jansen van Rensburg et al. 2007).

Non-heading Chinese cabbage is an annual, flowering vegetable with dark leaves supported by light green to white petioles that form a rosette (Van Averbeke et al. 2007a, b). Chinese cabbage has a short growing season, which takes 6–11 weeks from planting to the end of the vegetative stage (Jansen van Rensburg et al. 2007).

Fig. 4 Non-heading Chinese cabbage (*Brassica rapa* L. subsp. *chinensis*)

Depending on the cultivar, the plant may attain a height of 15–30 cm at the end of the vegetative stage (Jansen van Rensburg et al. 2007).

Chinese cabbage is a cool season crop, which requires adequate availability of soil water and plant nutrients for optimum growth (Grubben et al. 2004; Van Averbeke et al. 2007a, b) and it does not tolerate poorly drained conditions (Jansen van Rensburg et al. 2007). Chinese cabbage is frost tolerant and its seeds are able to germinate at a temperature as low as 5°C (Grubben et al. 2004). The seed takes 3–5 days to germinate under optimum soil moisture and temperature conditions. The optimum temperature for germination ranges between 20°C and 25°C (Grubben et al. 2004). In Southern Africa, Chinese cabbage grows as a weed on waste dumps and along road sides at altitudes ranging from 1,500 to 3,000 m (Grubben et al. 2004). Van Averbeke and Netshithuthuni (2010) reported that non-heading Chinese cabbage was very sensitive to water stress and concluded that this sensitivity was probably due to its shallow and poorly developed root system. Similar conclusions were reached by Nndwammbi (2009) and Larkom (2007).

Non-heading types of Chinese cabbage are harvested as single leaves or as complete plants (Matsumura 1981; Peirce 1987; Van Averbeke and Netshithuthuni 2010). The fresh and processed leaves are used to prepare a relish to accompany maize porridge and the flowers together with the leaves attached to the peduncle are used as medicine to treat high blood pressure (Van Averbeke et al. 2007a, b).

Fresh leaf yields obtained from non-heading Chinese cabbage vary. Tindall (1983) reported that yields ranged between 5 and 30 $t\,ha^{-1}$, depending on the prevailing environmental conditions, the cultivar and planting density used and that farmers' yields were typically about 15 $t\,ha^{-1}$. Grubben et al. (2004) reported that the optimum fresh leaf yield of non-heading Chinese cabbage was 25 $t\,ha^{-1}$. Chabalala and Van Averbeke (2011) recorded a maximum fresh leaf yield of 22.2 $t\,ha^{-1}$ but Van Averbeke and Netshithuthuni (2010) obtained a fresh leaf yield as high as 39.1 $t\,ha^{-1}$.

In a greenhouse pot experiment that involved the growth of Chinese cabbage with three types of animal manure, the addition of these organic nutrient sources at

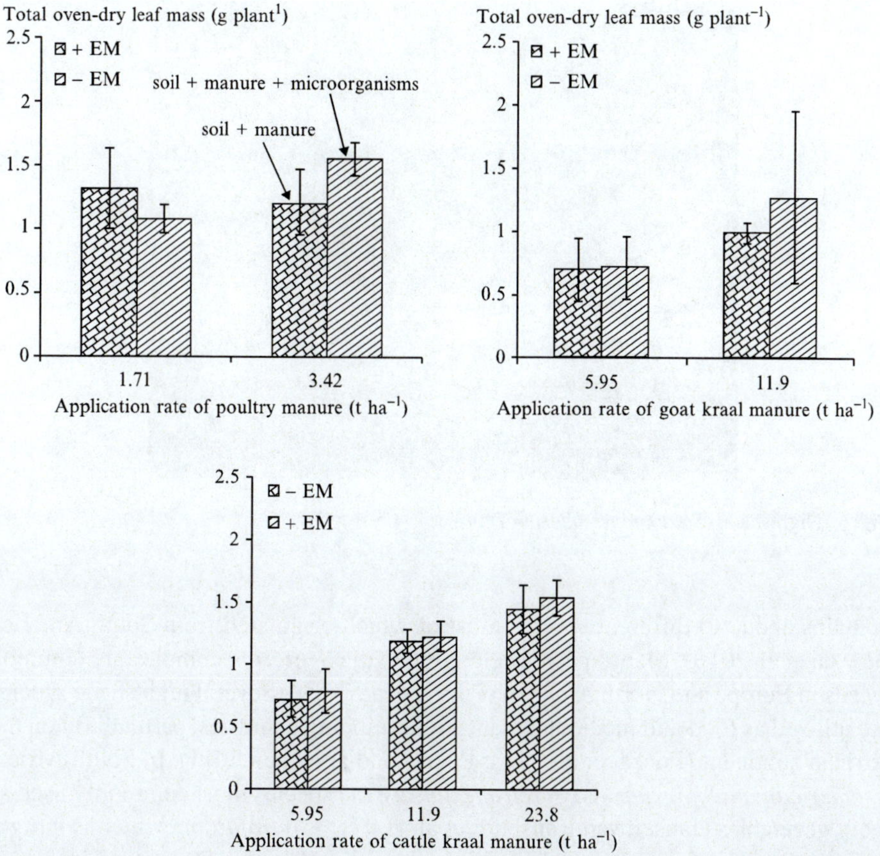

Fig. 5 Leaf yield response of non-heading Chinese cabbage to application rates of manure from poultry, goats and cattle *–EM* soil amended with manure only, *+EM* Soil amended with manure and effective microorganisms (after Okorogbona et al. 2011)

different levels of layer chicken manure, namely, Promis® applied at 1.71 and 3.42 t ha^{-1}, goat kraal manure (5.95 and 11.9 t ha^{-1}) and cattle kraal manure (5.95, 11.9, and 23.8 t ha^{-1}), Okorogbona et al. (2011) reported that biomass production of the vegetable responded positively to the application of these organic nutrient sources up to the highest rates of each of these organic materials incorporated into a low fertility soil. Evidence of this is indicated in Fig. 5 which shows the leaf yield response of non-heading Chinese cabbage to application rates of manure from poultry, goats and cattle.

8.3 Nightshade (Solanum retroflexum Dun.*)*

The Nightshade (*Solanum retroflexum* Dun.) that featured in this review is presented in Fig. 6. The *Solanum nigrum* L. complex, commonly known as black nightshade,

Fig. 6 Nightshade (*Solanum retroflexum* Dun.)

contains about 30 different species most of which originated from South America (Poczai et al. 2010). Many species belonging to the *S. nigrum* complex are common weeds in Europe and North America (Poczai et al. 2010). Several nightshade species are utilized as food and medicinal plants in developing countries, particularly on the African continent (Poczai et al. 2010; Poczai and Hyvonen 2011). In South Africa, *S. americanum*, *S nigrum* and *S. retroflexum* are the species most commonly used as leafy vegetables (Jansen van Rensburg et al. 2007). *S. retroflexum*, which is indigenous to South Africa, is widely cultivated in the Vhembe District of Limpopo Province of South Africa (Van Averbeke et al. 2007a, b).

Black nightshade species are erect, annual or biannual herbaceous plants that can attain a height of 75 cm. The leaves are alternate and have a bright green colour, which at times has purple pigmentation (Jansen van Rensburg et al. 2007). The small flowers are about 4–10 mm long with white petals conspicuously arranged in a drooping umbel-like inflorescence (Jansen van Rensburg et al. 2007). According to Jansen van Rensburg et al. (2007), *S. nigrum* species are mainly found in fairly humid environments with at least 500 mm of rain per annum. Their optimal temperature for growth ranges between 20°C and 30°C but most species are said to tolerate temperatures ranging between 15°C and 35°C (Jansen van Rensburg et al. 2007). When grown during winter, maximum growth and biomass production are obtained when the plants are exposed to full sunlight but during summer, shading up to 60% can be beneficial (Jansen van Rensburg et al. 2007). Poor germination is commonly encountered and this has been associated with inadequate removal of sugar and germination inhibitors present in the fruit during the extraction of the seed (Grubben et al. 2004). Ju et al. (2007) reported that *S. melongena* L. had a shallow root system but no information on the root system of *S. retroflexum* could be retrieved.

When used as a leafy vegetable, the leaves and tender shoots of nightshade are harvested and cooked (Van Averbeke et al. 2007a, b). The leaves are rich in proteins, fibers, vitamins, amino acids and the berries can be used as dye (Poczai and Hyvonen 2011). According to Jansen van Rensburg et al. (2007), leaves of *Solanum retroflexum* Dun. are also eaten raw and the ripe fruit is consumed fresh or as a preserve but the green fruit is poisonous. The limited information available on this crop suggests that the fresh leaf yield of *S. retroflexum* is subject to considerable variability. Jansen van Rensburg et al. (2007) indicated that under favourable conditions, cumulative fresh leaf yields of 20 t ha^{-1} could be achieved but Chabalala and Van Averbeke (2011) recorded fresh leaf yields as high as 81.8 t ha^{-1} from three shoot cuttings.

The biomass response of *S. retroflexum* Dun. to different application rates of manure from poultry, cattle and goats was investigated in a greenhouse pot experiment, where Azeez et al. (2010) found that biomass production increased as the rate of chicken manure application was raised up to the rate of 17 t ha^{-1} and decline in biomass production occurred when the rate was increased further to 34 t Promis® ha^{-1}. When cattle kraal manure was used as a fertilizer, for the growth of *S. retroflexum* Dun, Azeez et al. (2010) recorded increases in biomass production up to the highest application rate of 170 t ha^{-1}. The same trend was observed when goat kraal manure was used as a fertilizer, with the highest biomass being recorded at the highest application rate of 68.25 t ha^{-1} treatment employed in the experiment.

8.4 Pumpkin (*Cucurbita maxima* Duchesne)

The Pumpkin (*Cucurbita maxima* Duchesne) that featured in this review is presented in Fig. 7. The name pumpkin is used for different species of the genus *Cucurbita* (Schippers 2002; Ogar and Asiegbu 2005; Jansen van Rensburg et al. 2007). Pumpkin originated on the American continent (Harris et al. 2006) and was dispersed to other continents by transoceanic voyagers at the turn of the sixteenth century to become a familiar vegetable crop in many countries. Schippers (2002) pointed out that Cucurbitaceae are found mainly in the warm parts of all continents. Cucurbitaceae consists of 119 genera and about 825 species, many of which are eaten in one form or another (Schippers 2002). According to Harris et al. (2006), the three most widely grown *Cucurbita* species, *C. pepo* L., *C. maxima* Duchesne and *C. moschata* Duchesne, are characterized by polymorphism in fruit characteristics, fast growth rates and large sized plants.

Pumpkin is an annual plant of which the stems, leaf blades and petioles are covered with sharp, stiff, translucent hairs, which can irritate the human skin (Jansen van Rensburg et al. 2007). Pumpkin is a drought tolerant crop (Grubben et al. 2004; Jansen van Rensburg et al. 2007; Philip et al. 2009). Optimum mean daily temperature ranges from 18°C to 27°C (Grubben et al. 2004). The plant has an extensive fibrous root system (Grubben et al. 2004), which can extend to a soil depth of 1.8 m (Weaver and Bruner 1927).

Fig. 7 Pumpkin (*Cucurbita maxima* Duchesne)

When grown as a leafy vegetable, the leaves, flowers and young fruit are picked and cooked as a pot herb. The leaves are also used to treat anaemia and diabetes (Ogar and Asiegbu 2005; Jansen van Rensburg et al. 2007). Harvesting of the leafy parts of pumpkin is carried out at different growth stages but is usually done sequentially and incompletely to enable the plant to also produce fruit (Vorster 2007). Pumpkin seed is consumed the same way as nuts and contains about 30% protein and 47% oil (Ogar and Asiegbu 2005). The seed oil can be used for the preparation of margarine, pomade and to act as carrier for certain drugs (Ogar and Asiegbu 2005). Little information has been published on the leaf yield of pumpkin but Chabalala and Van Averbeke (2011) harvested 34 t fresh leaves ha^{-1} over a period of 52 days following planting, whilst Mbuli and Van Averbeke (2010, 2011) recorded fresh leaf yields of 24.5 and 31.6 $t\,ha^{-1}$, respectively. In a green house pot experiment, Azeez et al. (2010) found that biomass production of pumpkin increased as the rate of chicken manure application was raised up to the rate of 8.53 $t\,ha^{-1}$ but observed a drastic decline when the rate was increased to 17.1 t Promis® ha^{-1}. When manure from cattle and goat kraals were used as fertilizers, biomass production of pumpkin increased up to the highest application rate treatments used in the experiment which was 170 $t\,ha^{-1}$ in the case of cattle manure and 68.25 $t\,ha^{-1}$ in the case of goat manure.

This review elucidated that crops differed in terms of their nutrient requirements. Differences in nutrient requirements occurred from one crop species to the other and even among varieties of the same crop species. Amaranth, Chinese cabbage, nightshade and pumpkin emerged among the wild vegetables in South Africa that had potential to be developed for commercialization. The limited information retrieved on the use of animal manure indicated that amaranth thrived well and had considerable yield even when the crop was considered not to have attained its peak at application rates of 9.22 t poultry manure ha^{-1}, 10 $t\,ha^{-1}$ sheep manure, 10.30 t goat manure ha^{-1} and 11.7 cattle manure ha^{-1}. Chinese cabbage was yet to attain its biomass peak when layer chicken manure was applied to a low fertility soil at 3.42 $t\,ha^{-1}$, goat kraal manure at 11.9 $t\,ha^{-1}$ and cattle manure at 23.8 $t\,ha^{-1}$. Biomass production of nightshade continued to increase at 68.25 $t\,ha^{-1}$ of goat manure, 170 $t\,ha^{-1}$ cattle manure and peaked at 17 $t\,ha^{-1}$ of chicken manure. Pumpkin had its peak biomass production at 8.53 t chicken manure ha^{-1} while the peak biomass production was yet to be attained at 170 t cattle kraal manure ha^{-1} and 68.25 t goat kraal manure ha^{-1}.

9 Conclusion

The information compiled from various research works in this review, tentatively concluded that despite the availability of animal manure, the uncertainties about manure trends, limited its use by farmers due to the complexities surrounding the organic nutrient source. In South Africa, smallholder farming system was described to be dominated by black farmers whose aspirations included increasing the commercialization of their agricultural production activities in the course of providing a better livelihood. Unfortunately, limited availability of resources and a number of socio economic factors caused a draw-back in the smallholder goals.

African leafy vegetables were consumed mainly by black people who mainly produced the leafy vegetables by primarily using organic sources to fertilize their cropping land or harvested the crops from the wild. Among these organic sources of nutrients used for soil replenishment, ruminant manure and chicken manure were of paramount concern. The wide varieties of vegetables cultivated or harvested from the wild in South Africa provide advantages to the rural resource poor farmers from the economic, nutritional and medicinal perspectives. The need for the use of organic and inorganic fertilizers to complement natural supply of nutrients supplied to meet up with the increasing demand for food by the world increasing human population was also revealed. For reasons of affordability and accessibility, main soil nutrient source used to improve soil for increased crop yield by smallholders was animal manure obtained from different sources particularly from ruminant kraals and poultry houses. The review revealed that manure contained all essential nutrients required by plant even though these are not in same quantity as synthesized fertilizers. It was also revealed that animal manure is a heterogeneous material. Its quality as a fertilizer is affected by a wide range of factors and the review pointed out that ruminant manure tended to differ from chicken manure. Chicken manure typically had a higher nutrient concentration than ruminant manure, particularly nitrogen.

When used as a fertilizer in crop production, both types of manure were shown to have positive effects on crop growth and yield but in the case of chicken manure, the application rates at which the response curve tended to level off or decline, appeared to be considerably lower than in the case of ruminant manure. The review indicated that crops differed in terms of their nutrient requirements and the optimum application rate of animal manure is crop and manure specific. Amaranth, Chinese cabbage, nightshade and pumpkin were among the wild vegetables in South Africa that were mentioned to have the potentials of being developed for commercialization. The limited information retrieved on the use of animal manure for soil fertility in the production of leafy vegetables indicated that amaranth thrived well and had considerable yield while the crop was yet to attain its peak biomass production at application rates of 9.22 t poultry manure ha^{-1}, 10 tha^{-1} sheep manure, 10.30 t goat manure ha^{-1} and 11.7 cattle manure ha^{-1}. Chinese cabbage was yet to attain its biomass peak at the application rates of 3.42 t layer chicken manure ha^{-1}, 11.9 t goat kraal manure ha^{-1} and 23.8 t cattle manure ha^{-1}. Biomass production of nightshade continued to increase at 68.25 tha^{-1} of goat manure, 170 tha^{-1} cattle manure and peaked at 17 t chicken manure ha^{-1}. Pumpkin had its peak biomass production at 8.53 t chicken manure ha^{-1} while the peak biomass production was yet to be attained at 170 t cattle kraal manure ha^{-1} and 68.25 t goat kraal manure ha^{-1}.

The information retrieved in this work showed that a smallholder farmer is not basically the anonym of a commercial farmer but a resource poor farmer whose goal of increasing his or her agricultural production is hindered by lack of resources. Manure from animal origin cannot be classified as the same, as manures differ in composition and quality which are dependent on several factors. Animal manure cannot be used for soil amendment at the same application rates due to differences in the nutrient availability and nutrient release pattern in the varying organic nutrient sources. Vegetables that featured in this review responded differently to application rates of the different types of animal manure and as such, it could be recommended that farmers should endeavour to take into consideration the factors affecting the quality of the type of animal manure considered for use in growing or cultivating leafy vegetables.

Acknowledgement The financial support provided to the authors of this review work by Tshwane University of Technology (TUT) and the Water Research Commission (WRC) of South Africa towards the preparation of this article is hereby acknowledged. Opinions expressed and conclusions arrived at, are those of the authors and are not necessarily to be attributed to the WRC or TUT.

References

Abbasi MK, Hina M, Khalique A, Khan SR (2007) Mineralization of three organic manures used as nitrogen source in a soil incubated under laboratory conditions. Commun Soil Sci Plant Anal 38:1691–1711

Adey S (2006) A journey without maps: towards sustainable subsistence agriculture in South Africa. Ph.D. thesis, Rural Sociology Group, Wageningen University

Agehara S, Warcnke DD (2005) Soil moisture and temperature effects on nitrogen release from organic nitrogen sources. Soil Sci Soc Am J 69:1844–1855

Almeida AM (2011) The Damara in the context of Southern Africa fat-tailed sheep breeds. Trop Anim Health Prod 43:1427–1441

Amanullah MM (2007) 'N' release pattern in poultry manured soil. J Appl Sci Res 3(10):1094–1096

Amanullah MM, Sekar S, Muthukrishnan P (2010) Prospects and potentials of poultry manure. Asian J Plant Sci 9(4):172–182

Archer J (1988) Crop nutrition and fertilizer use, 2nd edn. Farming Press Ltd, Ipswich

Asfaw Z (1995) Conservation and use of traditional vegetables in Ethiopia. In: Attere AF (ed) Traditional African vegetables: proceedings of the IPGRI international workshop on genetic resources of traditional vegetables in Africa- conservation and use, held in Kenya, ICRAF-HQ, Nairobi, 29–31 Aug 1995

Aviagen (2002) Ross broiler management manual. Aviagen Limited, Newbridge

Azeez JO, Van Averbeke W, Okorogbona AOM (2010) Differential responses in yield of pumpkin (*Cucurbita maxima* L.) and nightshade (*Solanum retroflexum* Dun.) to the application of three animal manures. Bioresour Technol 101:2499–2505

Backeberg GR, Sanewe AJ (2010) Underutilised indigenous and traditional crops: why is research on water use important for South Africa?: foreword. S Afr J Plant Soil 27(4):291–292

Bationo A, Kihara J, Vanlauwe B, Waswa B, Kimetu J (2006) Soil organic carbon dynamics, functions and management in West African agro-ecosystems. Agric Syst. doi:10.1016/j.agsy.2005.08.011

Bowman DD (2009) Manure pathogens: manure management, regulations and water quality protection. WEF Press, Alexandria

Chabalala MP, Van Averbeke W (2011) Response of eight African leafy vegetables grown under field conditions to different rates of chemical fertilizer application. Centre of Organic and Smallholder Agriculture, Department of Crop Sciences, Tshwane University of Technology, Unpublished

Chang E, Chung R, Wang F (2008) Effect of different types of organic fertilizers on the chemical properties and enzymatic activities of an oxisol under intensive cultivation of vegetables for 4 years. Soil Sci Plant Nutr 54:587–599

Chardwick DR, John F, Pain BF, Chambers BJ, Williams J (2000) Plant uptake of nitrogen from the inorganic nitrogen fraction of animal manures: a laboratory experiment. J Agric Sci 134:158–168

Dakora FD, Keya SO (1997) Contribution of legume nitrogen fixation to sustainable agriculture in sub-Saharan Africa. Soil Biol Biochem 29(5/6):809–917

De Campos AB, Reichert JM, Huang C (2004) Near-surface hydrology, runoff production, and soil loss from surface-applied liquid swine manure. Conserving Soil and Water for Society: Sharing Solutions: 13th International Soil Conservation Organisation Conference, held in Australia on 4–8 July 2004. Queensland, Bribane

Department of Agriculture (2001) Annual report. The Government Printer, Pretoria

Diouf M, Gueye M, Faye B, Dieme O, Lo C (2007) The commodity systems of four indigenous leafy vegetables in Senegal. Water SA 33(3):343–348

Duada SN, Ajayi FA, Ndor E (2008) Growth and yield of water melon (*Citrullus lanatus*) as affected by poultry manure. J Agric Soc Sci 4:121–124

Edmeades DC (2003) The long term effects of manures and fertilizers on soil productivity and quality. Nutr Cycl Agroecosyst 66:165–180

Esminger ME (1971) Poultry management. In: Poultry Science, 1st edn. Interstate Printers and Publishers, Danville

Esse PC, Buerkert A, Hiernaux P, Assa A (2001) Decomposition of and nutrient release from ruminant manure on acid sandy soils in the Sahelian zone of Niger, West Africa. Agri Ecosyst Environ 83:55–63

Eyzaguirre P (1995) Conservation through increased use: complementary approaches to conserving African traditional vegetable. In: Guarino L (ed) Traditional African vegetables, Proceedings of the IPGRI international workshop on genetic resources of traditional vegetables in Africa: conservation and use, Nairobi, 29–31 August 1995

Faber M, Wenhold F (2007) Nutrition in contemporary South Africa. Water SA 33(3):393–400

Faber M, Van Jaarsveld PJ, Laubscher R (2007) The contribution of dark green leafy vegetables to total micro nutrient intake of two- to five-year-old children in a rural setting. Water SA 33(3):407–411

Faber M, Oelofse A, Van Jaarsveld PJ, Wenhold FAM, Jansen Van Rensburg WS (2010) African leafy vegetables consumed by households in the Limpopo and KwaZulu-Natal Provinces in South Africa. S Afr J Clin Nutr 23(1):30–38
Fageria NK (2009) The use of nutrients in crop plants. CRC Press, New York
Farren MT, Al-Assir IA, Sabra AS, Rubetz IG (1993) Layer and broiler poultry manure as nitrogen fertilizer sources for cabbage production. Comm Soil Sci Plant Anal 24(13 and 14):1583–1589
Fatunbi AO, Ncube L (2009) Activities of effective microorganism (EM) on the nutrient dynamics of different organic materials applied to soil. Am-Eur J Agron 2(1):26–35
Fening JO, Ewusi-Mensah N, Safo EY (2010) Improving the fertilizer value of cattle manure for sustaining small holder crop production in Ghana. J Agron 9(3):92–101
Food and Agriculture Organization (FAO) (1998) Guide to efficient plant nutrition management. Land and Water Development Division, Rome
Food and Agriculture Organization (FAO) (2005) Fertilizer use by crop in South Africa. Land and Water Development Division, Rome
Food and Agriculture Organization (FAO) (2010) The state of food security in the world: addressing food insecurity in protracted crisis. Food and Agriculture Organization of the United Nations, Rome
Fox FW, Norwood Young ME (1982) Food from the veld: edible wild plants of Southern Africa. Delta Books, Johannesburg
Fraser HM, Fleming RJ, O'Halloran IP, Van Eerd LL, Zandstra JW (2006) Non-Nutrient value of manure-literature review for Ontario Ministry of agriculture, food and rural affairs. College, University of Guelph, Ridge Town
FSSA (1989) Fertilizer handbook, 3rd edn. Fertilizer Society of Sub-Saharan Africa, International Livestock Centre for South Africa, Hennopsmeer
Fujiwara T, Murakami K, Yoshida E, Takahashi I (2009) Potential of near infrared spectroscopy to determine the lipid content of untreated garbage compost. Jpn J Soil Sci Plant Nutr 55(2):309–314
Germishuis H (2006) Republic of South Africa poultry and products annual report, USDA Foreign Agricultural Service, Washington, DC, 9 Jan, p 4
Gilley JE, Eghball B (2002) Residual effects of compost fertilizer application on nutrients in runoff. Trans Am Soc Agric Eng 45:1905–1910
Gilmour JT, Koehler MA, Cabrera ML, Szajdak L, Moore PA Jr (2004) Alum treatment of poultry litter: decomposition and nitrogen dynamics. J Environ Qual 33:402–405
Goh KM, Vityakon P (1983) Effects of fertilizers on vegetable production. 1 Yield of spinach and beetroot as affected by rates and forms of nitrogenous fertilizers. N Z J Agric Res 26:349–356
Gosh PK, Ramesh P, Bandyopadhyay KK, Tripathi AK, Hati KM, Misra AK, Acharya CL (2004) Comparative effectiveness of cattle manure, poultry manure, phosphocompost and fertilizer-NPK on three cropping systems in vertisols of semi-arid tropics. Bioresour Technol 95:77–84
Grattan SR, Grieve CM (1999) Salinity-mineral nutrient relations in horticultural crops. Sci Hortic 78:127–157
Grubben GJH, Denton OA, Messiaen CM, Schippers RR, Lemmens RHMJ, Oyen IPA (2004) Plant resources of tropical Africa 2: vegetables. Backhuys, Leiden
Gwaze RF, Chimonyo M, Dzama K (2010) Nutritionally-related blood metabolites and faecal egg counts in indigenous Nguni goats of South Africa. S Afr J Anim Sci 40(5):480–483
Harris PJ (1988) Ecology of the soil population. In: Wild A (ed) Russell's soil condition and plant growth, 11th edn. Longman Scientific and Technical, Harlow, pp 472–499
Harris F (2002) Management of manure in farming systems in semi-arid West Africa. Exp Agric 38:131–148
Harris SP, Marie-Christine D, Michel P, Jules J (2006) First known image of *Cucurbita* in Europe, 1503–1508. Ann Bot 98:41–47
Hebinck P, Monde N (2007) Production of crops in arable fields and home gardens. In: Hebinck P, Lent PC (eds) Livelihoods and landscapes: the people of Guquka and Koloni and their resources. Brill, Leiden, pp 181–220

Hossain ATMS, Rahman F, Saha PK, Solaiman ARM (2010) Effects of different aged poultry litter on the yield and nutrient balance in boro rice cultivation. Bangladesh J Agric Res 35(3):497–505

Jacobus DJ (2004) The influence of bedding material and collecting period on the feeding value of broiler and layer litter. M.Sc. dissertation, University of the Free State, Bloemfontein

Jang S, Hamayun M, Sohn E, Kang S, Choi K, Shin D, Lee I (2008) Growth and gibberellins level of two rice cultivars as influenced by different nitrogen containing compounds. J Crop Sci Biotechnol 11(4):223–226

Jansen Van Rensburg WS, Van Averbeke W, Slabbert R, Faber M, Van Jaarsveld P, Van Heerden I, Oelofse A (2007) African leafy vegetables in South Africa. Water SA 33(3):317–326

Johnston D, Chapman R, Masse D, Topp E (2002) Evaluation of commercial odour control agents for suppressing *Escherichia coli*. J Environ Qual 31:2120–2123

Ju XT, Gao Q, Christie P, Zhang FS (2007) Interception of residual nitrate from a calcareous alluvial soil profile on the North China Plain by deep-rooted crops: a ^{15}N tracer study. Environ Pollut 146:534–542

Khaliq T, Mahmood T, Kama J, Masood A (2004) Effectiveness of farmyard manure, poultry manure and nitrogen for corn (*Zea mays* L.) productivity. Int J Agric Biol 6(2):260–263

Kihanda FM, Warren GP, Attwal SS (2004) The influence of goat manure application on crop yield and soil nitrate variations in semi–arid eastern Kenya. In: Bationo A (ed) Managing nutrient cycles to sustain soil fertility in sub-Saharan Africa. Academy Science Publishers, Nairobi, pp 174–186

Kimani SK, Lekasi JK (2004) Managing manures throughout their production cycle enhances their usefulness as fertilizers: a review. In: Bationo A (ed) Managing nutrient cycles to sustain soil fertility in sub-Saharan Africa. Academy Science Publishers, Nairobi, pp 188–198

Kimetu JM, Mugendi DN, Palm CA, Mutuo PK, Gachengo CN, Bationo A, Nandwa S, Kungu JB (2004) Nitrogen fertilizer equivalencies of organics of differing quality and optimum combination with inorganic nitrogen sources in central Kenya. Nutr Cycl Agroecosyst 68:127–135

Kirsten JF, Van Zyl J (1998) Defining small-scale farmers in the South African context. Agrekon 37(4):560–571

Koelkebeck KW, Anderson KE (2007) Molting layers-alternative methods and their effectiveness. Poult Sci 86:1260–1264

Koelkebeck KW, Parsons CM, Biggs P, Utterback P (2006) Nonwithdrawal molting programs. J Appl Poul Res 15:483–491

Kotze WAG, Joubert MS (1992) Poultry manure as fertilizer for deciduous fruit production. Deci Fruit Gro 42(7):250–255

Kuun PJC, Bezuidenhout JD, Classens AS, Oberem PT (1997) Agricultural science in action, 5th edn. Rustica press, Ndabeni

Laker MC (2007) Introduction to special edition of Water SA on indigenous crops, water and human nutrition. Water SA 33(3):311–315

Larkom J (2007) Oriental vegetables. Frances Lincoln Ltd, London

Larney FJ, Buckley KE, Hao X, Mccaughey WP (2006) Fresh, stockpiled, and composted beef cattle feedlot manure: nutrient levels and mass balance estimates in Alberta and Manitoba. J Environ Qual 35:1844–1854

Law-Ogbomo KE, Ajayi SO (2009) Growth and yield performance of *Amaranthus cruentus* influenced by planting density and poultry manure application. Not Bot Hort Agro Cluj-Nap 37(1):195–199

Lazcano C, Gómez-Brandón M, Domínguez J (2008) Comparison of the effectiveness of composting and vermicomposting for the biological stabilization of cattle manure. Chemosphere 72:1013–1019

Lekasi JK, Tanner JC, Kimani SK, Harris PJC (2003) Cattle manure quality in Maragua district, central Kenya: effect of management practices and development of simple methods of assessment. Agric Ecosyst Environ 94:289–298

Lewu FB, Mavengahama S (2010) Wild vegetables in Northern KwaZulu Natal, South Africa: current status of production and research needs. Sci Res Essay 5(20):3044–3048

Lopez-Mosquera ME, Cabaleiro F, Sainz MJ, Lopez-Fabal A, Carral E (2008) Fertilizing value of broiler litter: effect of drying and pelletizing. Bioresour Technol 99:5626–5633

Lu N, Edwards JH (1994) Poultry litter quantity influences collard growth in pots and affects cabbage growth and nutrient uptake. J Hortic Sci 29(10):1143–1148

Ma BL, Dwyer LM, Gregorich EG (1999) Soil nitrogen amendment effects on seasonal nitrogen cycling in maize production. Agron J 91:1003–1009

Machete CL, Mollel NM, Ayisi K, Mashatola MB, Anim FDK, Vanasche, F (2004) Smallholder irrigation and agricultural development in the Olifants river basin of Limpopo Province: management transfer, productivity, profitability and food security issues, WRC Report No: 1050/1/04, Water Research Commission, Pretoria

Maerere AP, Kimbi GG, Nonga DLM (2001) Comparative effectiveness of animal manures on soil chemical properties, yield and growth of *Amaranthus* (*Amaranthus cruentus* L). Afr J Sci Technol 1(4):14–21

Mahimairaja S, Bolan NS, Hedley MJ (1995) Agronomic effectiveness of poultry manure composts. Commun Soil Sci Plant Anal 26(11and12):1843–1861

Makinde EA, Ayeni LS, Ojeniyi SO (2010) Morphological characteristics of *Amaranthus cruentus* L. as influenced by kola pod husk, organomineral and NPK fertilizers in southwestern Nigeria. N Y Sci J 3(5):130–134

Malherbe I De V (1964) Soil fertility, 5th edn. Oxford University Press, Cape Town

Marten GG, Abdoellah OS (1988) Crop diversity and nutrition in west Java. Ecol Food Nutr 21:17–43

Martirosyan DM, Miroshnichenko LA, Kulakova SN, Pojojeva AV, Zoloedov VI (2007) Amaranth oil application for coronary heart disease and hypertension. Lipid Health Dis. doi:10.1186/147/-511X-6-1

Materechera S (2010) Utilization and management practices of animal manure for replenishing soil fertility among small-scale crop farmers in semi-arid farming districts of the North West Province, South Africa. Nutr Cycl Agroecosyst 87(3):415–428

Materechera SA, Mkhabela TS (2002) The effectiveness of lime, chicken manure and leaf litter ash in ameliorating acidity in a soil previously under black wattle (*Acacia mearnsii*) plantation. Bioresour Technol 85:9–16

Materechera SA, Modiakotla LN (2006) Cattle manure increases soil weed population and species diversity in a semi-arid environment. S Afr J Plant Soil 23(1):21–28

Matsumura T (1981) The present status of Chinese cabbage growing in Japan. In: Talker NS, Griggs TD (eds) International symposium on Chinese cabbage, Proceedings, Tsukuba, Asian Vegetable Research and Development Center, Shanhua, 31 March–5 April 1980, pp 1–10

Maunder EMR, Meaker JL (2007) The current and potential contribution of home-grown vegetables to diets in South Africa. Water SA 33(3):401–406

Maundu PM (1995) The status of traditional vegetable utilization in Kenya. In: Attere AF (ed) Traditional African vegetables: proceedings of the IPGRI international workshop on genetic resources of traditional vegetables in Africa: conservation and use, held in Nairobi, International Centre for Research in Agroforestry–Headquarters, Nairobi, 29–31 Aug 1995, pp 75–88

Mbuli SS, Van Averbeke W (2010) Intercropping of maize: green maize and pumpkins: 2009 progress report, Centre of Organic and Smallholder Agriculture, Department of Crop Sciences, Tshwane University of Technology, Pretoria, Unpublished

Mbuli SS, Van Averbeke W (2011) Intercropping of maize: green maize and pumpkins: 2010 progress report. Centre of Organic and Smallholder Agriculture, Department of Crop Sciences, Tshwane University of Technology, Pretoria, Unpublished

Mhlonto S, Muchaonyerwa P, Mnkeni PNS (2007) Effects of sheep kraal manure on growth, dry matter yield and leaf nutrient composition of a local *Amaranthus* accession in the central region of the Eastern Cape Province, South Africa. Water SA 33(3):363–368

Mkhabela TS (2004) Substitution of fertilizer with poultry manure. Agrekon 43(3):347–408

Mkhabela TS (2006) A review of the use of manure in small-scale crop production system in South Africa. J Plant Nutr 29:1157–1185

Mkhabela TS, Materechera SA (2003) Factors influencing the utilization of cattle and chicken manure for soil fertility management by emergent farmers in the moist midland of KwaZulu-Natal Province, South Africa. Nutr Cycl Agroecosyst 65:151–162
Mkhabela MS, Mkhabela MS, Shikulu JP (2001) Response of maize (*Zea mays* L.) cultivars to different levels of nitrogen application in Swaziland: seventh eastern and Africa regional maize conference, held in Kenya, KCB institute, Nairobi, 11–15 Feb 2001
Mkile Z (2001) The use and agronomic effectiveness of kraal manures in the Transkei region of the Eastern Cape, South Africa. M.Sc. dissertation, University of Fort Hare, Alice
Mnzava NA (1995) The status of traditional vegetable utilization in Kenya. In: Attere AF (ed) Traditional African vegetables: proceedings of the IPGRI international workshop on genetic resources of traditional vegetables in Africa: conservation and use, held in Nairobi, International Centre for Research in Agroforestry–Headquarters, Nairobi, 29–31 Aug 1995, pp 8–22
Modi AT (2007) Growth temperature and plant age influence on nutritional quality of *Amaranthus* leaves and seed germination capacity. Water SA 33(3):369–375
Mphaphuli NS (2003) Development of an organic growth medium for the propagation of vegetable seedlings in containers using pine needles and poultry manure. M.Tech. dissertation, Tshwane University of Technology, Pretoria
Muller CJC, Cloete SWP, Olivier JJ, Botha JA, de Waal H (2006) Heritability of live weight and condition score in a Holstein herd and correlations with milk traits – preliminary estimates. S Afr J Anim Sci 36(2):79–88
Niewoudt WL (2000) An economic evaluation of a crop insurance programme for small-scale commercial farmers in South Africa. Agrekon 39(3):269–291
Nndwammbi LG (2009) Determination of plant and soil irrigation parameters for different crops at Dzindi irrigation scheme, South Africa. M.Sc. dissertation, University of Limpopo, Polokwane
Nyamangara J, Mudhara M, Giller KE (2005) Effectiveness of cattle manure and nitrogen fertilizer application on the agronomic and economic performance of maize. S Afr J Plant Soil 22(1):59–63
Odhav B, Beekrum S, Akula U, Bainat HH (2007) Preliminary assessment of nutritional value of traditional leafy vegetables in KwaZulu-Natal, South Africa. J Food Comp Anal 20:430–435
Ogar EA, Asiegbu JE (2005) Effects of fertilizer rates and cutting frequency on the marketable vegetable and pod yields in fluted pumpkin in southeastern Nigeria. J Agric Food Environ Ext 4(1):66–69
Ogbona PE, Umar-Shaaba YG (2011) Yield responses of sesame (*Sesamium indicum* L) to rates of poultry manure application and time of planting in a derived savannah ecology of south eastern Nigeria. Afr J Biotechnol 10(66):14881–14887
Ogbonna PE, Obi UI (2007) Effect of time of planting and poultry manure application on growth and yield of egusi melon *(Colocynthis citrullus L.)* in a derived savannah agro-ecology. J Agric Food Environ Ext 6(2):33–39
Okorogbona AOM (2011) Changes in the quality of slaughtered cattle gastro-intestinal wastes as affected by period of decomposition and the effect of treatments on decomposers of the abattoir wastes. Global Vets 6(2):140–151
Okorogbona AOM, Van Averbeke W, Ramunsadiwa TD (2011) Growth and yield response of Chinese cabbage as affected by nutrient availability in air-dried and pulverised different types of animal manure using low biological activity soil. World J Agric Sci 7(1):1–12
Okwu DE, Ukanwa NS (2007) Nutritive value and phytochemical contents of fluted pumpkin (*Telfaria occidentalis* Hook F.) vegetable grown with different levels of turkey dropping: African crop science proceedings, held in Egypt, Al Minya, El- Minia.
Omolayo FO, Ayodele OJ, Fasina AS, Godonu K (2011) Effects of poultry manure from different sources on the growth and marketable yield of leaf amaranth (*Amaranthus cruentus (hybridus)* L.) amaranthaceae. Int Res J Agric Sci 1(2):29–34
Peacock C (1996) Goat production in the tropics: a manual for development workers. Oxfam/Farm-Africa Publications, Oxford

Peirce LC (1987) Vegetables: characteristics, production and marketing. Wiley, New York
Philip B, Shittu AM, Aiyelaagbe IOO, Adedokun T (2009) Economic potentials of plantain and fluted pumpkin intercropping as a poverty reduction strategy in south-western Nigeria. World J Agric Sci 5(5):525–534
Place G, Bowman D, Burton M, Pufty T (2008) Root penetration through a high bulk density soil layer: differential response of a crop and weed species. Plant Soil 307:179–190
Poczai P, Hyvonen J (2011) On the origin of *Solanum nigrum*: can networks help? Mol Biol Rep 38:1171–1185
Poczai P, Matyas K, Taller J, Szabo I (2010) Study of the origin of the rarely cultivated edible *Solanum* species: morphological and molecular data. Biol Plant 54(3):543–546
Pospisil A, Pospisil M, Varga B, Svecnjak Z (2006) Grain yield and protein concentration of two amaranth species (*Amaranthus* spp.) as influenced by nitrogen fertilization. Eur J Agron 25:250–253
Postgate J (1998) Nitrogen fixation, 3rd edn. Cambridge University Press, Cambridge
Pote PPT (2008) Technical constraints to smallholder agriculture: case study of Nkonkobe Municipality, Eastern Cape, South Africa. M.Sc. dissertation, University of Fort Hare, Alice
Powell JM, Broderick GA, Grabber JH, Hymes-Fecht UC (2009) Technical note: effects of forage protein-binding polyphenols on chemistry of dairy excreta. J Dairy Sci 92:1765–1769
Prasad R, Power JF (1997) Soil fertility management for sustainable agriculture. CRC Press, Boca Raton
Ramrao WY, Tiwari SP, Singh J (2006) Crop-livestock integrated farming system for the marginal farmers in fed regions of Chhattisgarh in Central India. Livest Res Rural Dev 18(7):1–4
Rautaray SK, Ghosh BC, Mittra BN (2003) Effects of fly ash, organic wastes and chemical fertilizers on yield, nutrient uptake, heavy metal contents and residual fertility in a rice-mustard cropping sequence under acid lateritic soils. Bioresour Technol 90(3):275–283
Reddy DD, Rao AS, Rupa TR (2000) Effects of continuous use of cattle manure and fertilizer phosphorus on crop yields and soil organic phosphorus in a vertisol. Bioresour Technol 75(2):113–118
Reinten E, Coetzee JH (2002) Commercialization of South African indigenous crops: aspects of research and cultivation of products. In: Janick J, Whipkey A (eds) Trends in new crops and new uses. ASHS Press, Alexandria, pp 76–80
Rufino MC, Rowe EC, Delve RJ, Giller KE (2006) Nitrogen cycling efficiencies through resource-poor African crop–livestock systems. Agric Ecosyst Environ 112:261–282
Sanchez PA, Swaminathan MS (2005) Hunger in Africa: the link between unhealthy people and unhealthy soils. Lancet 365:442–444
Sanchez JE, Willson TC, Kizilkaya K, Parker E, Harwood RR (2001) Enhancing the mineralizable nitrogen pool through substrate diversity in long term cropping systems. Soil Sci Soc Am J 65:1442–1447
Schippers RR (2002) African indigenous vegetables: an overview of the cultivated species, Revised edn. Chatham: National Resources Institute/ACP-EU Technical Centre for Agricultural and Rural Cooperation
Sharifi RS, Taghizadeh R (2009) Response of maize (*Zea mays* L.) cultivars to different levels of nitrogen fertilizer. J Food Agric Environ 7(3and4):519–521
Shini S (2003) Physiological responses of laying hens to the alternative housing systems. Int J Poult Sci 2(5):357–360
Shoenian S (2010) Nutrient management on sheep farms: a beginner's guide to raising sheep: study guide Sheep 201. University of Maryland, Keedysville
Shortall JG, Liebhardt WC (1975) Yield and growth of corn as affected by poultry manure. J Environ Qual 4(2):186–191
Silva PSL, Silva JD, Oliveira HT, Sousa AKF, Gustavo PD (2006) Residual effect of cattle manure application on green ear yield and corn grain yield. Hortic Bras 24:166–169
Sims JT, Bitzer CC (1988) Estimating the availability of poultry manure through laboratory and field studies. J Environ Qual 17(1):47–54

Singh G, Kawatra A, Sehgal S (2001) Nutritional composition of selected green leafy vegetables, herbs and carrots. Plant Food Hum Nutr 56:359–364

Stacey G, Burris RH, Evans HJ (1992) Biological nitrogen fixation. Chapman and Hall Inc, New York

Synman MA (2010) Influence of body weight, age and management system on reproduction of South African Angora goat does. S Afr J Anim Sci 40(1):41–43

Tikhonovich IA, Provorov NA, Ramonov VI, Newton WE (1995) Nitrogen fixation- fundamentals and applications: proceedings of the 10th international congress on nitrogen fixation held in Russia in May 1995, St Petersburg

Tindall HD (1983) Vegetables in the tropics. Macmillan Press, London

Tshikalange TE (2007) Response of *Brassica rapa* L. subsp. *chinensis* to nitrogen, phosphorus and potassium in pots. M.Tech. dissertation, Tshwane University of Technology, Pretoria

Tyson SC, Cabrera ML (1993) Nitrogen mineralization in soil amended with composted and uncomposted poultry litter. Commun Soil Sci Plant Anal 24(17 and 18):2361–2374

Van Averbeke W (2008) Initial report on recommendations for crops and resource requirements/ production system/technology for implementation based on experimental results and situation analysis of end-user resource availability: plant nutrition, Centre of Organic and Smallholder Agriculture, Tshwane University of Technology, Pretoria, Unpublished

Van Averbeke W, Mohammed SS (2006) Smallholder farming styles and development policy in South Africa: the case of Dzindi Irrigation Scheme. Agrekon 45(2):136–157

Van Averbeke W, Netshithuthuni C (2010) Effect of irrigation scheduling on leaf yield of non-heading Chinese cabbage (*Brassica rapa* L. subsp. *chinensis*). S Afr J Plant Soil 27(4):322–327

Van Averbeke W, Yoganathan S (1997) Using kraal manure as a fertilizer. Government Printer, National Department of Agriculture, Directorate Communication, Pretoria

Van Averbeke W, Juma KA, Tshikalange TE (2007a) Yield response of African leafy vegetables to nitrogen, phosphorus and potassium: the case of *Brassica rapa* L. subsp. *chinensis* and *Solanum retroflexum* Dun. Water SA 33(3):355–362

Van Averbeke W, Tshikalange TE, Juma KA (2007b) The commodity system of *Brassica rapa* L. subsp. *chinensis* and *Solanum retroflexum* Dun. In Vhembe District of Limpopo Province, South Africa. Water SA 33(3):349–354

Van Averbeke W, Mnkeni PSN, Harris PJC, Mkile Z (2008) Fertility status of cropped soils in smallholder farming systems of the Transkei region, Eastern Cape Province, South Africa. In: Hanekklaus S, Hera C, Reitz RM, Schnug E (eds) Fertilizer and fertilization for sustainability in agriculture: the first world meets the third world challenges for the future, Pretoria, "Terra Nostra" Iasi, Bucharest, pp 513–526

Van den Heever E (1995) The use and conservation of indigenous leafy vegetables in South Africa. In: Attere AF (ed) Traditional African vegetables: proceedings of the IPGRI international workshop on genetic resources of traditional vegetables in Africa: conservation and use, held in Nairobi, International Centre for Research in Agroforestry–Headquarters, Nairobi, 29–31 Aug 1995, pp 186–190

Van den Heever E, Coertze AF (1996) Indigenous leaf crops A1–*Amaranthus* (marog). Agricultural Research Council Leaflet, ARC-VOPI, Pretoria

Van Wyk B (2005) Food plants of the world: identification, culinary uses and nutritional value. Briza, Pretoria

Vitosh ML, Davis JF, Knezek BD (1973) Long-term effects of manure, fertilizer and plough depth on chemical properties of soils and nutrient movement in a mono-culture corn system. J Environ Qual 2(2):296–299

Vorster HJ (2007) The role and production of traditional leafy vegetables in three rural communities in South Africa. M.Sc. dissertation, University of Pretoria

Vorster HJ, Stevens JB, Steyns GJ (2008) Production systems of traditional leafy vegetables: challenges for research and extension. S Afr J Agric Ext 37:85–96

Weaver JE, Bruner WE (1927) Root development of vegetable crops. MacGraw-Hill, New York

Wenhold F, Faber M (2008) Nutritional status of South Africans: links to agriculture and water, WRC Report TT362/P/08, Water Research Commission of South Africa, Gezina

Wild A (1988) Plant nutrients in soil. In: Wild A (ed) Russell's soil condition and plant growth, 11th edn. Longman Scientific and Technical, Harlow, pp 652–779

Wilkerson VA, Mertens DR, Casper DP (1997) Prediction of excretion of manure and nitrogen by Holstein dairy cattle. J Dairy Sci 80(12):3193–3204

Wilkinson SR (1979) Plant nutrient and economic value of animal manures. J Anim Sci 48(1):123–133

Yoganathan S, Sotana MM, Van Averbeke W, Mandiringana OT, Materechera S, Harris PCJ, Mnkeni PSN (1998) Kraal manure as a fertilizer in small scale crop production in central Eastern Cape, South Africa. In: 15th international symposium of the AFSRE, proceedings, Pretoria, 29 Nov–4 Dec 1998, vol 1, pp 361–368

Vermicompost and Soil Quality

Supradip Saha, Debashis Dutta, Deb Prasad Ray, and Rajib Karmakar

Abstract Sustainability is a major issue for policy makers, researchers and extension workers worldwide. Achievement towards sustainable agriculture has not been satisfactory so far. Negative impacts of industrial agriculture are threatening biodiversity and biodiversity role in maintaining functional biosphere. Vermicomposting is a promising solution for the loss of biodiversity due to recycling of natural resources. Though almost all soil processes are regulated by soil microbes, the role of vermicompost in maintaining microbial diversity and soil functions is not fully understood. Here we review the major advances and benefits of vermicomposting. The major points are the following. Extracellular enzyme activity is increased upon application of vermicompost in soil. Initial enhancement of microbial growth is also observed and explained by the initial activation of the microbial enzymes and intracellular enzyme activity. Upon aging of vermicomposting the enzymatic activity decreases. Higher microbial population in vermin-cast is observed versus the surrounding soil. Major changes in bacterial and fungal communities are observed. Improvement in mineralization of nutrients is reported in most studies. Specifically, C and N mineralization is highly changed by the application of vermicompost. Enhancement of crop yields achieved on soil amended with vermicompost is explained by better mineralization of nutrients.

S. Saha (✉)
Division of Agricultural Chemicals, Indian Agricultural Research Institute, New Delhi, India
e-mail: s_supradip@yahoo.com

D. Dutta • R. Karmakar
Department of Agriculture, Government of West Bengal, Kolkata, India

D.P. Ray
National Institute for Research on Jute and Allied Fibre Technology, Kolkata, India

E. Lichtfouse (ed.), *Farming for Food and Water Security*, Sustainable Agriculture Reviews 10, DOI 10.1007/978-94-007-4500-1_10,

Keywords Vermicompost • Soil biology • Soil functions • Mineralization • Enzymatic activity • C:N ratio • Macronutrients • Micronutrients • Soil quality • Sustainability

1 Introduction

Modern agriculture is one of the greatest threats worldwide and it has led to growing concern about conserving biodiversity and its role in maintaining functional biosphere. Negative local consequences due to agricultural intensification such as increased erosion, lower soil fertility, and reduced biodiversity; negative regional consequences, such as pollution of ground water and eutrophication of rivers and lakes; and negative global consequences, including impacts on atmospheric constituents and climate is in alarming condition. Concerns about the ability to maintain long-term intensive agriculture are also growing. There has been a growing movement to decrease rates of inorganic fertilizer applications to soils by using soil nutrients more efficiently and by the increased use of organic matter.

Biotic interactions in soil are known to influence soil fertility and plant growth by influencing soil nutrient dynamics and rhizospheric environment (Wardle 2002). Belowground communities include a large variety of organisms showing highly complex interactions across trophic or non-trophic groups (Coleman 2008). Among the great diversity of soil biota, earthworms plays key role in this interaction (Six et al. 2002).

Earthworms are keystone soil organisms in regulating nutrient cycling through: (i) their own metabolism that leads to high availability of carbon (C) and nitrogen (N) from metabolic wastes (ii) the dispersal and the stimulation of soil microorganism activity associated with passage through the intestinal tract and (iii) the distribution and the mixing of organic matter and soil mineral particles (Lee 1985; Edwards and Bohlen 1996; Lavelle and Spain 2001). Earthworm play important roles by regulating mineralization and humification processes and in turn the soil organic matter dynamics (Lavelle and Martin 1992; Brown et al. 2000; Kizilkaya and Hepsen 2007). It plays a dynamic role in improving physical and chemical properties of soil as well (Lee 1985; Curry and Byrne 1997).

Positive interaction of earthworm with other biotic organisms can be achieved by using them in the composting process. Vermicomposting is the biooxidation and stabilization of the organic waste resources, involving the joint action of earthworms and associated micro-flora. Composted product helps in improving nutrient mineralization and thus it helps in plant growth. Among the mechanisms by which earthworms modify plant growth at the individual or community levels (Fig. 1; Scheu 2003; Brown et al. 2004), five have been claimed to be responsible for the positive effect noted on plant production: (i) increased mineralization of soil organic matter, which increases nutrient availability (Lavelle and Martin 1992; Subler et al. 1997), especially for nitrogen (N), the major limiting nutrient in terrestrial ecosystems; (ii) modification of soil porosity and aggregation (Blanchart et al. 1999; Shipitalo

Fig. 1 Vermicomposting pit at village level

and Le Bayon 2004), which induces changes in water and oxygen availability for plants (Doube et al. 1997; Allaire-Leung et al. 2000); (iii) production of plant growth regulators via the stimulation of microbial activity (Nardi et al. 2002); (iv) biocontrol of pests and parasites (Blouin et al. 2005); (v) stimulation of symbionts (Furlong et al. 2002).

Thus the composting process will influence the soil functions, soil biological activity and interaction with microflora. This article is a reappraisal of various investigations involving vermicomposting process and its influence on soil biology and functions.

2 Vermicomposting Process

The term "vermicomposting" refers to an aerobic, bioxidation and stabilization non-thermophilic process of organic waste decomposition that depends upon earthworms to fragments, mix and promote microbial activity (Gunadi et al. 2002). Vermicomposting as a principle originates from the fact that earthworms in the process of feeding fragment the substrate thereby increasing its surface area for further microbial colonization (Chan and Griffiths 1988). During this process, the important plant nutrients such as nitrogen, potassium, phosphorus and calcium present in

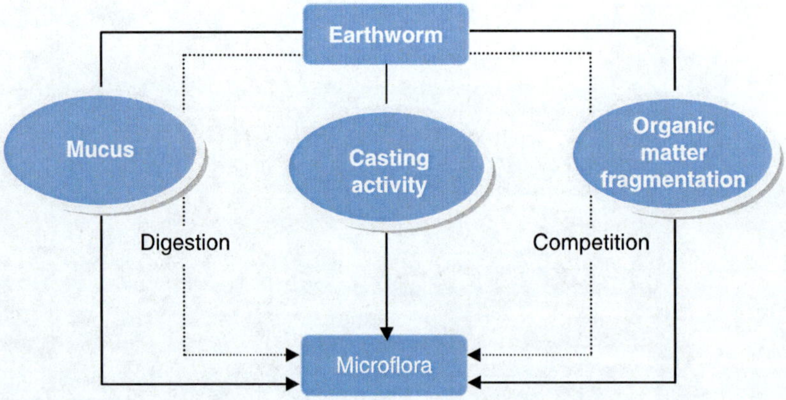

Fig. 2 Positive (—) and negative (········) effects of earthworms on microbial biomass and activity (Adapted from Dominguez et al. 2010)

the feed material are converted through microbial action into forms that are much more soluble and available to the plants than those in the parent substrate (Ndegwa and Thompson 2001).

"Ecosystem engineer", earthworms are voracious feeders on organic waste and while utilizing only a small portion for their body synthesis they excrete a large part of these consumed waste material in a half digested form. Since the intestine of earthworms harbour wide range of microorganisms, enzymes hormones, etc., these half digested substrate decomposes rapidly and are transformed into a form of vermicompost with in a short time (Lavelle 1988). This process takes place in the mesophilic temperature range (35–40°C). Earthworm prepares organic manures, through their characteristic functions of breaking up organic matter and combines it with soil particles. The final product is a stabilized, well humidified, organic fertilizer, with adhesive effects for the soil and stimulator for plant growth and most suitable for agricultural application and favourable environmentally (Fig. 2). The action of earthworms in this process is both physical/mechanical and biochemical.

Recently concluded investigation confirmed that decrease in total organic carbon (about 35%), C:N ratio from 31.2 to 12.3 and biochemical parameters like hydrolytic enzymes averagely 40% and dehydrogenase 23% indicated the progress of mineralization and the stabilization of organic matter at the end of the vermicomposting process (Macci et al. 2010). Furthermore, the increase in humification rate viz. pyrophosphate extractable carbon (PEC) from 17.6 to 33.3 mg g^{-1}, and PEC: water-soluble carbon from 1.76 to 2.97 confirms the progress of mineralization.

3 Soil Biological Activity

Due to the involvement in the biological transformations of nutrients in soil, enzyme activities have been widely used as an index of soil fertility or ecosystem status (Tate 2000; Saha 2010). Physicochemical and biochemical transformations

(Dominguez 2004) in short time (Aira et al. 2002) make vermicomposting a unique system (Aira et al. 2006a). Vermicomposting involves the bio-oxidation and stabilization of organic matter through the joint action of earthworms and microorganisms and thus it influences the biotransformation of nutrients.

Dehydrogenase activity is an indicator for potential non-specific intracellular enzyme activity of the total microbial biomass (Ladd 1978). Enzymes of the C cycle, such as β-glucosidase and cellulase is important because they have been associated with mass loss and therefore with the turnover of carbon in a wide range of ecosystems (Sinsabaugh and Moorhead 1994). Enzymes related to N cycle, protease activity is responsible in depolymerization of organic nitrogen (Paul and Clark 1996), is proven to be a critical point in the N cycle (Schimel and Bennett 2004) as polymers are not accessible to microorganisms (Chapin et al. 2002). Alkaline phosphatase is involved with the P cycle, hydrolysing organic P esters to its available form, inorganic phosphorus (Alef et al. 1995). Therefore, studying the decomposition rates of organic matter is prime important since these enzymes metabolize large organic polymers into smaller ones.

Enzymatic changes during the vermicomposting is studied in details and reported in literature. Aira et al. (2007) found that extracellular enzyme activity increased with rate of organic sources. Presence of earthworms found to be stimulating in microbial growth which is related to the initial activation of the microbial enzymes through enhancement of microbial biomass and intracellular enzyme activity. Upon aging of vermicomposting the enzymatic activity tends to be reducing. Macci et al. (2010) also of similar conclusion in their investigation. Stimulatory effect of earthworms on microbial biomass could be explained by mucus and casts production. Mucus is known to be a source of easy assimilable carbon for microorganisms (Doube and Brown 1998; Aira et al. 2007). Casts are nutrient enriched structures with available forms of C, N and P (Scheu 1987; Aira et al. 2003). Upon aging of vermicompost, there was a decrease of microbial biomass and activity of β-glucosidase, cellulase and protease. After initial increase, rapid decrease in alkaline phosphatase was also recorded (Aira et al. 2006b).

During studying the kinetics of dehydrogenase enzymes upon application of vermicompost, it was concluded that use of earthworm cast did not alter the enzyme-substrate affinity, while mineral fertilizer reduced this affinity or changed the composition and activity of soil microbiota (Masciandaro et al. 2000). Application of vermicompost increased V_{max}, and K_m was observed as same as control. While, it doubled in organo-mineral and mineral treatments. V_{max} and K_m represents a measurement of the quantity of enzyme and enzyme-substrate affinity.

Positive influence of Vermicomposts on soil biological properties is well documented (Aira et al. 2006a; Tejada and Gonzalez 2009; Tejada et al. 2009; Pramanik et al. 2010; Arancon et al. 2006). Benitez et al. (2000) studied the rhizospheric enzyme dynamics and the study found that few fold increase in dehydrogenase activity upon incorporation of vermicompost as mulch. Though it suggests that microbial numbers and potential activity in the rhizosphere were related to the addition of organic materials, but interestingly it was recorded that urease activity in the rhizosphere was found to be threefold inhibited. Phosphatase activity was also stimulated by the presence of vermicompost. Soil dehydrogenase, urease, β-glucosidase, phosphatase, and arylsulfatase activities along with microbial biomass increased more in the vermicompost produced from animal manure than plant source (Tejada and Gonzalez 2009; Tejada 2010). Amylase, protease, urease and acid phosphatase activities were also significantly higher in vermicompost treated soils compared with the control (Pramanik et al. 2010). Fernandez-Gomez et al. (2010) successfully vermicomposted tomato-fruit wastes and the compost was found to have positive role in dehydrogenase, β-glucosidase, protease and urease activity. Aira et al. (2006b) reported in enhancement of cellulase and β-glucosidase along with microbial biomass. In another interesting study, extracellular β-glucosidase, phosphatase, urease and protease activities were found to be always higher or equal to that measured at the beginning of the vermicomposting process, suggesting that the enzymes bound to humic matter resisted biological attack and environmental stress (Macci et al. 2010). The application of vermicompost enhanced dehydrogenase enzyme activity over time but altered soil urease activity to a very limited extent (Romero et al. 2010). The results showed that both urease and phosphatase activity of the compost increased significantly due to earthworm activity while there was no significant increase in microbial biomass and cellulase activity. In contrast, Albiach et al. (2000) concluded that no changes in soil biological activity after application of vermicompost.

It was clear that increase in the amount of vermicompost produced greater increase in enzymatic activity related P cycle than unamended control but less than composted manure (Saha et al. 2008a). Similar conclusion was drawn for other soil enzymes related to C and N cycle (Gopinath et al. 2008; Saha et al. 2008b). In another study, application of lime and inoculation of *Bacillus polymyxa* enhanced phosphatases and urease activities of vermicompost applied soil (Pramanik et al. 2007).

Composting is the predominant method of processing raw manure and other waste materials. Vermicomposting has proven to be high in potential in nutrient transformation. Thus it's quite obvious that comparison of vermicomposting vs composting comes into picture. Several studies were undertaken for estimating enzymatic activities which describe organic matter decomposition in two microbial-driven processes, composting and vermicomposting (Garcia et al. 1994, 1995; Benitez et al. 2002; Haritha Devi et al. 2009).

In a comparative study, vermicompost was found better in enhancement of microbial populations size and activity, and produced higher crop yields, which is attributed to higher nutrient concentrations in vermicomposted soil (Tognetti et al. 2005). Haritha Devi et al. (2009) studied the extracellular enzyme activities and microbial populations were studied during the normal composting and vermicomposting of fruit pulp, vegetable waste, groundnut husk and cow dung. They found that in vermicompost, the maximum enzyme activities (cellulase, amylase, invertase, protease and urease) were observed during 21–35 days. The cellulase, amylase and protease activities of vermicompost reached the maximum values by 28th day. Similarly the invertase and urease activities reached to maximum on 35th day. Nogales and Benitez (2007) compared the impact of application of the compost and vermicompost and it was found that initial stimulation of dehydrogenase activity, β-glucosidase, and urease activities followed by decline. Enhancement of enzymatic activities was attributed to increased amounts of humic acids in soil. In another comparative study, Tejada and Gonzalez (2008) reported that the soil microbial biomass and dehydrogenase, urease, β-glucosidase, phosphatase, and arylsulfatase activities increased more in the cow dung amended vermicompost soils compared with the green forage amended vermicompost soils. Therefore, these results suggest that application of vermicomposts rich in humic acids increase the soil enzymatic activity with time.

Toxicity of waste material is another concern in viewpoint of sustainable development and ecological concerns. Vermicompost was found to be very effective in reduction of toxicity of waste material (Benitez et al. 2004). Vermicomposted olive-mill solid waste increased the original soil dehydrogenase activity by fivefold, indicating a loss of toxicity of the waste during the vermicomposting process. Further enhancement of the enzyme activity upon second time application of the compost, further confirms the toxicity reduction phenomenon. Arancon et al. (2005) also reported the enhancement of microbial biomass and dehydrogenase activity after application of vermicomposts produced from food wastes and paper wastes. Feedstocks also converted to vermicompost for its utilization as source of nutrient supply (Warman and Anglopez 2010). Vermicompost was not only recorded for its influence on reduction of toxicity of source but also it has significant influential role in herbicide dissipation (Romero et al. 2010). Probable reason is quite obvious that increased microbial biomass, due to application of vermicompost, helps in faster degradation of the xenobiotics. Application of vermicompost amended soil decreased diuron persistence. These additions, under different soil management conditions, minimize the bioavailability and persistence of diuron and consequently the risk of leaching and seepage into aquifers (Romero et al. 2010). These experiments demonstrate that vermicomposting can be an alternate technology for the recycling and environmentally safe disposal/management of textile mill sludge using an epigeic earthworm, *E. foetida* (Garg et al. 2006).

Earthworms were responsible for additional alkaline phosphatases, some of them being produced in the worm gut and are then excreted through cast deposition (Satchell and Martin 1984; Ranganathan and Vinotha 1998; Vinotha et al. 2000). As the parent soil used was slightly acidic, this is in accordance with the observations of Eivazi and Tabatabi (1977) that showed a greater acid phosphatase activity in acid soils and a greater alkaline phosphatase in alkaline soils. The increased basic acid phosphatase activity we observed in casts extended earlier observations made with casts of temperate epigeic worms, such as *Eisenia fetida*, *Dendrobaena veneta*, *L. rubellus* (Satchell and Martin 1984) and *D. octaedra* (Flegel and Schrader 2000).

So, in general, extracellular enzyme activity was increased upon application of vermicompost in soil. Enhancement is attributed to quality of vermicompost which in turn dependant on organic sources. Initial enhancement of microbial growth was also observed which is related to the initial activation of the microbial enzymes and intracellular enzyme activity. Upon aging of vermicomposting the enzymatic activity tends to be reducing. Due to involvement of two different type of composting mechanism, enzymatic activity varied with time scale.

4 Interaction with Microflora

Much of the work regarding earthworm effects on other organisms has focused on the functional significance of microbial-earthworm interactions, and little is known on the effects of earthworms on microfloral and faunal diversity. Earthworms can affect soil microflora and fauna populations directly and indirectly by three main mechanisms: (1) fragmentation, burrowing and casting; (2) grazing; (3) dispersal. These activities change the soil physico-chemical and biological status and may cause drastic shifts in the density, diversity, structure and activity of microbial and faunal communities within the drilosphere. Due to limited movement in soil, small organisms, particularly microflora and microfauna, may be benefitted from the long ranging movements of earthworms (Fig. 3).

The digestive system of earthworms consists of a pharynx, oesophagus and gizzard followed by an anterior intestine that secretes enzymes and a posterior intestine that absorbs nutrients. During progress through digestive system there is a dramatic increase in number of micro organisms of upto 1,000 times. The earthworms were found to be feeding directly upon the cells of certain micro-organisms. Other species were found to be toxic to *E. fetida*. The seeding of vermiculture beds with

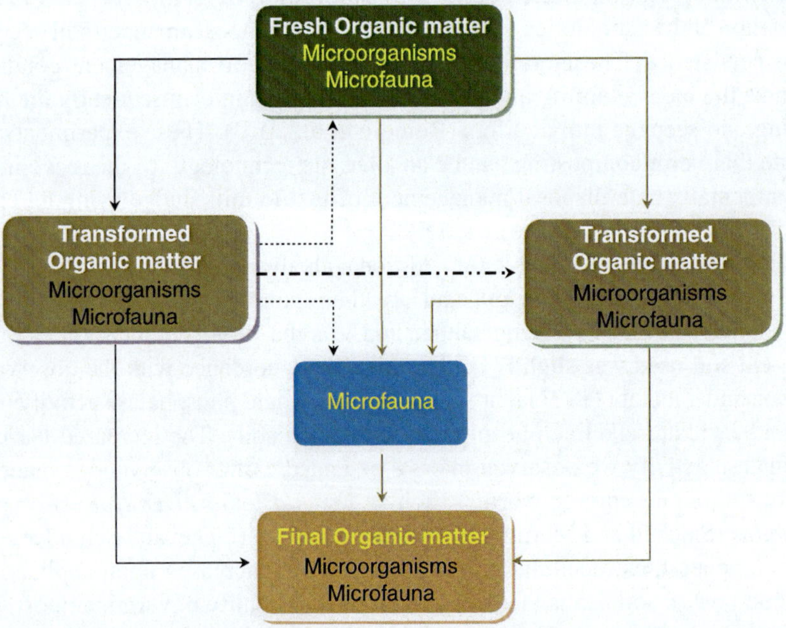

Fig. 3 Role of earthworm and microbes in the decomposition of organic matter during vermicomposting process (Modified from Dominguez et al. 2010). *Black lines* indicate process governed by earthworm and *grey lines* depicts microbial pathway. *Solid* and *broken lines* indicate direct and indirect effects. Direct effects include digging, digestion and casting by earthworms, which initially modify the organic matter, microorganisms and microfauna. Aging and mixing of casts with fresh organic matter or transport of microorganisms are indirect effects

the bacterium *Acinetobacter calcoaceticus* stimulated earthworm growth and consumption of the substrate (Hand et al. 1988), while no difference was observed for *Acetobacter diazotrophicus* inoculation over the worm reproductivity.

The burrowing and casting activities of earthworms contribute to the activity of soil micro organisms (Edwards and Bohlen 1996) and nutrient enriched earthworm casts are good media supporting microbial growth (Lee 1985). Many authors have studied the microbial community in the gut of earthworms (Fischer et al. 1995). It is well known that Gram-negative bacteria are common inhabitants of the intestinal canal of earthworms. Number of *Vibro* sp. and *Aeromonas hydrophila* were reported to frequent in the gut of earthworms *Eisenia lucens* and *Pheretima* sp. respectively. Danne et al. (1998) found, by scanning electron microscopy, that there were numerous rod shaped bacteria in egg capsule of *E. fetida*, and suggested a mutualistic association.

Vermicompost fertilization resulted in significant enhancement of microbial biomass, available phosphorus, and nitrogen content of wheat soil (Gaind and Nain 2007). Surprisingly, as fungi may be part of the diet of earthworms, the activity of *E. fetida* triggered fungal growth during vermicomposting (Aira et al. 2006a, b). Pramanik et al. (2010) found that application of vermicompost increased the proportion of fungal biomass in total soil microorganisms. Earthworm activity greatly decreased the bacterial growth rate and did not affect the fungal growth rate after 1 month of vermicomposting. Animal manure is microbiologically rich environments in which bacteria constitutes the largest fraction, with fungi mainly presents as spores (Garrett 1981; Aira et al. 2007). Moreover the first stages of decomposition in these organic wastes are mainly dominated by bacteria because of availability of water and easily decomposable substrate. Hence the activity of earthworm is expected to influence the bacterial growth more profusely than fungal growth. Earthworm casts and adjacent uningested soil from thirty different locations were compared to determine the abundance and diversity of fungal species. The casts contained larger fungal populations (g^{-1} dry soil weight) and numbers of fungal species than the soil. Fungal populations and the number of fungal species in casts and soil also varied significantly between samples from different locations. A total of 27 fungal species were recorded from the casts and soil. Indices of dominance (0.084 casts; 0.14 soil) and general diversity (2.53 casts; 2.02 soil) demonstrated that the casts displayed more diverse fungal flora than the soil. The diversity of fungal species increased in earthworm casts after passing through the earthworm gut (Tiwari and Mishra 1993). Most of the enzymes showed correlation with change in number and types of different microbial groups like bacteria, fungi and actinomycetes during vermicomposting with maximum number of 126×10^6, 28×10^4 and 93×10^5 colony forming units (CFU) g^{-1} sample respectively (Haritha Devi et al. 2009).

The casts contained higher microbial populations and enzyme activities than the soil. Except for fungal populations, statistically significant increases were found in all other parameters. Microbial populations and enzyme activities showed similar temporal trends with higher values in spring and summer and lower values in winter. Selective feeding by earthworms on organically rich substrates, which break down during passage through the gut, is likely to be responsible for the higher microbial populations and greater enzyme activity in the casts (Tiwari et al. 1989). As the vermicast become aged, there is a reduction in the moisture level leading to reduction in microbial population and activity. Consequently reduced microbial activity leads to reduced enzyme activity and NPK contents (Parthasarathi and Ranganathan 1999).

Studies on incidence of cellulolytic and lignolytic organisms in earthworm worked soils showed that symbiotic microflora of worms are involved in lignin degradation. The total microbial load in the different regions of the gut of worms has also shown more intense colonization of microbes in the anterior part of the intestine than the other region. Bisesi (1990) found that application of earthworm biotechnology in conjunction with indegenous microbial activity, under ambient conditions of temperature and seasonal changes enhance the rate of stabilization and turnover of biological sludge. Earthworm has been shown to selectively consume different type of plant material, and to select different fungal species when offered on filter paper disc (Cooke 1983). The presence of fungal propagules in the earthworm gut, and in cast material, has been known for some time (Parle 1963) and earthworms have been implicated in both the reduction and dispersal of soil- borne animal and plant fungal disease and the spread of beneficial group such as mycorrhizal fungi. Parle (1963) reported that population of yeast and fungi did not proliferate during passage through the gut, although actinomycetes and bacteria did.

Recent studies document North American earthworm invasions and their profound effects on the structure of the soil profile, which is the habitat for soil microorganisms mainly fungi and bacteria. Dramatic alterations made to these layers during earthworm invasion significantly change microbial community structure and therefore microbial activities such as C transformations. Understanding the impacts of earthworm invasion on the microbes themselves will give insight into earthworm effects on microbial activities. Bacterial and actinomycete communities in earthworm guts and casts have not been studied in environments recently invaded by earthworms. Earthworm invasion tended to decrease fungal species density and fungal species diversity and richness. The presence of earthworms decreased zygomycete species abundance probably due to disruption of fungal hyphae. Physical disruption of hyphae may also explain decreased mycorrhizal colonization rates, decreased mycorrhizal abundance and altered mycorrhizal morphology in the presence of earthworms. Mixing of organic layers into mineral soil during earthworm invasion tended to decrease microbial biomass in forest floor materials while increasing it in mineral soil. In newly invaded forest soils, microbial respiration and the metabolic quotient tended to decline. In forests where either the microbial community has had time to adapt to earthworm activities, or where the destruction of the forest floor is complete, as in invasions by the Asian *Amynthas hawayanus*, the presence of earthworms tends to increase the metabolic quotient indicating a shift to a smaller, more active microbial community (McLean et al. 2006).

Both basal and substrate-induced microbial respiration, microbial biomass C and N and fungal population in lateritic soil were increased due to vermicompost application. Ergosterol and chitin content were significantly higher in vermicompost treated soils over the control (Pramanik et al. 2010). Significant increase in the colonisation of these microbes in the vermicompost treated soil was observed. The symbiotic association of mycorrhizae in the roots showed a remarkable difference in infection. Except for actinomycetes, the colonies of the other microbes assessed 2 months after the harvest of the crop in the drained plots, showed significantly higher counts in the experimental plots. The stubbles in experimental plots retained higher counts of mycorrhizae than those in the control plots. It could be deduced that the vermicompost application has enhanced the activity of these selected microbes in the soil system. There was high level of Total N in the experimental plot which comparatively received less quantity of fertilisers. This may be due to the higher count of N-fixers observed than control.

In general, higher microbial population in vermin-cast was observed than the surrounding soil. Feeding and break down of organically rich substrates by earthworms during passage through the gut, is likely to be responsible for the higher microbial populations and greater enzyme activity in the casts. Significant changes in bacterial and fungal communities were observed.

5 Soil Functions

Measurement of soil functions was generally done by determining the rates of microbial process without knowing the microbial species effectively involved in the measured process. The links between microbial diversity and soil functioning are unknown and it is difficult to measure microbial diversity (Nannipieri et al. 2003). The central problem of the link between microbial diversity and soil function is to understand the relations between genetic diversity and community structure and between community structure and function.

In a recent study it was concluded that microbial biomass and diversity were reduced but enhancement was observed in microbial activity and, in turn in carbon mineralization (Gomez-Brandon et al. 2011). Optimization of the vermicomposting process could be possible because low values of microbial biomass that are indicative of stabilized materials were reached after 2 weeks of vermicomposting. However, a period of between 18 and 21 weeks was needed to achieve a more stabilized substrate in relation to the microbial activity.

Impact of vermicompost on soil functions was well documented in literature (Fig. 4). Vermicompost application significantly increased the concentration of organic C, mineralizable N, available P and exchangeable K in soil (Pramanik et al. 2010; Tejada et al. 2010). Arancon et al. (2006) concluded that enhancement of dehydrogenase activity and microbial biomass-N was correlated positively with the increased amounts of NH_4-N, NO_3-N and orthophosphates in the vermicompost-treated plots than in the controls. Thus the result suggests better mineralization of nutrients in vermicompost applied soil.

Suthar (2006) found that the conjugation treatment of inorganic fertilizers with vermicompost provides better plant growth by supplying essential micronutrients required for plant growth through vermicompost other than plant nutrients. The presence of some plant growth promoters such as phytohormones in

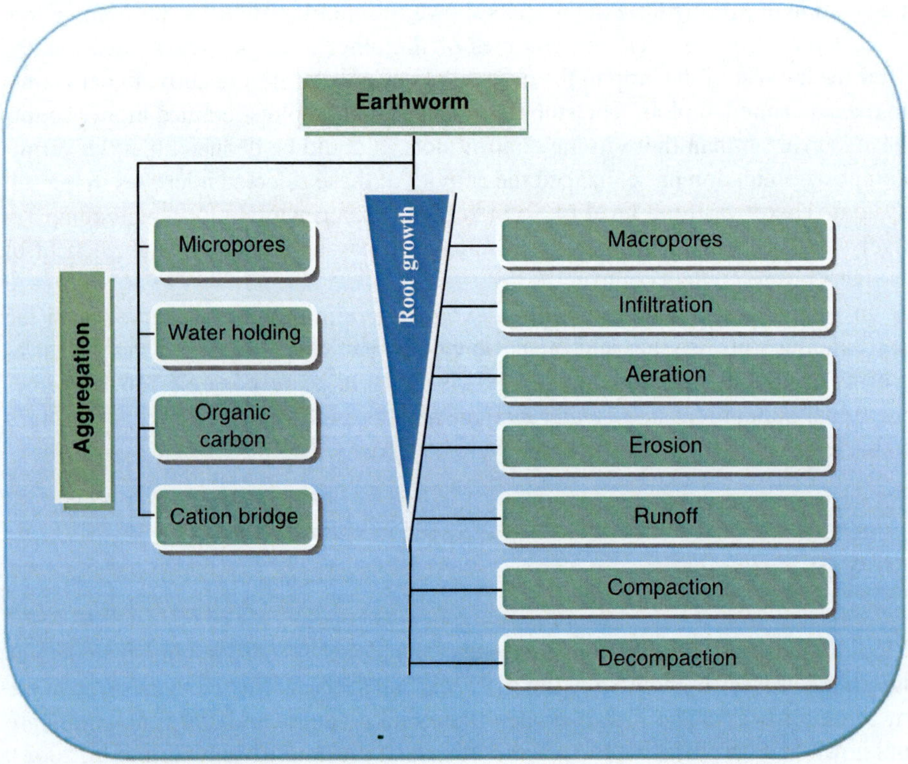

Fig. 4 Influence of earthworm in drilospehere soils (Adapted from Kavdir and Ilay 2010)

vermicompost further promotes the plant production. Moreover, organic manure input supports suitable microclimatic condition for rapid mineralization and transformation of plant nutrients, proliferated by vermicompost inhabiting soil organisms for sustainable land production. Suthar (2009) conducted a experiment with Vermicomposting showed a decrease in pH, organic C and C:N ratio, but increase total N, available P and exchangeable K at the end. C:N ratio of end material (vermicompost) was within the agronomic acceptable limit (<20). The high level of NPK in worm-processed material indicates the candidature of this species for waste management operations. The earthworm also showed an excellent growth in different wastes. Results thus indicate that *A. parva* appeared a potential tool for conversion of organic wastes into value added products for sustainable land restoration practices.

Kaushik and Garg (2003) conducted a experiment with vermicomposting of mixed solid textile mill sludge and cow dung and found that Vermicomposting resulted in significant reduction in C:N ratio and increase in N content. Total K and Ca were lower in the final cast than the initial feed mixture. Total P was higher in the final product than the initial feed mixture. Total heavy metal contents were lower in the final product than initial feed mixture. Solid textile mill sludge can be potentially useful as raw substrate in vermicomposting if mixed with up to 30% cow dung (on dry weight basis).

Suthar (2009) reported that use of vermicompost has advantages of infield crop production, but here such effects could be attributed to the nutritional status of vermicompost and to a variety of other factors (soil microbial structure and activity, mineralization, soil enzymatic factors and presence of some phyto-hormones in worm-processed materials). The higher concentration of soil enzymes, soil organic matter and soil microorganisms in worm casts creates suitable microclimatic conditions in soils for rapid mineralization and transformation of plant nutrients in soil. An important feature of vermicompost is that, during the processing of the various organic wastes by earthworms, many of the nutrients that it contents are change to forms that are more readily taken by plants such as nitrate or ammonium nitrate, exchangeable phosphorous and soluble potassium, calcium and magnesium (Suthar and Singh 2008). Suthar (2006) demonstrated that during the vermicomposting of some crop residues mixed with cattle dung resulted in an increase in total N (91–144%), available P (63–105%), and exchangeable K (45–90%) content of it. Therefore, ready vermicompost relatively contains more exchangeable plant nutrient than those by other plant growth media. Azarmi et al. (2009) reported that addition of vermicompost at rate of 15 $t ha^{-1}$ significantly increased contents of soil total organic carbon, total N, P, K, Ca, Zn and Mn substantially compared with control plots. The soils treated with vermicompost had significantly more electrical conductivity in comparison to unamended plots. The addition of vermicompost in soil resulted in decrease of soil pH. The physical properties such as bulk density and total porosity in soil amended with vermicompost were improved. The results of this experiment revealed that addition of vermicompost had significant positive effects on the soil chemical, physical properties. Kaushik and Garg (2004) reported that vermicomposting resulted in significant reduction in C:N ratio and increase in available NPK and Ca after 77 days of worm activity in all the feeds. Vermicomposting can be an alternate technology for the management of textile mill sludge if mixed with cow dung in appropriate quantities.

Tajbakhsh et al. (2011) reported that vermicompost is an excellent product being homogeneous, retaining most of the original nutrients and reduced level of organic contaminants, and can be applied to soil to increase soil organic matter content which can be released upon decomposition, improve soil structure, and increase cation exchange capacity. Vermicomposting seems to be more appropriate and an efficient technology to convert organic waste to a valuable community resources at low input basis.

The combined treatment of composting and vermicomposting was the most effective in terms of stabilizing the cattle manure. Moreover, earthworms promoted the retention of nitrogen and gradual release of P, as well as a reduction in electrical conductivity, thereby producing improved substrates for agricultural use (Lazcano et al. 2008). In another study, two earthworm species were compared for their mineralization rate and it was found that *Perionyx sansibaricus* was better efficient in mineralization and decomposition rate than that of *P. excavates* (Suthar and Singh 2008).

Dominguez et al. (1997) also compared vermicomposting and conventional processes for composting organic waste. Products of vermicomposting were shown to be superior to ordinary compost, particularly in (1) the degree of nitrogen mineralization and incorporation into biohumus, (2) the ability to kill pathogens, and (3) the low content of heavy metals.

Increased growth of rice plants in the presence of earthworms (*Millsonia anomala*, Megascolecidae) and demonstrated that enhanced nitrogen release (generally considered as the principal mechanism involved in earthworm positive effect on plants) was not responsible for this result (Blouin et al. 2006). We cannot, however, discard the hypothesis that microbial generalist parasites or symbionts may have been controlled or stimulated by *M. anomala*. This leaves the production of plant growth regulators (Muscolo et al. 1998; Nardi et al. 2002; Quaggiotti et al. 2004) as the probable explanation of the stimulatory effect of *M. anomala* on rice in our experiment. The possibility and the detailed mechanisms of the control of plant physiology via phytohormones secreted into soils by the bacteria associated with earthworms activities should be studied thoroughly. In earthworm rich soil, microbial biomass decreased, and there was a concomitant increase of available nutrients. There was no difference between the glucose-sensitive microbial biomass and the control but the respiratory quotient was greater. The fungal-to-bacterial ratio was slightly higher in that soil than in uningested soil (Zhang et al. 2000).

Yagi et al. (2003) reported that the vermicompost increased the levels of Ca, pH, organic matter and cation exchange capacity more than the manure. C-humic acids decreased and C-humin increased with vermicompost application. Plaza et al. (2008) reported that vermicomposting able to promote organic matter humification in cattle manure, thus enhancing the quality of these materials as soil organic amendments. Atiyeh et al. (2000) reported that the ash and total nitrogen contents increased greatly for a few weeks after the introduction of earthworms, reflecting a rapid breakdown of carbon compounds and mineralization of nitrogen by the earthworms. CO_2 evolution decreased rapidly 1 week after the introduction of earthworms, and continued at a lower rate throughout the 17 weeks, indicating increasing stability of the organic matter. Earthworms reduced microbial biomass early in the process, but enhanced nitrogen mineralization and increased the rates of conversion of ammonium-nitrogen into nitrate. The major general effect of earthworms on the organic wastes was to accelerate the maturation of the organic wastes as demonstrated by enhanced growth of lettuce and tomato seedlings.

Tejada and Gonzalez (2008) reported that vermicomposts have a positive effect on the soil biological properties and rice quality and yield parameters with respect to the no-vermicompost-amended soil. The fact that soil microbial biomass-C and soil respiration were higher in vermicompost with higher fulvic acids content may have been due to a greater labile fraction of organic matter in the former product. The labile fraction of organic matter is the most degradable and, therefore, the most susceptible to mineralization (Cook and Allan 1992), acting as an immediate energy source for microorganisms. Wisniewska and Kalembasa (2009) studied the transformation of carbon and nitrogen compounds in waste organic materials during their decomposition in sandy soil. It was found that amount of nitrogen released in the mineral forms $N\text{-}NH_4$, $N\text{-}NO_3$ increased initially and carbon mineralization is higher in fresh, composted and vermicomposted farm yard manure.

In another study Atiyeh et al. (2001) found that ammonium ions can be adsorbed on to the negative charges of the substrates, leached, taken up by the plants, or converted to nitrates via nitrification. It is very likely that the ammonium-nitrogen was converted rapidly to nitrates because of the increased microbial activity in the vermicompost-substituted substrates, thereby providing a slow release of nitrates. Chaoui et al. (2003) conduct a bioassay with earthworm cast in wheat to assess the amendment effects on plant growth and nutrient uptake and to validate the nutrient release results from the incubation study. Both microbial respiration and biomass were significantly greater in the conventional compost treatment compared to earth warm cast treatment for the initial 35 days of incubation followed by similar respiration rates and biomass to the end of the study at 70 days of incubation. Soil NO_3^- increased rapidly in the earth warm cast and conventional compost treatments in the initial 30 days of incubation, attaining 290 and 400 mg N kg^{-1} soil, respectively. Nitrate in the earth warm cast treatment then declined to 120 mg N kg^{-1} soil by day 70, while nitrate in the conventional compost treatment remained high. While ammonium levels decreased in the conventional compost treatment as nitrate level increased with increasing incubation time, a low level of ammonium was maintained in the earth warm cast treatment throughout the incubation. All cast and compost amendments significantly increased wheat P and K uptake compared to either the non-amended control or the mineral fertilizer treatment. The results show that casts are an efficient source of plant nutrients and that they are less likely to produce salinity stress in container as compared to compost and synthetic fertilizers.

Mashaee et al. (2008) investigated nitrogen mineralization rate in soils amended with compost, vermicompost and cattle manure, and found that nitrogen mineralization is higher in order Cattle manure>Vermicompost>Compost and found that the mineralization perfectly follow the mixed first-and zero-order kinetics. The effect of different combinations of vermicompost and coir dust on microbial respiration and nitrogen mineralization in soil was studied under laboratory conditions by Walpola and Wanniarachchi (2009). They found that treatment with 75% vermicompost and 25% coir dust demonstrated the highest carbon mineralization and NH_4^+-N contents followed by treatment 100% vermicompost and 50% vermicompost and 50% coir dust. The highest NO_3-N content was observed in the treatment of 100% vermicompost. Kalembasa (2006) found that the biggest amounts of NO_3^- and NH_4^+ ions were released during mineralization process in soil materials mixed with vermicomposts produced from waste activated sludge with addition of peat and droppings. The amounts of CO_2 liberated from analysed objects were lower for vermicomposts produced from waste activated sludge than for the fresh, composted and vermicomposted FYM.

Study by Chaoui et al. (2003) suggested that earthworm cast resulted in higher plant biomass production due to a slower rate of nitrogen mineralization that was more synchronized with the plant requirements. Similar observations were made by Cox (1993). The plant biomass and shoot elemental content data show that casts are an efficient source of plant nutrients and that the slower rate of N release gives it an advantage as compared to compost and synthetic fertilizers.

Vermicompost was found better than farmyard manure in faster P transformation (Saha et al. 2008b). Organic matter through the application of vermicompost increases the bioavailability of phosphorus in the soil effecting plant growth in potato cropping (Erich et al. 2002). Compost application also effects nitrogen mineralization in soil (Debosz et al. 2002). Most important feature of vermicompost is that, during processing of the various organic wastes by earthworms, many of the nutrients are changed to available forms that are more readily taken up by plants (Chaudhuri 2005). Ghosh et al. (1999) studied the magnitude of the transformation of phosphorus from the organic to inorganic state, and thereby into available forms was found to be considerably higher in the case of earthworm-inoculated organic wastes, showing that vermicomposting may prove to be an efficient technology for providing better P nutrition from different organic wastes. The transformation of P into three major inorganically bound forms; Al-P, Fe-P and Ca-P also tended to change substantially during the process of vermicomposting. Pramanik et al. (2009) found that higher rate of P-mineralization was recorded in soil treated with rock-phosphate as compared to other treatments. But after 60 days of incubation, available P content was declined in this soil, and finally it (14.36 mg kg^{-1}) became statistically at par with available P-content of soils received vermicompost prepared from grass (13.66 mg kg^{-1}) and cow dung (13.43 mg kg^{-1}). Application of vermicompost in soil increased pH, organic carbon, mineralizable nitrogen and exchangeable potassium content in soil as compared to the application of rock phosphate alone. Nourbakhsh (2007) found in his experiment that decomposable fraction of solid wastes decreased due to vermicomposting. Gaind and Nain (2007) reported that vermicompost fertilization resulted in highest microbial biomass, available phosphorus, and nitrogen content of wheat soil. It was also found effective in minimizing the alkalinity of soil compared to other treatments as indicated by pH change. Orozco et al. (1996) found that after ingestion of the coffee pulp by the earthworms, an increase in available P, Ca, and Mg but a decrease in K were detected.

Many studies have examined impacts of earthworm on C and N fluxes in soils (Binet and Trehen 1992; Lavelle and Martin 1992; Bohlen et al. 2004; Lavelle et al. 1997; Whalen and Janzen 2002); however, less attention has been paid to how and to the extent to which earthworms influence the dynamics of soil phosphorous (P). Earthworm casts collected directly from pasture were found to contain three times more water extractable P than the surrounding soil (Sharpley and Syers 1976). In addition, the ingestion and thorough mixing of soil in the intestinal tract of *Lumbricus rubellus* and *Aporrectodea caliginosa* favor the dissolution of phosphate rock and thus the availability of the derived-P in the soil (Mackay and Kladivko 1985). Lopez-Hernandez et al. (1989) also reported that water soluble P increased by 2.7 times in fresh casts of a tropical geophageous earthworm *Pontoscolex corethrurus*.

Inoculation of epigeic species of earthworms to the organic wastes during composting helps to enhance the transformation of organic P into mineral forms, and also to keep the magnitude of fixation of released P into insoluble inorganic forms at low levels, thus increasing the availability of P in these compounds (Ghosh et al. 1999).

Enhancement of microbial population and activity, NPK content and enzyme activities in the fresh casts are due to enhanced mineralization of nutrients, high substrate concentrations and high moisture level. Dryness of aged vermicasts implies the loss of N through volatilization, leaching and denitrification. Reduction in P is due to lesser conversion of organic to inorganic P due to reduced microbial activity. Loss of K input is due to leaching (Parthasarathi and Ranganathan 1999).

Improvement in mineralization of nutrients was reported by majority of studies. Specifically C and N mineralization was most influenced by the application of vermicompost. This might be attributed to higher microbial biomass in the vermicompost.

6 Conclusion

Huge potential of earthworms, keystone in the decomposition of organic matter and efficient nutrient recycling, in the form vermicomposting technology is need to be harnessed for sustainable agriculture system. The powerful tool can be exploited in wide range of sources effectively. Comparatively slower rate of nitrogen and phosphorous mineralization, synchronized with plant requirements enhance the value of the nutrient source. Comparable improvement in soil biological attributes as of compost made vermicompost a significant alternative. So definite improvement of the biological properties as well as soil functions and the sustained supply of macro as well as micronutrients will significantly improve the condition of soil for better harvest.

Acknowledgements Thanks are to be given to many scientists who have contributed information and help.

References

Aira M, Monroy F, Dominguez J, Mato S (2002) How earthworm density affects microbial biomass and activity in pig manure. Eur J Soil Biol 38:7–10

Aira M, Monroy F, Domínguez J (2003) Effects of two species of earthworms (*Allolobophora* spp.) on soil systems: a microfaunal and biochemical analysis. Pedobiologia 47:877–881

Aira M, Monroy F, Dominguez J (2006a) *Eisenia fetida* (Oligochaeta, Lumbricidae) activates fungal growth, triggering cellulose decomposition during vermicomposting. Microbial Ecol 52:738–747

Aira M, Monroy F, Dominguez J (2006b) Microbial biomass governs enzyme activity decay during aging of worm-worked substrates through vermicomposting. J Environ Qual 36:448–452

Aira M, Monroy F, Dominguez J (2007) Earthworms strongly modify microbial biomass and activity triggering enzymatic activities during vermicomposting independently of the application rates of pig slurry. Sci Total Environ 385:252–261

Albiach R, Canet R, Pomares F, Ingelmo F (2000) Microbial biomass content and enzymatic activities after the application of organic amendments to a horticultural soil. Bioresour Technol 75:43–48

Alef K, Nannipieri P, Trazar-Cepeda C (1995) Phosphatase activity. In: Alef K, Nannipieri P (eds) Methods in applied soil microbiology and biochemistry. Academic, London, pp 335–344

Allaire-Leung SE, Gupta SC, Moncrief JF (2000) Water and solute movement in soil as influenced by macropore characteristics. 1. Macropore continuity. J Contam Hydrol 41:283–301

Arancon NQ, Edwards CA, Bierman P, Metzger JD, Lucht C (2005) Effects of vermicomposts produced from cattle manure, food waste and paper waste on the growth and yield of peppers in the field. Pedobiologia 49:297–306

Arancon NQ, Edwards CA, Bierman P (2006) Influences of vermicomposts on field strawberries: part 2. Effects on soil microbiological and chemical properties. Bioresour Technol 97:831–840

Atiyeh RM, Arancon NQ, Edwards CA, Metzger JD (2000) Influence of earthworm-processed pig manure on the growth and yield of greenhouse tomatoes. Bioresour Technol 75:175–180

Atiyeh RM, Edwards CA, Subler S, Metzger JD (2001) Pig manure vermicompost as a component of a horticultural bedding plant medium: effects on physicochemical properties and plant growth. Bioresour Technol 78:11–20

Azarmi R, Giglou MT, Hajieghrari B (2009) The effect of sheep-manure vermicompost on quantitative and qualitative properties of cucumber (*Cucumis sativus* L.) grown in the greenhouse. Afr J Biotech 8:4953–4957

Binet F, Trehen P (1992) Experimental microcosm study of the role of *Lumbricus terrestris* (Oligochaeta, Lumbricidae) on nitrogen dynamics in cultivated soils. Soil Biol Biochem 24:1501–1506

Benitez E, Melgar R, Sainz H, Gomez M, Nogales R (2000) Enzyme activities in the rhizosphere of pepper (*Capsicum annuum*, L.) grown with olive cake mulches. Soil Biol Biochem 32:1829–1835

Benitez E, Saizn H, Melayar R, Nogales R (2002) Vermicomposting of a lignocellulosic waste from olive oil industry: a pilot scale study. Waste Manage Res 20:134–142

Benitez E, Melgar R, Nogales R (2004) Estimating soil resilience to a toxic organic waste by measuring enzyme activities. Soil Biol Biochem 36:1615–1623

Bisesi MS (1990) Vermial and microbial management of biological sludges under dynamic conditions of temperature and seasonable changes. Biol waste 32:99–109

Blanchart E, Albrecht A, Alegre J, Duboisset A, Gilot C, Pashanasi B, Lavelle P, Brussaard L (1999) Effects of earthworms on soil structure and physical properties. In: Lavelle P, Brussaard L, Hendrix P (eds) Earthworm management in tropical agroecosystems. CAB International, Wallingford, pp 149–172

Blouin M, Zuily-Fodil Y, Pham-Thi AT, Laffray D, Reversat G, Pando A, Tondoh J, Lavelle P (2005) Belowground organism activities affect plant aboveground phenotype, inducing plant tolerance to parasites. Ecol Lett 8:202–208

Blouin M, Barot S, Lavelle P (2006) Earthworms (*Millsonia anomala*, Megascolecidae) do not increase rice growth through enhanced nitrogen mineralization. Soil Biol Biochem 38:2063–2068

Bohlen PJ, Pelletier D, Groffman PM, Fahey TJ, Fisk MC (2004) Influence of earthworm invasion on redistribution and retention of soil carbon and nitrogen in northern temperate forests. Ecosystems 7:13–27

Brown GG, Barois I, Lavele P (2000) Regulation of soil organic matter dynamics and microbial activity in the drilosphere and the role of interactions with other edaphic functional domain. Eur J Soil Biol 36:177–198

Brown GG, Edwards CA, Brussaard L (2004) How earthworms affect plant growth: burrowing into the mechanisms. In: Edwards CA (ed) Earthworm ecology. CRC Press, Boca Raton, pp 13–49

Chan PLS, Griffiths DA (1988) The vermicomposting of pre-treated pig manure. Biol wastes 24:57–69

Chaoui HI, Zibilske LM, Ohno T (2003) Effects of earthworm casts and compost on soil microbial activity and plant nutrient availability. Soil Biol Biochem 35:295–302

Chapin FS III, Matson PA, Mooney HA (2002) Principles of terrestrial ecosystem ecology. Springer, New York

Chaudhuri PS (2005) Vermiculture and vermicomposting as biotechnology for conversion of organic wastes into animal protein and organic fertilizer. Asian J Microbiol Biotech Env Sci 7:359–370

Coleman DC (2008) From peds to paradoxes: linkage between soil biota and their influences on ecological processes. Soil Biol Biochem 40:271–289

Cook BD, Allan DL (1992) Dissolved organic matter in old field soils: total amounts as a measure of available resources for soil mineralization. Soil Biol Biochem 24:585–594

Cooke A (1983) The effects of fungi on food selection by *Lumbricus terrestris* L. In: Satchell JE (ed) Earthworm ecology. Chapman & Hall, London, pp 365–373

Cox DA (1993) Reducing nitrogen leaching-losses from containerized plants: the effectiveness of controlled-release fertilizers. J Pl Nutr 16:533–545

Curry JP, Byrne D (1997) Role of earthworms in straw decomposition in a winter cereal field. Soil Biol Biochem 29:555–558

Danne LL, Molina JAE, Sadowsky M (1998) Scanning electron microscopy of the microflora in egg capsules of the earthworm *Eisenia foetida*. Pedobiologia J 42:78–87

Dominguez J (2004) State of the art and new perspectives on vermicomposting research. In: Edwards CA (ed) Earthworm ecology, 2nd edn. CRC Press, Boca Raton, pp 401–424

Dominguez J, Briones MJ, Mato S (1997) Effect of diet on growth and reproduction of *Eisenia andrei* (Oligochaeta, Lumbricidae). Pedobiologia 41:566–576

Dominguez J, Aira M, Gomez-Brandon M (2010) Vermicomposting: earthworm enhances the work of microbes. In: Insam H, Frank-Whittle I, Goberna M (eds) Microbes at work: from waste to resources. Springer, Berlin/Heidelberg, pp 93–110

Doube BM, Brown GG (1998) Life in a complex community: functional interactions between earthworms, organic matter, microorganisms, and plant growth. In: Edwards CA (ed) Earthworm ecology. St. Lucie Press, Boca Raton, pp 179–211

Doube BM, Williams PML, Willmott PJ (1997) The influence of two species of earthworm (*Aporrectodea trapezoides* and *Aporrectoedea rosea*) on the growth of wheat, barley and faba beans in three soil types in the greenhouse. Soil Biol Biochem 29:503–509

Edwards CA, Bohlen PJ (1996) Biology and ecology of earthworms. Chapman & Hall, London, p. 426

Eivazi F, Tabatabi MA (1977) Phosphatases in soil. Soil Biol Biochem 9:167–172

Erich MS, Fitzgerald CB, Porter GA (2002) The effect of organic amendment on phosphorus chemistry in a potato cropping system. Agr Ecosyst Environ 88: 79–88

Fernandez-Gomez MJ, Nogales R, Insam H, Romero E, Goberna M (2010) Continuous-feeding vermicomposting as a recycling management method to revalue tomato-fruit wastes from greenhouse crops. Waste Manage 30:2461–2468

Fischer K, Hahn D, Amann RI, Daniel O, Zeyer J (1995) In situ analysis of the bacterial community in the gut of the earthworm *Lumbricus terrestris* L., by whole-cell hybridization. Can J Microbiol 41:666–673

Flegel M, Schrader S (2000) Importance of food quality on selected enzyme activities in earthworm casts (*Dendrobaena octaedra*, Lumbricidae). Soil Biol Biochem 32:1191–1196

Furlong MA, Singleton DR, Coleman DC, Whitman WB (2002) Molecular and culture-based analyses of prokaryotic communities from an agricultural soil and the burrows and casts of the earthworm *Lumbricus rubellus*. Appl Environ Microbiol 68:1265–1279

Gaind S, Nain L (2007) Chemical and biological properties of wheat soil in response to paddy straw incorporation and its biodegradation by fungal inoculants. Biodegra 18:495–503

Garcia C, Hernandez MT, Costa F, Ceccanti B (1994) Biochemical parameters in soils regenerated by addition of organic wastes. Waste Manage Res 12:457–466

Garcia C, Ceccanti C, Masciandaro G, Hernandez T (1995) Phosphatase and β-glucosidase activities in humic substances from animal wastes. Bioresour Technol 53:79–87

Garg VK, Kaushik P, Dilbaghi N (2006) Vermiconversion of wastewater sludge from textile mill mixed with anaerobically digested biogas plant slurry employing *Eisenia foetida*. Ecotox Environ Safe 65:412–419

Garrett SD (1981) Soil fungi and soil fertility. MacMillan Company/Pergamon Press, New York

Ghosh M, Chattopadhyay GN, Baral K (1999) Transformation of phosphorus during vermicomposting. Bioresour Technol 69:149–154

Gomez-Brandon M, Aira M, Lores M, Dominguez J (2011) Changes in microbial community structure and function during vermicomposting of pig slurry. Bioresour Technol 102:4171–4178

Gopinath KA, Saha S, Mina BL, Kundu S, Pande H, Gupta HS (2008) Impact of organic amendments n growth, yield, and quality of wheat and on soil properties during transition to organic production. Nutr Cycl Agroecosys 82:51–60

Gunadi B, Edwards CA, Arancon Q (2002) Changes in trophic structure of soil arthropods after the application of vermicomposts. Eur J Soil Biol 381:61–165

Hand P, Hayes WA, Frankland JC, Satchell JE (1988) Vermicomposting of cow slurry. Pedobiologia 31:199–209

Haritha Devi S, Vijayalakshmi K, Pavana Jyotsna K, Shaheen SK, Jyothi K, Surekha Rani M (2009) Comparative assessment in enzyme activities and microbial populations during normal and vermicomposting. J Environ Biol 30:1013–1017

Kalembasa D (2006) Content of selected nitrogen compounds (NO_3 AND NH_4^+) during mineralization process of vermicomposts in sandy soil materials. Adv Agric Sci (Zeszyty Problemowe Postepow Nauk Rolniczych) 513:181–189

Kaushik P, Garg VK (2003) Vermicomposting of mixed solid textile mill sludge and cow dung with the epigeic earthworm *Eisenia foetida*. Bioresour Technol 90:311–316

Kaushik P, Garg VK (2004) Dynamics of biological and chemical parameters during vermicomposting of solid textile mill sludge mixed with cow dung and agricultural residues. Bioresour Technol 94:203–209

Kavdir Y, Ilay R (2010) Earthworms and soil structure. In: Karaca A (ed) Biology of earthworm. Springer, New York, pp 39–50

Kizilkaya R, Hepsen S (2007) Microbiological properties in earthworm cast and surrounding soil amended with various organic wastes. Comm Soil Sci Pl Anal 38:2861–2876

Ladd JN (1978) Origin and range of enzymes in soil. In: Burns RG (ed) Soil enzymes. Academic, London, pp 51–96

Lavelle P (1988) Earthworm activity and the soil system. Biol Eert Soils 6:237–251

Lavelle P, Martin A (1992) Small-scale and large-scale effects of endogeic earthworms on soil organic matter dynamics in soils of the humid tropics. Soil Biol Biochem 24:1491–1498

Lavelle P, Bignell D, Lepage M, Wolters V, Roger P, Ineson P, Heal OW, Dhillion S (1997) Soil function in a changing world—the role of invertebrate ecosystem engineers. Eur J Soil Biol 33:159–193

Lavelle P, Spain A (2001) Soil ecology. Kluwer Academic, Dordrecht/Boston/London, 654 pp

Lazcano C, Gomez-Brandon M, Dominguez J (2008) Comparison of the effectiveness of composting and vermicomposting for the biological stabilization of cattle manure. Chemosphere 72:1013–1019

Lee KE (1985) Earthworms: their ecology and relationship with soils and land use. Academic, Sydney

López-Hernández D, Nino M, Nannipieri P, Fardeau JC (1989) Phosphatase activity in *Nasutitermes ephratae* termite nests. Biol Fertil Soils 7:134–7

Macci C, Masciandaro G, Ceccanti B (2010) Vermicomposting of olive oil mill wastewaters. Waste Manage Res 28:738–747

Mackay AD, Kladivko EJ (1985) Earthworms and rate of breakdown of soybean and maize residues in soil. Soil Biol Biochem 17:851–857

Masciandaro G, Ceccanti B, Ronchi V, Bauer C (2000) Kinetic parameters of dehydrogenase in the assessment of the response of soil to vermicompost and inorganic fertilisers. Biol Fert Soils 32:479–483

Mashaee SS, Aliasgharzade N, Ostan S (2008) Kinetics of nitrogen mineralization in soils amended with compost, vermicompost and cattle manure. J Sci Technol Agr Nat Resour 11:405–415

McLean MA, Migge-Kleian S, Parkinson D (2006) Earthworm invasions of ecosystems devoid of earthworms: effects on soil microbes. Biol Invasions 8:1257–1273

Muscolo A, Cutrupi S, Nardi S (1998) IAA detection in humic substances. Soil Biol Biochem 30:1199–1201

Nannipieri P, Ascher J, Ceccherini MT, Landi L, Pietramellara G, Renella G (2003) Microbial diversity and soil functions. Eur J Soil Sci 54:655–670

Nardi S, Pizzeghello D, Muscolo A, Vianello A (2002) Physiological effects of humic substances on higher plants. Soil Biol Biochem 34:1527–1536

Ndegwa PM, Thompson SA (2001) Integrating composting and vermicomposting in the treatment of bioconversion of biosolids. Bioresour Technol 76:107–112

Nogales R, Benítez E (2007) Effect of olive-derived organic amendments on Lead, Zinc, and biochemical parameters of an artificially contaminated soil. Comm Soil Sci Pl Anal 38:795–811

Nourbakhsh F (2007) Influence of vermicomposting on soil waste decomposition kinetics in soils. J Zhejiarg Univ Sci 8:725–730

Orozco FH, Cegarra J, Trvjillo LM, Roig A (1996) Vermicomposting of coffee pulp using the earthworm *Eisenia foetida*: effects on C and N contents and the availability of nutrients. Biol Fert Soil 22:162–166

Parle JN (1963) Microorganisms in the intestines of earthworms. J Gen Microbiol 31:1–13

Parthasarathi K, Ranganathan LS (1999) Longevity of microbial and enzyme activity and their influence on NPK content in pressmud vermicasts. Eur J Soil Biol 35:107–113

Paul EA, Clark FE (1996) Soil microbiology and biochemistry, 2nd edn. Academic, San Diego

Plaza C, Nogales R, Senesi N, Benítez E, Polo A (2008) Organic matter humification by vermicomposting of cattle manure alone and mixed with two-phase olive pomace. Bioresour Technol 99:5085–5089

Pramanik P, Ghosh GK, Ghosal PK, Banik P (2007) Changes in organic – C, N, P and K and enzyme activities in vermicompost of biodegradable organic wastes under liming and microbial inoculants. Bioresour Technol 98:2485–2494

Pramanik P, Bhattacharya S, Bhattacharyya P, Banik P (2009) Phosphorous solubilization from rock phosphate in presence of vermicomposts in Aqualfs. Geoderma 152:16–22

Pramanik P, Ghosh GK, Chung YR (2010) Changes in nutrient content, enzymatic activities and microbial properties of lateritic soil due to application of different vermicomposts: a comparative study of ergosterol and chitin to determine fungal biomass in soil. Soil Use Manage 26:508–515

Quaggiotti S, Ruperti B, Pizzeghello D, Francioso O, Tugnoli V, Nardi S (2004) Effect of low molecular size humic substances on nitrate uptake and expression of genes involved in nitrate transport in maize (*Zea mays* L.). J Exp Bot 55:803–813

Ranganathan LS, Vinotha SP (1998) Influence of pressmud on the enzymatic variations in the different reproductive stages of *Eudrilus eugeniae* (Kinberg). Curr Sci 74:634–635

Romero E, Fernandez-Bayo J, Diaz JMC, Nogales R (2010) Enzyme activities and diuron persistence in soil amended with vermicompost derived from spent grape marc and treated with urea. Appl Soil Ecol 44:198–204

Saha S (2010) Soil functions and diversity in organic and conventional farming. In: Lichtfouse E (ed) Sociology, organic farming, climate change and soil science, vol 3, Sustainable agriculture reviews. Springer, New York, pp 275–301

Saha S, Mina BL, Gopinath KA, Kundu S, Gupta HS (2008a) Relative changes in phosphatase activities as influenced by source and application rate of organic composts in field crops. Bioresour Technol 99:1750–1757

Saha S, Mina BL, Gopinath KA, Kundu S, Gupta HS (2008b) Organic amendments affect biochemical properties of a sub temperate soil of Indian Himalayas. Nutr Cycl Agroecosys 80:233–242

Satchell JE, Martin K (1984) Phosphatase activity in earthworm faeces. Soil Biol Biochem 2:191–194

Scheu S (1987) Microbial activity and nutrient dynamics in earthworm cast (Lumbricidae). Biol Fertil Soils 5:230–234

Scheu S (2003) Effect of earthworms on plant growth: patterns and perspectives. Pedobiologia 47:846–856

Schimel JP, Bennett J (2004) Nitrogen mineralization: challenges of a changing paradigm. Ecology 85:591–602

Sharpley AN, Syers JK (1976) Potential role of earthworm casts for the phosphorus enrichment of run-off waters. Soil Biol Biochem 8:341–346

Shipitalo MJ, Le Bayon RC (2004) Quantifying the effects of earthworms on soil aggregation and porosity. In: Edwards CA (ed) Earthworm ecology. CRC Press, Boca Raton, p 441

Sinsabaugh RL, Moorhead DL (1994) Resource allocation to extracellular enzyme production: a model for nitrogen and phosphorous control of litter decomposition. Soil Biol Biochem 26:1305–1311

Six J, Feller C, Denef K, Ogle SM, Sa JCD, Albrecht A (2002) Soil organic matter, biota and aggregation in temperate and tropical soils – effects of no-tillage. Agronomie 22:755–775

Subler S, Baranski CM, Edwards CA (1997) Earthworm additions increased short-term nitrogen availability and leaching in two graincrop agroecosystems. Soil Biol Biochem 29:413–421

Suthar S (2006) Potential utilization of guar gum industrial waste in vermicompost production. Bioresour Technol 97:2474–2477

Suthar S (2009) Vermicomposting of vegetable-market solid waste using *Eisenia fetida*: impact of bulking material on earthworm growth and decomposition rate. Ecol Eng 35:914–920

Suthar S, Singh S (2008) Vermicomposting of domestic waste by using two epigeic earthworms (*Perionyx excavatus* and *Perionyx sansibaricus*). Int J Env Sci Tech 5:99–106

Tajbakhsh J, Goltapeh EM, Varma A (2011) Vermicompost as a Biological Soil Amendment. Biology of Earthworms 24:215–228

Tate RL (2000) Soil microbiology, 2nd edn. Wiley, New York

Tejada M (2010) Organic amendment based on vermicompost and compost: differences on soil properties and maize yield. Waste Manage Res. doi:10.1177/0734242X10383622

Tejada M, Gonzalez JL (2008) Influence of two organic amendments on the soil physical properties, soil losses, sediments, and runoff water quality. Geoderma 145:325–334

Tejada M, Gonzalez JL (2009) Application of two vermicomposts on a rice crop: effects on soil biological properties and rice quality and yield. Agron J 101:336–344

Tejada M, Garcia-Mrtinez AM, Parrado J (2009) Effects of a vermicompost composted with beet vinasse on soil properties, soil losses and soil restoration. Catena 77:238–247

Tejada M, Gomez I, Hernandez T, Garcia C (2010) Utilization of vermicomposts in soil restoration: effects on soil biological properties. Soil Sci Soc Am J 74:525–532

Tiwari SC, Mishra RR (1993) Fungal abundance and diversity in earthworm casts and in uningested soil. Biol Fert Soils 16:131–134

Tiwari SC, Tiwari BK, Mishra RR (1989) Microbial populations, enzyme activities and nitrogen-phosphorus-potassium enrichment in earthworm casts and in the surrounding soil of a pineapple plantation. Biol Fert Soils 8:178–182

Tognetti C, Laos F, Mazzarino MJ, Hernandez MT (2005) Composting vs. vermicomposting: a comparison of end product quality. Compost Sci Util 13:6–13

Vinotha SP, Parthasarathi K, Ranganathan LS (2000) Enhanced phosphatase activity in earthworm casts is more of microbial origin. Curr Sci 79:1158–1159

Walpola BC, Wanniarachchi SD (2009) Microbial respiration and nitrogen mineralization in soil amended with different proportions of vermicompost and coir dust. Bangladesh J Ag Res 34:537–543

Wardle DA (2002) Communities and ecosystems: linking the aboveground and belowground components. Princeton University Press, Princeton, 400 p

Warman PR, AngLopez MJ (2010) Vermicompost derived from different feedstocks as a plant growth medium. Bioresour Technol 101:4479–4483

Whalen JK, Janzen HH (2002) Labeling earthworms uniformly with ^{13}C and ^{15}N: implications for monitoring nutrient fluxes. Soil Biol Biochem 34:1913–1918

Wisniewska B, Kalembasa D (2009) The mineralisation process of carbon and nitrogen compounds during the decomposition of organic materials in sandy soil. Zeszyty Problemowe Postepow Nauk Rolniczych 537:369–379

Yagi R, Ferreira ME, Pessoa Da Cruz MC, Barbosa JC (2003) Organic matter fractions and soil fertility under the influence of liming, vermicompost and cattle manure. Scientia Agricola 60:549–557

Zhang B, Li G, Shen T, Wang J, Su Z (2000) Changes in microbial biomass C, N, and P and enzyme activities in soil incubated with the earthworms *Metaphire guillelmi* or *Eisenia fetida*. Soil Biol Biochem 32:2055–2062

Index

E. Lichtfouse (ed.), *Farming for Food and Water Security*, Sustainable Agriculture Reviews 10, DOI 10.1007/978-94-007-4500-1,